WITHDRAWN FROM
THE LIBRARY

UNIVERSITY OF
WINCHESTER

D0301803

POPE AND THE EARLY
EIGHTEENTH-CENTURY BOOK TRADE

POPE

AND THE

EARLY
EIGHTEENTH-CENTURY
BOOK TRADE

DAVID FOXON

The Lyell Lectures, Oxford 1975–1976

Revised and edited by
JAMES McLAVERTY

CLARENDON PRESS · OXFORD
1991

Oxford University Press, Walton Street, Oxford OX2 6DP
Oxford New York Toronto
Delhi Bombay Calcutta Madras Karachi
Petaling Jaya Singapore Hong Kong Tokyo
Nairobi Dar es Salaam Cape Town
Melbourne Auckland
and associated companies in
Berlin Ibadan

Oxford is a trade mark of Oxford University Press

Published in the United States
by Oxford University Press, New York

© *David Foxon and James McLaverty 1991*

All rights reserved. No part of this publication may be reproduced,
stored in a retrieval system, or transmitted, in any form or by any means,
electronic, mechanical, photocopying, recording, or otherwise, without
the prior permission of Oxford University Press

British Library Cataloguing in Publication Data
Foxon, David
Pope and the early eighteenth-century book trade : the
Lyell lectures, Oxford 1975–1976.—(Lyell lectures in
bibliography; 1975–6).
1. Poetry in English. Pope, Alexander, 1688–1744
I. Title   II. McLaverty, J.
821.5
ISBN 0–19–818402–6

Library of Congress Cataloging in Publication Data
Foxon, David F. (David Fairweather)
Pope and the early eighteenth-century book trade : Lyell lectures
in bibliography, 1975–1976 / David Foxon ; revised and edited by
James McLaverty.
Includes bibliographical references.
1. Pope, Alexander, 1688–1744—Knowledge—Book arts and sciences.
2. Book industries and trade—Great Britain—History—18th century.
3. Authors and publishers—Great Britain—History—18th century.
4. Literature publishing—Great Britain—History—18th century.
5. Pope, Alexander, 1688–1744—Criticism, Textual. 6. Pope,
Alexander, 1688–1744—Bibliography. 7. Pope, Alexander, 1688–1744—
Publishers. I. McLaverty, J. II. Title.
PR3637.B58F69   1990
821'.5—dc20   90–35571
ISBN 0–19–818402–6

Typeset by Eta Service (Typesetters) Ltd, Beccles, Suffolk
Printed in Great Britain by
The Alden Press, Oxford

KING ALFRED'S COLLEGE
WINCHESTER

821.5
POP/   01279042

FOX

CONTENTS

LIST OF ABBREVIATIONS

BC	*Book collector*
BL	British Library
Bodl.	Bodleian
BJECS	*British journal for eighteenth century studies*
Correspondence	*The correspondence of Alexander Pope*, ed. George Sherburn, 5 vols. (Oxford, 1956)
DUJ	*Durham University journal*
Early career	Sherburn, George, *The early career of Alexander Pope* (Oxford, 1934)
ECL	*Eighteenth-century life*
ECS	*Eighteenth-century studies*
Foxon	Foxon, David F., *English verse 1701–1750*, 2 vols. (Cambridge, 1975)
Griffith	Griffith, Reginald H., *Alexander Pope: A bibliography*, 2 vols. (Austin, 1922–7)
HLB	*Harvard Library bulletin*
HLQ	*Huntington Library quarterly*
JEGP	*Journal of English and Germanic philology*
MLR	*Modern language review*
MP	*Modern philology*
NCBEL	*New Cambridge bibliography of English literature*, ed. George Watson, 5 vols. (Cambridge, 1969–77)
NQ	*Notes and queries*
PBSA	*Papers of the Bibliographical Society of America*
PQ	*Philological quarterly*
PRO	Public Record Office
PubH	*Publishing history*
PULC	*Princeton University Library chronicle*
RES	*Review of English studies*
SB	*Studies in bibliography*
SELit	*Studies in English literature*
SP	*Studies in philology*
Twickenham	*The Twickenham edition of the poems of Alexander Pope*, ed. John Butt et al., 11 vols. (London and New Haven, 1939–69)

LIST OF ILLUSTRATIONS

The following list gives the source of the illustration and the library in which the book or manuscript is held (with the call-mark, when I have it). The dimensions in millimetres are given with the illustrations themselves (the measurements are of the page, or, less precisely, of the opening, except in the case of imprints, where the measurement is of the type area).

The illustrations are reproduced by kind permission of the following libraries:

Beinecke Rare Book and Manuscript Library, Yale University
Bibliothèque de l'Arsenal
Bodleian Library
British Library
Cambridge University Library
Folger Shakespeare Library
Houghton Library, Harvard University
Huntington Library, San Marino, California
Keele University Library
National Library of Scotland
Pierpont Morgan Library, New York

LIST OF TABLES

EDITOR'S INTRODUCTION

David Foxon's Lyell Lectures, delivered in Oxford in March 1976, are important for two reasons: they provide the first full account of Pope's publishing career and they offer a new and challenging view of his text. As a supplement to Pope biography they examine his complicated relations with printers and booksellers and the ways in which he achieved financial security, while the analysis of Pope's text focuses on elements that are generally neglected—the design and format of books and their typography—and finds that in these areas Pope was an innovator. The two strands in the lectures are bound together by an emphasis on the importance of historical investigation and understanding; it is only through a knowledge of the workings of the contemporary book trade that Pope's career as an author, both in its financial and book-making aspects, can be understood.

The five chapters that are printed here correspond to the first five of six lectures. The first three give a roughly chronological account of Pope's literary career against the background of the book trade. His first works were produced by Jacob Tonson and John Watts, the leading literary publisher and his partner, who was perhaps the most elegant printer of the period. The translations of Homer, which made Pope's fortune, were published by subscription at the expense of Tonson's rival, Bernard Lintot; Pope was given complete control over type, paper, and illustration, and this enabled him to set new fashions in book production and typography. In publishing his subsequent works Pope kept control over his copyrights and aimed to maximize his profits, using less influential members of the trade to whom he could dictate his own terms; this produced its own problems.

The two final chapters begin the exploration of the problems faced by an editor of Pope. Underpinning the discussion are two surviving volumes which contain Pope's corrections: the proofs of the second volume of the quarto *Iliad* at the Bibliothèque de l'Arsenal and a copy of the octavo *Works* II (1736) in the British Library. Both these show Pope making detailed changes to accidentals (punctuation, capitalization, and italics), and an analysis of the progress of selected poems from manuscript through their various printed editions shows the same pattern recurring throughout Pope's working life. The general direction of Pope's changes is towards a classical typography, without capitals or italics, but complications are introduced by his own inconsistencies and by his feeling that a different typographical style should distinguish editions intended for the general public from the volumes intended for a select readership of subscribers.

There was originally a final lecture which dealt with the theory of textual criticism and rapidly surveyed some authors with a pronounced interest in typography: William Drummond and his friend William Alexander, Earl of Stirling, Jonson, Drayton, Benlowes, Cowley, Prior, Thomson, Akenside. These authors, like Pope, adopted styles of capitalization and italicization different from their printers' normal practice, and revised these accidentals in later editions, producing similar problems for an editor. At an early stage in planning publication it was decided to confine this book to Pope, and this final lecture is not included here, but it forms part of the typescript of the lectures, copies of which are deposited in the Bodleian, British, Beinecke, and Clark Libraries.

David Foxon had originally planned to prepare the lectures for publication himself, intending to incorporate further research, and he carried on collecting materials until the early 1980s, when it became clear that his health would not permit him to finish the work. As I had worked on Pope under his supervision at the time he first planned the lectures, and as we had kept in close touch since then, it seemed natural that I should take over the task of seeing the lectures into print. It was not our plan that I should try to bring the major lines of enquiry opened up by the lectures to a conclusion; publication had always been planned as a report on work in progress and one aim of this book is to draw these new approaches to the attention of other scholars. My task has been to revise and, where necessary, supplement the lectures as they were prepared and delivered. The first stage involved using the actual scripts of the lectures to correct the book-length typescript which has been available through deposit in selected libraries. The scripts include some corrections, some abridgements, and some redistribution of material; the lectures on Pope's text are particularly heavily revised, gaining in detail and sophistication. The next task was to complete the annotation, tie up any loose ends, and take account of subsequent scholarship. I have tried to do this as lightly as possible and those anxious to spot the sorcerer's apprentice at work will find he has taken charge of only two sections: I have completed Chapter 3 by dealing with Pope's relations with Warburton, and I have supplied the copyright appendix. I have not tried to take account of developments in the theory of textual criticism since 1976, but have allowed the lecturer's views to stand as given then. That year saw the publication of an important paper by Fredson Bowers, 'Scholarship and editing', *PBSA* lxx (1976), 161–88, which also rejected a narrow and dogmatic interpretation of Greg's rationale of copy-text, and development has been rapid since then.

Both author and editor are grateful to the Lyell electors, who issued the initial invitation to give the series of lectures and subsequently supported their publication. A fellowship at the William Andrews Clark Library enabled David to write the bulk of the lectures in congenial and helpful surroundings. I am not able to acknowledge here the many scholars who helped him in his work; indeed it would be impossible to do so, because the lectures represent a lifetime's study and discussion. I am conscious of the contributions of Michael Treadwell, Richard Noble, and John Chalmers. I also wish to acknowledge the work of D. F. McKenzie, whose 1976 Sandars Lectures, 'The London book trade in the later seventeenth century', shared many of the Lyell Lectures' concerns and form an admirable companion to them. I am grateful to the Bibliographical Soiciety for permission to make use of Table IV from Michael Treadwell's 'London trade publishers 1675–1750', *Library*, 6th ser. iv (1982), 109, and to the Grolier Club, New York, for permission to quote from the Bowyer printing ledgers.

I am conscious of the many debts I have incurred in preparing the lectures for the press. David Fleeman encouraged the project from an early stage and on a visit to the United States in 1987 checked materials that would otherwise have been unavailable to me; I owe much to his good sense and kindness. Maynard Mack responded to my queries with a generosity familiar to those who have worked on Pope, and the importance and distinction of his own work is apparent in the way it is drawn on throughout this study. I have also received generous advice or help from Kim Scott Walwyn, Frances Whistler, Alice Park, and Robert Peden at the Press, and from Michael Treadwell, Richard Goulden, Mervyn Jannetta, Don Nichol, Kathleen McKilligan, Frank Doherty, Richard Rouse, Pat Voiels, and David Young.

The identification and assembling of the illustrations has been a major undertaking in the preparation of this book, and it could not have been accomplished without generous help from librarians. Christine Fyfe of Keele University Library provided typically shrewd advice, while the staff of the Bodleian showed exemplary patience and good humour, even though for much of the period of preparation one floor or other of the stacks was closed for renovation. I am grateful to the librarians who have taken trouble in answering my queries about photographs: Vicki Denby, Roger Stoddard, and Jennie Rathbun (Houghton); Karen Kearns and Thomas V. Lange (Huntington); Julia Blanchard (Pierpont Morgan); Vincent Giroud (Beinecke); Anne Muchoney (Folger); Maureen Townley (National Library of Scotland).

Finally my thanks go to David Foxon, who expressed his anxieties about collaboration in the introduction to his catalogue of *English verse 1701–50*, but has given me every possible assistance and proved himself the most sympathetic and helpful of collaborators.

I

POPE'S EARLY RELATIONS WITH THE BOOK TRADE

INTRODUCTION: THE MEANING OF THE IMPRINT

When I started to plan these lectures I thought I would have to begin by apologizing for spending so much time talking about the book trade rather than about Pope. But, as I have worked on Pope, I have discovered that with all the documentary evidence that survives, added to the bibliographical facts, it has been possible to find a whole series of patterns that are of interest. A good deal of hypothesis inevitably enters into such work, and I make no claim that all I say will stand up to further research, nor can I give all the evidence in these lectures. But even a superficial account leaves little time for talking about the general practices of the book trade except in relation to particular aspects of Pope's work.

I must, however, start by saying something about the meaning of the imprint. By the Licensing Act of 1662 and, when that lapsed, by an ordinance of the Stationers' Company of 1681, every piece of printing had to bear 'the name either of the printer or of a bookseller with a shop in London or the suburbs'.[1] How closely this was enforced I do not know, but I have the impression that a similar provision of the Stamp Act of 1712 led to tighter control. This Act, quite apart from its requirement that newspapers and broadsides should be printed on stamped paper, laid a duty of 2s. a sheet on all pamphlets, which it defined as works not exceeding 20 sheets in folio, 12 in quarto, or 6 in octavo—roughly, works under 100 pages in length. These duties remained in force into the nineteenth century and were not repealed until 1833. All pamphlets had to be taken to the Stamp Office within a week and the duty paid, and an imprint was clearly necessary for enforcement; a penalty of £20 was laid on anyone selling a work without one.[2] I suspect this legislation, by putting the matter in the hands of the

---

[1] C. Blagden, *The Stationers' Company* (1960), p. 163. Dates correspond to a year beginning 1 Jan.; where a document uses the old style (beginning 25 Mar.), both years are given.

[2] For an account of the Stamp Act and its consequences, see D. F. Foxon's 1978 Sandars Lectures, 'The Stamp Act of 1712', deposited in the Cambridge University and British Libraries. The Stamp Duty Ledgers were destroyed a century ago, thus depriving bibliographers of an invaluable source.

excise, was far more consistently enforced than previous regulations. From now on, any work with an imprint as unspecific as 'London: printed in the year 1715' must be suspected of being surreptitious or printed for private circulation.

The fullest form of an imprint is one which names three people, or groups of people:

London: printed by *X* (the printer), for *Y* (the bookseller who owned the copyright), and sold by *Z*.

In the eighteenth century the printer's name is rarely given, at least in works printed in London, and the form is more commonly:

London: printed for *Y*, and sold by *Z*.

Very often in this period, and particularly for pamphlets, it is further abbreviated to:

London: printed and sold by *Z*.

It is this last form which is my present concern. *Z* is usually what the eighteenth century called 'a publisher',[3] or one who distributes books and pamphlets without having any other responsibility—he does not own the copyright or employ a printer, or even know the author. This final form of the imprint is often misleading because in the process of abbreviation the punctuation makes it appear that *Z* is also the printer; in the provinces this may be so, and occasionally with minor London printers, but in the regular London trade *Z* is never the printer.[4]

We can distinguish four cases in which the formula 'and sold by' is used, though they may sometimes overlap. The first, illustrated from Thomas Creech's translation of *De rerum natura* (Figure 1), is, I think, the outward sign of a practice which much more often lies concealed. Cyprian Blagden has shown that in addition to the practice of inviting public subscriptions to a book, which we shall deal with in relation to Pope's *Iliad* and *Odyssey*, there had grown up by the early eighteenth century a practice by which a bookseller invited other members of the trade to subscribe for blocks of copies of a new or newly reprinted book before its publication, a practice which still goes on.[5] The booksellers who subscribed paid less than the normal trade price, while with any luck the bookseller who made the proposal would receive enough money from the subscribers to meet his bills for paper and printing when they fell due; the copies that remained to be disposed of over the years would then represent clear profit. Usually there is no sign of this arrangement in the imprint, but I think Creech's Lucretius may be an example of

[3] M. Treadwell has given an excellent account of these publishers, 'London trade publishers 1675–1750', *Library*, 6th ser. iv (1982), 99–134. As he explains in a preliminary note, there was once a plan that he should collaborate with Mr Foxon and Dr D. F. McKenzie on a paper on this topic.

[4] Sometimes the imprint is clarified by the punctuation, 'printed: and sold by'. John Watts, who played both roles, uses the imprint 'London: printed by John Watts, and to be sold at the Printing Office'.

[5] C. Blagden, 'The memorandum book of Henry Rhodes, 1695–1720 II', *BC* iii (1954), 110–13. I think Blagden is mistaken in thinking that those binders (like Cholmley, Fayram, and Woodward in his examples) who collected subscriptions were acting as publishers.

LONDON:
Printed by J. MATTHEWS for G. SAWBRIDGE, at the *Three Golden Flower de Luces* in *Little Britain*; and fold by *J. Churchill* and *W. Taylor* in *Pater-Nofter-Row*; *J. Wyat*, and *R. Knaplock* in St. *Paul's* Church Yard; *R. Parker, G. Straban*, and *J. Phillips* near the *Royal Exchange*; *B. Tooke* and *R. Goflin* in *Fleetfireet*; *J. Brown* without *Temple Bar*; *J. Tonfon* in the *Strand*; *W. Lewis* in *Covent-Garden*; *J. Harding* in St. *Martin's Lane*; and *J. Graves*, next Door to *White's* Chocolate Houfe, St. *James's*. MDCCXIV.

FIG. 1. Imprint listing subscribing booksellers: ordinary-paper copy of Lucretius, *Of the nature of things*, trans. Thomas Creech (1714), i. title-page (Bodl. Vet. A4 e. 1935; 30 × 89)

LONDON:
Printed by J. MATTHEWS, for G. Sawbridge, at the *Three Golden Flower de Luces* in *Little Britain*. MDCCXIV.

FIG. 2. Imprint listing printer and copyright-holder only: large-paper copy of Lucretius, *Of the nature of things*, trans. Thomas Creech (1714), i. title-page (National Library of Scotland; approx. 20 × 95)

OXFORD,
Printed for *Steph. Fletcher* Bookfeller in *Oxford*, and to be Sold by *H. Clements* at the *Half-Moon* in S. *Paul's* Church Yard, *London*. 1716.

FIG. 3. Imprint listing provincial bookseller and London agent: R. Newton, *A sermon preach'd at the consecration of the Hart-Hall chapell in Oxford* (1716), title-page (Bodl. 12 θ 1780(10); 24 × 85)

LONDON:
Printed for R. DODSLEY at *Tully's Head* in *Pall-mall*, and fold by *T. Cooper* in *Pater-nofter-row*, MDCCXXXVII.　　　　　[*Pric* 1 *s.*]

FIG. 4. Dodsley's customary imprint: Pope, *First epistle of the first book of Horace* (1737), title-page (Bodl. Vet. A4 c. 290; 36 × 142)

LONDON,
Printed: And Sold by *J. Roberts* at the *Oxford-Arms* in *Warwick-Lane*, 1726.　　Price 6 *d.*

FIG. 5. Publisher's imprint: Swift, *Cadenus and Vanessa* (1726), title-page (Bodl. G. Pamph. 1285(13); 19 × 120)

LONDON:
Printed for J. ROBERTS near the *Oxford Arms* in *Warwick-lane*. 1715.

FIG. 6. Publisher's imprint: Pope, *Key to the lock* (1715), title-page (Bodl. 12 θ 1905; 15 × 86)

Sold by *J. Roberts, J. Morphew, R. Burleigh, J. Baker*, and *S. Popping*. Price Three Pence.

FIG. 7. Imprint listing London publishers: Pope, *A full and true account of a horrid and barbarous revenge by poison, on the body of Mr. Edmund Curll, bookseller* (1716), title-page (BL C. 59. i. 5; 12 × 136)

the subscribers' names being included—the fine-paper copies (Figure 2) have the names of Matthews and Sawbridge alone.[6]

From this example of a trade practice that will concern us later, we can turn to publishing proper. The imprint in Figure 3 makes an obvious point: a provincial printer or bookseller had to have a London agent to supply not only the London

---

[6] My case is weakened by the fact that although we know that Henry Clements subscribed for Creech's trans. of Lucretius (12 copies on 15 Mar. 1714, and 50 more on 26 Jan. 1715), his name is not on the list. For examples where he subscribed and his name is present, see M. Hole, *A practical exposition of the . . . Catechism* (subscribed 27 Jan. 1715); J. Ogilby, *The traveller's guide* (30 Nov. 1711); and C. Wheatly, *The Church of England man's companion* (16 Mar. 1714), in *The notebook of Thomas Bennet and Henry Clements*, ed. N. Hodgson and C. Blagden (Oxford, 1956), pp. 74–5, 119–94. Perhaps in the case of the Lucretius, Clements's initial order of 12 was not big enough to give him a place in the imprint.

trade but also the provincial booksellers who normally depended on their London agents for supplies. The almost invariable Dodsley imprint (Figure 4) is a variation on this case; I doubt if Dodsley had warehouse space in Pall Mall for his stock, but it was in any case out of the way for the book trade, which was centred round St Paul's. Accordingly he used a major publisher in the city to hold and sell his books to the trade in London and the country.[7] Publishing of this sort was carried out for books by many of the leading booksellers, whose names appeared in the imprint as a guide to where one should go for supplies.

By contrast our third case, 'Printed: And Sold by J. Roberts' in Swift's *Cadenus and Vanessa* (Figure 5), brings us to those who specialized as publishers and dealt mainly with pamphlets and periodicals.[8] I think that there were two principal reasons for their existence. Whereas the regular trade in books was geared to a slow return as an edition was sold and exchanged through the trade, these works were ephemeral and largely sold through the pamphlet shops and by hawkers; as with our newsagents, much of the business was done on the basis of sale or return. In writing of publishers, John Dunton lays stress on the reliability of their bookkeeping, since they would normally have an edition delivered to them by the printer and subsequently account to him or to the proprietor for the copies sold.[9] As far as we can see, the publisher was normally paid a flat fee of say 1s. 6d. a hundred copies, and the pamphlets were charged to him at a trade price three-quarters of the retail price. The system and the discount seem to be exactly the same as for the publishing of magazines a century later.

The second reason for the existence of publishers was the anonymity they provided. In the days when the booksellers who did what we call publishing had shops, a customer could call on, for instance, Dodsley and ask who wrote such and such an anonymous work which bore his imprint—and if he were a good or influential customer, it would be hard for Dodsley to refuse an answer.[10] If it did not have his imprint, this situation could not arise. There is abundant evidence in the printers' ledgers of an author's controversial works being printed either at his own expense or by the order of his usual bookseller but issued through a publisher. This anonymity could also be a protection against legal action. John Morphew gave evidence in August 1714 that it was

a very usual thing for persons to leave books and papers at his House and at the Houses of other Publishers, and a long time after to call for the value thereof without making themselves known to the said Publishers, and if the Government makes enquiry concerning the Authors of any books or papers so left, in order to bring them to punishment, it often happens that nobody comes to make any demand for the value of the said Books.[11]

---

[7] The Coopers, like Dodsley, have a role to play in Pope's story. Mary Cooper took over from her husband Thomas on his death, 9 Feb. 1743.

[8] For an early discussion of these imprints in their various forms, see A. T. Hazen, 'The meaning of the imprint', *Library*, 5th ser. vi (1951), 120–3.

[9] Dunton, *The life and errors of John Dunton* (1705), p. 298 (Mr Nut) and pp. 342–3 (Mrs Baldwin).

[10] R. Straus, *Robert Dodsley, poet, publisher & playwright* (1919), pp. 135, 174.

[11] PRO SPD 35/1/28(29), 28 Aug. 1714, cited by J. Sutherland, *The Restoration newspaper and its development* (Cambridge, 1986), p. 219, and by L. Hanson, *Government and the press 1695–1763* (1936), p. 51.

The best definition of a publisher that I know, and one which summarizes these various points, is that given in J. Collier's *The parents and guardians directory* (1761), p. 233:

The Publisher advertises Pamphlets, enters them at the Stamp-office, folds, stitches them, and publishes them for such gentlemen as print them at their own expence, and for the booksellers who do not chuse to set their names to them. They also sometimes publish bound books for the booksellers, and some News-papers; they likewise purchase copies, and publish them at their own expence; but seldom or never take apprentices.

It follows from the nature of the publisher that any attempt to use his imprint in a pamphlet as evidence to identify an author is vain, since the imprint is designed to make that impossible. What is misleading to any clear-thinking bibliographer is that his imprint is likely to be 'Printed for J. Roberts' as in Pope's *Key to the lock* (1715) in Figure 6, instead of 'sold by J. Roberts'. Since publishers could own copyrights, the natural assumption is that a work with this imprint is one of those cases, but it is not so. In this case Lintot had paid fifteen guineas for the copyright, and was using Roberts to conceal Pope's identity and his own relationship with the work.[12] But this form of imprint was not reserved for controversial works: Edward Young's series of poems *The universal passion* (1725–8) were entered to Jacob Tonson in the Stationers' Register by Roberts but published by Roberts as 'printed for J. Roberts'. The only way to avoid being misled is to regard any imprint which says a book is printed for a publisher as meaning it is sold by him.

To do this we have to know who the main publishers were, and fortunately they were few in number. If you have the capacities of a Michael Treadwell, it becomes comparatively easy to trace the main firms involved in the trade, and the results are set out in Table 1. The important figures to notice are James Roberts and the Coopers, Thomas and Mary, who all published for Pope. You will see that the late 1720s produced a shortage of publishers, which was filled by my next category, the mercuries. But between 1714 and 1717 there were five publishers, and they are all listed on Pope's squib *A full and true account of a horrid and barbarous revenge by poison, on the body of Mr. Edmund Curll, bookseller* (1716) in Figure 7. Curll's riposte, John Oldmixon's *The Catholick poet* (1716), lists the same five but 'printed for'—adding 'and sold by all the booksellers in England, Dominion of Wales, and Town of Berwick upon Tweed' (Figure 8); I don't know why Scotland is omitted. Pope's *A further account of the most deplorable condition of Mr. Edmund Curll, bookseller* (1716) uses the generalized (and presumably illegal) phrase, 'publishers, mercuries, and hawkers' (Figure 9). It was these mercuries who took the place of the publishers in the 1720s; their names can be seen in Figure 10. Their chief centres were the Royal Exchange and Temple Bar, with westerly outposts at Charing Cross and Westminster Hall. As their name suggests, mercuries started as distributors of newspapers to the hawkers; in 1666, at

---

12 The 2nd and 3rd edns. of Swift's *Cadenus and Vanessa* (1726) change 'sold by J. Roberts' to 'printed for', and the latter seems to be the most common form of Roberts's imprint. D. F. Foxon, *English verse 1701–1750* (Cambridge, 1975), records 216 instances of 'printed for', as opposed to 61 of 'sold by'.

TABLE 1: Early Eighteenth-Century Publishers

Lines represent publishing businesses. A publisher's period running the business extends from the time marked by his / her name to that marked by the next name or the ending of the business.

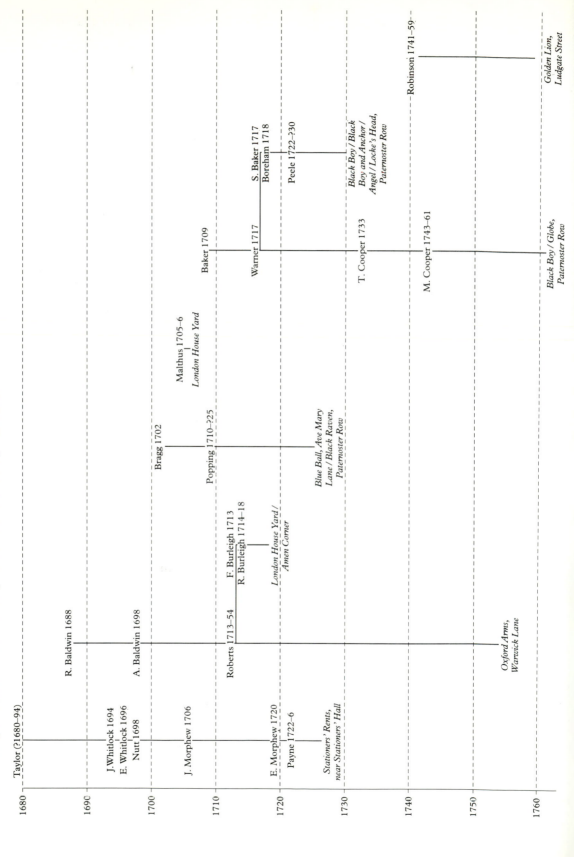

LONDON: Printed for *J. Morphew*, *J. Roberts*, *R. Burleigh*, *J. Baker*, and *S. Popping* ; and sold by all the Bookfellers in *England*, Dominion of *Wales*, and Town of *Berwick upon Tweed*. 1716.
(Price Three-pence.)

FIG. 8. Imprint listing London publishers: John Old-mixon, *The Catholick poet* (1716), p. 6 (Beinecke; approx. 16 × 129)

LONDON
Printed, and Sold by all the Pub-lifhers, Mercuries, and Hawkers, within the Bills of Mortality.
1716.

FIG. 9. Generalized publishers and mercuries imprint: Pope, *A further account of the most deplorable condition of Mr. Edmund Curll, bookseller* (1716), title-page (BL Ashley 1298; 27 × 70)

DUBLIN Printed:

LONDON Re-printed, and Sold by Mrs. NUTT, Mrs. COOK, and Mrs. CHARL-TON, at the *Royal Exchange*; Mrs. DODD at the *Peacock*, Mrs. CHARLTON at the *Golden Ball*, and Mr. SLOW, over-againft *Devereux Court*, without *Temple-Bar*; Mr. TAYLOR, Bookfeller, at the *Meufe Back-Gate*, over-againft *Duke's-Court*; and at the Pamphlet-Shops at *Charing-Crofs*, and *Weftminfter-Hall*. M.DCC.XXXV.

FIG. 10. Imprint listing London mercuries: *True taste: or, female philosophy* (1735), title-page (Bodl. Firth c. 9(8); 35 × 172)

LONDON:
Printed for *A. MOOR* near St. *Paul's Church*, and Sold by the Bookfellers. 1720.

FIG. 11. Pseudonymous imprint: Pope and James Craggs, *Duke upon Duke* (1720), title-page (Bodl. Pamph. 357(11); 45 × 135)

the very beginning of the newspaper trade, Mrs Elizabeth Andrews took 'a third to a quarter' of the copies of the *London gazette* that Newcombe printed.[13] Some continued to carry on a very large business and act in parallel to the publishers. In the legal proceedings about the publication of the *London journal* for 12 August 1721, one William Hewitt said he had served 'Mr. Dodd about a year and a half, the said Mr. Dodd's Wife being a Retailler of News Papers & Pamphlets commonly called a Mercury'. He was sent to fetch the papers and 'did for the use of his said Mistress fetch from Mr. Wilkins's Printing-House in Little Brittain a hundred and Eight Quires' or some two thousand copies of the paper.[14] The distinguishing feature of the mercuries was that they were usually women, often the wives, widows, or daughters of printers;[15] they seem to merge into the category of stall-holders or those who kept pamphlet shops, as indeed our present wholesale newsagents own shops and also supply the paper rounds. Thomas Mortimer writes in *A new and complete dictionary of trade and commerce* (1766):

[13] There are several references in the State Papers to the bookwoman for the *Gazette*; see J. Greenwood, *Newspapers and the Post Office 1635–1834* (1971), pp. 19, 22, 23 (numbering from beginning of ch. 1; pages unnumbered).

[14] J. R. Sutherland, 'The circulation of newspapers and literary periodicals, 1700–30', *Library*, 4th ser. xv (1934–35), 117. The *London journal* was a sheet and a half, and it is not clear if Hewitt uses the term 'quires' precisely. He could mean 2,700 copies or 2,700 sheets and 1,800 copies. Richard Goulden tells me there were two Anne Dodds, mother and daughter; the mother died in 1739. I suspect neither should be identified with the Mrs Dodd in Thomas Gent's *Life*, ed. J. Hunter (1832), pp. 134, 136–7.

[15] Note that Dunton groups Abigail Baldwin with his 'Honest (MERCURIAL) Women' (*Life and errors*, p. 316). Elizabeth Nutt by contrast stepped down the scale after her husband's death, but he had handed over the publisher's business to John Morphew.

This business is principally carried on by women, who keep little shops in the most popu-
lous parts of the town; as at the Royal Exchange, Temple-Bar, Charing-Cross, &c. They
sell all kinds of news-papers, king's-speeches, Votes of the House of Commons, single
plays, pamphlets of all sorts, almanacks, memorandum books, or indeed any book, if bespoke; though they keep but few bound ones by them.

One distinction between publishers and mercuries is that the latter did not as a
rule own copyrights; though I suppose that the widow of a stationer might have
that right, the trade sales catalogues specifically exclude stall-holders from those
who might attend them. It is accordingly impossible for the imprint 'Printed
for A. Dodd' to indicate that she is the copyright holder in her first appearance,
on Swift's *Part of the seventh epistle of the first book of Horace imitated* (1713). In
fact we know that the copyright had been entered to John Barber in the
Stationers' Register by the publisher John Morphew. Why he passed it on to Mrs
Dodd is another question. The same imprint appears in the early editions of the
Dunciad, and can mean no more than it does in the imprints of James Roberts—
sold by.

The anonymity conferred by the use of publishers and mercuries was only
achieved at the expense of their being frequently arrested and interrogated, and it
is perhaps for that reason that around 1714 a new form of imprint is found which
should perhaps form a separate category of pseudonymous imprints. Figure 11 is
from *Duke upon Duke*, a highly libellous ballad written by Pope and James
Craggs and 'Printed for A. Moor'. As Bookweight, the bookseller in Fielding's
The author's farce (1729), says,

The study of bookselling is as difficult as the law: and there are as many tricks in the one
as the other. Sometimes we give a foreign name to our own labours, and sometimes we
put our names to the labours of others. Then as the lawyers have John-a-Nokes and
Tom-a-Stiles, so we have Messieurs More near St. Paul's, and Smith near the Royal
Exchange.[16]

I suspect these names have some origin in reality, but there is no doubt that they
were freely used to satisfy the requirement of the Stamp Act that there should be
a name in the imprint. We shall find further examples of misleading imprints and
misleading copyright entries in the Stationers' Register, but they are exceptional;
those we have been looking at are the rule.

Imprints Illustrated: The Career of
William Lewis

We can usefully illustrate some of the variations and ambiguities of the imprint
from the career of William Lewis, Pope's Roman Catholic school-fellow, to

---

[16] *The complete works of Henry Fielding* (1903), viii. 222. For examples of Smith and Moore in imprints,
see Foxon, P708 (1727) 'sold by Messieurs Smith & Moore', and S717 (1727) 'sold by J. Smith, & A.
Moore'.

whom he entrusted his first separate publication, the *Essay on criticism* of 1711.[17] He was a retail bookseller in Russell Street, Covent Garden, and his only obvious connection with Pope was his publication of the *Essay*, which is the only work he ever entered in the Stationers' Register, but we can find evidence that the relationship between them was deeper than this. Lewis's name first occurs in imprints in 1709, when we find it appearing on Robert Gould's *Works*, brought out in two volumes by his widow, and on four republished plays: Behn's *The rover*, Vanbrugh's *The provok'd wife*, Wycherley's *The plain-dealer*, and Jonson's *The alchemist*. His name is coupled in the imprints of these plays with Tonson's or Wellington's, 'printed for Richard Wellington: and sold by William Lewis', and his role is clear: all four plays had been recently acted at the Theatre Royal in Drury Lane and Lewis, with a shop just off Drury Lane, was in the ideal place to sell them. The connection with the theatre is a consistent strand in Lewis's career and it doubtless combined with his relationship with Pope to bring him into contact with John Watts, who, as we shall see, was Pope's first printer. Four of Watts's earliest ventures as a proprietor, the *Letters of Abelard and Helouise* (1713), *The Cid*, *Electra*, and George Bell's *The divinity of our Lord Jesus Christ* (all 1714), have the imprint 'printed for J. Watts [or J. W.]; and sold by W. Lewis' with the names of between five and eight other vendors added. The imprint shows that Watts was the originator of these projects, but Lewis's name can only have been placed before those of senior members of the trade because he had a key role as retailer.

In 1710 we find Lewis's name among those for whom the great *Historical and critical dictionary* of Pierre Bayle was printed. This was one of the major projects of Jacob Tonson—of whom more will be said shortly—and it was published in four large folio volumes with thirteen names on the imprint, arranged as usual by their seniority in the trade.[18] The last two are Bernard Lintot (later to become Pope's bookseller) and William Lewis, much the youngest partner. By contrast with Creech's Lucretius, which I illustrated above as sold by fourteen booksellers, including Lewis, this was not a work to which the trade subscribed before publication but one in which the partners took shares in the copyright and the costs of production—no doubt refunding their share of the heavy expenses Tonson had incurred over the years. Accordingly the imprint is 'printed for' the thirteen. I wonder whether Lewis's share in this major investment which brought him into contact with senior members of the trade was not the result of Pope's influence with Tonson.

In the following year, four weeks before he issued the *Essay on criticism*, Lewis's name appears as one who sold the *Spectator*,[19] another sign of his accept-

[17] J. Nichols, *Literary anecdotes of the eighteenth century* (1812–14), iii. 6; viii. 168. Lewis's correspondence with Hearne suggests he may have had a special interest in antiquarian books (Bodl. MS Rawl. lett. 110, fos. 102–6).

[18] C. Harper (freed 1667); D. Brown (1670); J. Tonson (1677); A. & J. Churchill (1681 and ?1687); T. Horne (?1689); T. Goodwin (1682); R. Knaplock (1689); J. Taylor (?); A. Bell (1695); B. Tooke (1695); D. Midwinter (1698); B. Lintot (1700); W. Lewis (?).

[19] *Spectator* for 17 Apr. 1711. Lewis's address is 'under Tom's Coffee House, Covent Garden'.

ance in literary circles. The *Essay* itself has two versions of the imprint (Figure 12). Bibliographers have argued about their priority, but they are merely variants in standing type. Though one must have been printed earlier than the other in the press run, there is no reason to believe there was any priority in their publication. As a new and little-known bookseller, it was clearly wise for Lewis to join better established booksellers with him: Taylor and Osborn were well known, and John Graves, next door to White's Chocolate House in St James's Street, was well placed for the growing west-end trade later served by Pope's booksellers, Dodsley, in Pall Mall, and Brindley, in Bond Street. I am tempted to believe that these three subscribed for copies, or at least agreed to publish, buying them at a cheap rate; it might explain how Narcissus Luttrell was able to mark the price of his copy (BL 161. m. 24) 6*d.* as opposed to the advertised price of 1*s.* The earliest advertisement in the *Spectator* of 15 May 1711 (and a later advertisement on 1 January 1712 which omits Taylor, Osborn, and Graves) adds to the imprint 'sold by . . . J. Morphew, near Stationers-Hall'. Morphew, that is, was the real publisher who distributed copies to the pamphlet shops, and also fulfilled his function by entering the copyright to Lewis in the Stationers' Register on 11 May 1711. Not surprisingly (as with Creech's Lucretius, 1714) the only fine-paper copy I have found has the short imprint which would be used in the copies sold by Lewis himself.

In the following year Lewis published a poem by Pope's friend, the Catholic priest Thomas Southcote, *Monsieur Boileau's epistle to his gardiner.*[20] Thereafter he published occasional poems, plays and treatises, and in 1723 he became the regular bookseller for the poet Hildebrand Jacob; but, except for this association, all but one of the poems appearing with his name on the imprint after 1723 have some connection with Pope. Of the first we would know nothing but for Curll, who in *Mr. Congreve's last will and testament* (1729) reprints Congreve's *Letter to Viscount Cobham* and notes, 'The following EPISTLE . . . is here printed from a Manuscript of the Author . . . The Public having been notoriously abused, by a very erroneous Copy, surreptitiously obtained by one LEWIS in *Covent-Garden*, and vended under the cover of A. DOD and E. NUTT' (p. 24). This is a good illustration of the use of mercuries to conceal the source of a work; anyone seeing Lewis's name might well associate him with Pope, and I think that it is at Pope that Curll is pointing. Pope could well have been the source of the manuscript copy of Congreve's poem, and would not have been pleased to have it called erroneous. In 1736 Lewis published *Sannazarius on the birth of our saviour*, which according to Maurice Johnson's copy[21] was written by Edward Walpole of Dunston and 'perused & corrected by Alexr Pope esqr'; at this time Pope was on bad terms with his bookseller Gilliver, and it is likely that he suggested Lewis should be used.

---

[20] Pope had a longstanding friendship with Southcote; in 1728 he helped arrange for him to be given an abbey in France. See J. Spence, *Observations, anecdotes, and characters of books and men*, ed. J. M. Osborn (Oxford, 1966), anecdotes 69 and 70, and app., p. 615.
[21] Sold at Sotheby's, 23 Mar. 1970, lot 132; I have not seen it.

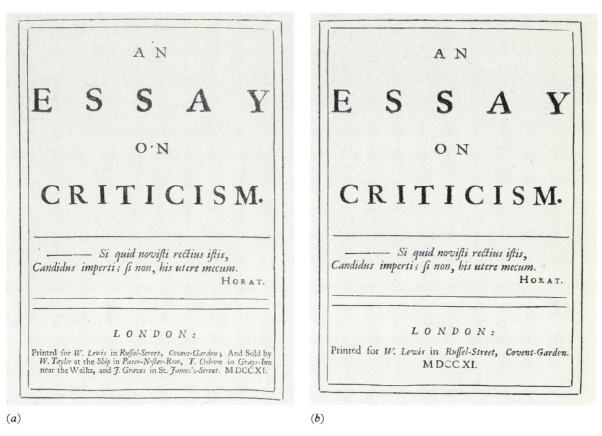

(a) *(b)*

FIG. 12. Variant title-pages of *Essay on criticism* (1711): *(a)* with imprint listing publishers or subscribing booksellers (Bodl. Don. e. 70; 219 × 164); *(b)* with imprint listing copyright-holder only (Beinecke; approx. 219 × 169)

Finally, we have a case where we might suspect a bookseller of giving (in Fielding's words) 'a foreign name to our own labours'. The first two editions of Cibber's famous attack, *A letter from Mr. Cibber to Mr. Pope*, of July 1742, have the imprint 'London, printed: and sold by W. Lewis in Russel-Street, Covent-Garden'; Cibber's *Second letter* of February 1743 is 'Printed for A. Dodd' but *Another occasional letter from Mr. Cibber to Mr. Pope* of January 1744 is again 'sold by W. Lewis'.[22] This could be a joke at the expense of Pope and his bookseller friend, but Cibber turns out to have been one of Lewis's theatrical clients. Late in his career he started to publish at his own expense: he entered his *Apology for the life of Mr. Colley Cibber* in the Register himself on 19 March 1740, and it was published with the imprint 'printed by John Watts for the author: and sold

[22] The undated 3rd edn. of the original letter and a reprint of this appeared with the imprint 'Glasgow: printed for W. Macpherson' which is clearly false.

by W. Lewis'; *The egoist* was similarly entered in the Register (on 10 January 1742) and bears the imprint 'London printed: and sold by W. Lewis'. Three more late Cibber works are also connected with Lewis by their imprints, and I conclude he was Cibber's choice to handle these publications. Cibber had spent much of his career at the Drury Lane theatre, where he had been one of the actor managers, and Lewis, because of the position of his shop, may well have been the bookseller he knew best. These changes of allegiance may leave us with a sense of insecurity, but I would like to think that *Pope's ghost: A ballad. To the tune of William and Margaret*, of October 1744, in which Pope's ghost returns to reproach Cibber in imitation of David Mallet's ballad, was indeed printed for W. Lewis as its imprint says.

TONSON AND WATTS

In the first half of his publishing career, up to the *Dunciad* of 1728, Pope dealt with established booksellers and printers, Tonson or Lintot, Watts or Bowyer, and for the first five years he depended on selling his copyrights in order to gain an income from his writing. For his Homer, starting with the *Iliad* from 1715 to 1720, he turned to publication by subscription, and thereby made his fortune; but even in the very earliest years his dealings with the trade show him taking an active and innovatory role.

The printers and booksellers of this early period are all well known with the exception of John Watts. Of him, Nichols wrote,

The fame of Mr. John Watts for excellently good printing will endure as long as any public library shall exist. The duodecimo editions of Maittaire's Classicks, 'ex officina Jacobi Tonson & Johannis Watts', would alone have been sufficient to have immortalized his memory, both for correctness and neatness. But there are many works of still higher importance; Clarke's Caesar for example; and several beautiful volumes of English Classicks, in quarto. (*Anecdotes*, i. 292 n.)

All the works mentioned have a place in our story; but, despite the claim that Watts's fame will endure, his reputation is lost. The books he should be famous for are discussed by Updike under the name of his partner Jacob Tonson;[23] and even the Bodleian has relegated his great Caesar from the distinction of an Arch class-mark to a mere number. His early career remains in obscurity, despite three recent biographies of his partner Tonson.[24]

It must be said that many of the books published by Tonson in the seventeenth century are undistinguished specimens of English printing, no doubt because the

---

[23] D. B. Updike, *Printing types*, 2nd edn. (Cambridge, 1937), ii. 135.

[24] G. Papali, *Jacob Tonson, publisher* (Auckland, 1968), a revision of a doctoral thesis of 1933; H. M. Geduld, *Prince of publishers* (Bloomington and London, 1969); K. M. Lynch, *Jacob Tonson, Kit-Cat publisher* (Knoxville, 1971).

standards of the trade were low.[25] But in 1698 Tonson commissioned the first work of the Cambridge University Press, the quarto editions of Horace, Virgil, Terence, and Catullus, Tibullus, and Propertius.[26] These were published by subscription in 1699 and 1702, and set a new standard. We do not, of course, know whether they represent the fulfilment of Tonson's own long held wishes, or result from ideas pressed upon him by the influential writers and noblemen with whom he was now associating in the Kit-Cat Club. What is clear is that his ambition for the making of fine books was fired, and he began to plan his edition of Caesar, of which Addison wrote to Leibniz, 'He intends to spare no cost in the Edition of this book which will probably be the noblest Volume that ever came from the English press'.[27] I think that the establishment of a printing house with John Watts can be associated with this project, though it may be that Tonson was equally aware of the advantages of having a printing business now that he had influential sponsors who could obtain government work and appointments for him—as indeed they did.

Tonson's Caesar was finally published in 1712, but work must have started on it a long time before that. It could well be that Tonson's trip to Holland with Congreve and Charles Main in August 1700 was concerned with preliminary work: each plate of the Caesar was to be dedicated to a subscriber, in keeping with the tradition of Ogilby's and Dryden's translations of Virgil, and the plates were to be engraved in Holland. We know subscriptions were being collected as early as the winter of 1702–3, and work started on the engravings when Tonson returned to Amsterdam in May 1703, as is clear from the correspondence of Addison and Vanbrugh.[28]

If we look at the books printed for Tonson in the years 1702 and 1703, seven dated 1703 use distinctive condensed swash italic capitals for titling, and this style is a feature of John Watts's printing in his celebrated period. Figure 13 shows the title-page of a copy of Rowe's *Fair penitent* (1703) and, for comparison, the section-title for *The shepherd's week* from Gay's *Poems on several occasions* (1720), which we know was printed by Watts. This type with its lower case and roman is used by other London printers,[29] but what is distinctive about Watts is

---

[25] Papali, *Jacob Tonson*, p. 51, identifies Elizabeth and Miles Flesher, Thomas Newcomb, Henry Hills, Robert Everingham, J. Heptinstall, Thomas Warren, and William Bowyer as printers who worked for Tonson. John Chalmers privately adds the names of Mary Clark, T. Hodgkin, I. Dawks, and Edward Jones.

[26] See D. F. McKenzie, *The Cambridge University Press* (Cambridge, 1966), i. 15.

[27] *Letters of Joseph Addison*, ed. W. Graham (Oxford, 1941), p. 43 (10 July 1703). It was originally to have been dedicated to James Butler, Duke of Ormonde, but in Dec. 1709 the Kit-Cat Club ordered Tonson to make Marlborough the dedicatee (Lynch, *Jacob Tonson*, p. 47).

[28] Lynch, *Jacob Tonson*, pp. 109–10. Nichols says that Tonson was also procuring paper for the edn. in 1703 (*Anecdotes*, i. 295), while E. R. Mores records a letter by Thomas Jones of 27 July 1710 suggesting Tonson was buying type 'not long ago' (*A dissertation upon English typographical founders and foundries* (*1778*), ed. H. Carter and C. Ricks (Oxford, 1961), p. 54). Tonson visited Amsterdam in 1700, 1703, and 1707; for this last visit, see I. H. van Eeghen, *De Amsterdamse boekhandel 1680–1725* (Amsterdam, 1960–78), i. 124.

[29] By Henry Hills in an undated broadside, *A new years gift. Sir Matthew Hale ... The sum of religion,*

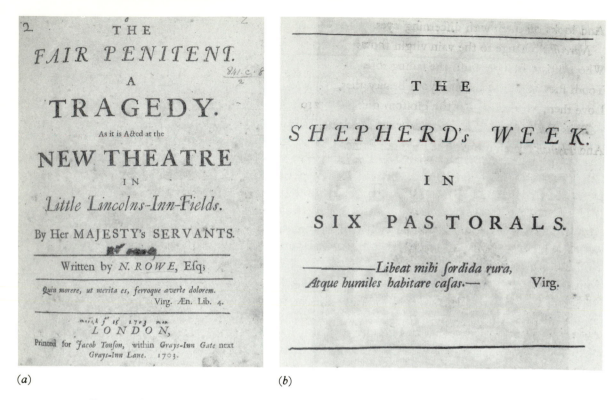

(a)　　　　　　　　　　　　　　　(b)

FIG. 13. Condensed swash italic capitals typical of John Watts: (*a*) Rowe, *The fair penitent* (1703), title-page (BL 841. c. 8(2); 218 × 155); (*b*) Gay, *Poems on several occasions* (1720), i. 65 (BL 83. k. 21; 290 × 222)

his use of the letterspaced italic capitals alone in titles, and I think we can identify his printing for Tonson at the end of 1702 by this means. There may, of course, have been occasional imitators, and I know of one interesting instance: the words 'Alexander Pope' are printed in this type by Bowyer on the title of his *Works* of 1717, though they are not found elsewhere in the volume; possibly Pope himself asked for this.

The books of 1703 in which these distinctive italic titles are found are Blackmore's *A hymn to the light of the world*, Congreve's *The mourning bride* (third edition) and *The tears of Amarylis for Amyntas*, Rowe's *The fair penitent* and *Tamerlane* (second edition), and volumes IV and V of *Plutarch's lives* (volumes II and III are the work of two other printers). Of these, the earliest is Blackmore's

and by Richard Janeway in [James Smallwood], *A congratulatory poem to his grace the Duke of Marlborough* (1704). The italic in *The fair penitent* appears to be the companion to the Two-line Great-Primer no. 2, in Mores's specimen of John James's foundry (*A dissertation*, p. ²29). Carter in his notes says that it is based on Garamond's Gros Canon and suggests it came from the Grovers' foundry (p. 118); Mores in his list of matrices belonging to the Grovers gives under 'Title letters and irregulars' 2-line Great Primer Roman (86) and italic (68) (p. 47). For a clear presentation of the list, see T. B. Reed, *A history of the old English letter foundries*, rev. and enlarged A. F. Johnson (1952), p. 191.

poem, which was dated in the previous year, 7 December 1702, by Narcissus Luttrell (Foxon B252); it is perhaps an inauspicious beginning, since the drop-head title on p. 16 'The cartons of Raphael Urbani, in the gallery at Hampton-Court' originally read 'chartones', and a printed cancel-slip had to be pasted over the word—it could, of course, have been Blackmore's error.

There is a problem, though, in dating the start of Watts's association with Tonson in late 1702. John Watts is almost certainly the son of William Watts, deceased victualler of St Martin-in-the-Fields, who was apprenticed to Robert Everingham on 3 October 1698.[30] Everingham was one of Tonson's regular printers, so there is an easy explanation of how Tonson spotted a promising young man to start a printing house for him. But in late 1702 John Watts had served only four years of his seven-year apprenticeship; he was not made free—by his elder brother William, Everingham having died in 1705—until 9 June 1707, two years after his service had been completed, though this delay does not seem to have been out of the ordinary.[31] If we accept the type as evidence of Watts at work, we have to ask whether he started to work for Tonson in Everingham's shop with his own supply of types or whether he was set up in business separately while still formally apprenticed, perhaps under the auspices of his brother. The latter solution would be the more probable if Nichols is correct in saying John Watts was eighty-five when he died on 26 September 1763 (*Anecdotes*, iii. 739), which would mean that he was not apprenticed until he was twenty and that he was twenty-four in 1702.

What we do know is that Tonson was listed as a printer in Bow Street, Covent Garden in October 1705,[32] when Watts's apprenticeship should have been complete, and from 1706 onwards we find the names of Tonson and Watts in the rate books of St Paul's, Covent Garden as occupants of a house where Bow Street police station now is. It must have been a large house, for it is one of the three on the east side with the highest rate of 6s.—another of the three was that of Grinling Gibbons four doors up.[33] There is no doubt that this house in Bow Street was a home for Tonson as well as a printing office, for a poem by Nicholas Rowe, 'The reconcilement between Jacob Tonson and Mr. Congreve', which appeared in Oldmixon's *Muses Mercury* for March 1707, makes Tonson say,

[30] D. F. McKenzie (ed.), *Stationers' Company apprentices 1641–1700* (Oxford, 1974), no. 1435.

[31] Michael Treadwell tells me that probate was granted to Everingham's widow on 7 Jan. 1706 (PRO PROB 6/82/7). Watts bound his first apprentice, Benjamin Young, on 9 Feb. 1708 (D. F. McKenzie (ed.), *Stationers' Company apprentices 1701–1800* (Oxford, 1978), no. 8599).

[32] H. Snyder, 'The reports of a press spy for Robert Harley', *Library*, 5th ser. xxii (1967), 337, quotes Clare's report for 10–17 Oct. 1705; the name is given as 'Mr. Thompson', but John Chalmers's suggestion that this is an error for Tonson must be correct; there is no Thompson in Bow Street.

[33] The rate books show the house under the name of Thomas Dutton in 1704; empty for half a year; under Samuel Walker for 1705–7. But in 1706 the collector has substituted John Watts's name for Walker's, and in 1707 Mr Tonson's. In 1708 Watts's name is inserted in place of Walker's in the ledger, and from then until 1717 Watts's name appears in all the records. The documents are in the Victoria Library, City of Westminster: overseer's accounts (H482–95); poor-rate collector's books and ledger (H5–25).

> What if from *Van*'s [Vanbrugh's] dear Arms I shou'd retire,
> And once more warm my *Bunnians* at your Fire;
> If I to *Bowstreet* shou'd invite you home,
> And Set a Bed up in my dining Room,
> Tell me, dear Mr. *Cong*[*reve*], wou'd you come?[34]

It may not be a mere coincidence that Will's Coffee House was on the opposite or west side of the street, on the corner of Russell Street, and though it had declined from its former status by the time Dryden died, there seems no doubt that there Pope met such wits of Dryden's circle as Wycherley, Walsh, and Congreve.[35] About 1712 Addison established Button's Coffee House on the opposite corner and the wits moved across the road. It was here that Steele set up the Lion's Mouth, which received contributions to the *Guardian*—very well placed for the printing house up the road; conversely, Tonson may have been glad to have his authors near at hand, if only to read proof. Beside the house was a passage to Earl's Court (later Duke's Court) and an announcement in the *Daily courant* in May 1710 refers to Mr Salkeld's 'next Door to Mr Tonson's Printing House in Earls Court in Bow-Street near Covent Garden'.[36] There is no trace of a separate establishment in Earl's Court in the rate books of St Martin-in-the-Fields (the parish boundary ran immediately behind the Bow Street houses), and the printing probably occupied much of the house and the yard behind.

Our first evidence of Watts apart from the rate books and his types (and from 1708 the ornaments which appear in his books[37]) comes from the pivotal years of his career, 1711 to 1713. On 9 April 1711 he appeared in the Stationers' Register for the first time, entering the copyright of Prior's *To the the right honourable Mr. Harley, wounded by Guiscard* to Jacob Tonson (elder or younger not specified). On 19 July Pope wrote to Caryll that 'Tonson's printer told me he drew off a thousand copies of his first impression' of the *Essay on criticism*; this suggests that Watts was in a subordinate position still, particularly by contrast with Pope's letter to the younger Tonson in February 1724 asking 'that you will use your interest with Mr Watts' about the printing of the *Odyssey*.[38] But the situation may have been about to alter soon. In the following year, 1712, the great Caesar was published 'sumptibus et typis Jacob Tonson', and it could be that with this work accomplished Watts had earned a partnership, for the Maittaire classics from 1713 bear the imprint 'ex officina Jacobi Tonson & Johannis Watts'. In the same year, 1713, Watts was called to the livery of the Stationers' Company.

This partnership in Bow Street lasted for four years; then in 1717 Jacob Tonson the elder retired to the country. On 29 October Jacob Tonson junior exe-

---

[34] For the full poem and further discussion, see M. Treadwell, 'Congreve, Tonson, and Rowe's "Reconcilement"', *NQ* ccxx (1975), 265–9.

[35] *Correspondence of Alexander Pope*, ed. G. Sherburn (Oxford, 1956), i. 11 (Pope to Wycherley): 'I have now chang'd the Scene from the Town to the Country; from *Will's* Coffee-House to *Windsor* Forest.'

[36] Quoted in *The correspondence of Richard Steele*, ed. R. Blanchard (Oxford, 1941), p. 258.

[37] First seen in John Philips, *Cyder* (1708).

[38] *Correspondence*, ii. 217; the date, which is Sherburn's, may be a little early.

cuted a contract with Henry Huddle and Thomas Perry of London, carpenters, for 'compleatly repairing in good & Workmanlike manner a house Scituate in Bow Street lately in the possession of Jacob Tonson the Eldr & John Watts'.[39] It seems clear that the work involved restoring the building to a fashionable dwelling house; much of the work is to be done 'in a handsom manner'. It corresponds with an entry in the rate collector's book for 1717, where John Watts is recorded as 'Gone'; in 1718 there is a note '2 quarters Empty', and it was then taken over by Mr Charles Winyates. The reason for the change is no doubt the elder Tonson's retirement to the country; he transferred all his business and his copyrights to his nephew on 17 September 1718. The younger Tonson lived in Bow Street later, but that seems to be another story.[40]

Jacob Tonson the younger complicates our attempts to work out the relationship between Watts and Tonson senior. It is clear that Watts and the elder Tonson occupied the Bow Street house, and it is reasonable to surmise that Tonson set him up in business there and later took him into partnership; hence their joint ownership of the house by 1717. But it was the younger Tonson who 'printed' the *London gazette*, taking over from the widow of Edward Jones of the Savoy on 1 March 1708, since Steele wrote 'on behalf of my friend Jacob Tonson Junr Printer of the Gazette' to James Brydges on 25 July 1711 in a vain attempt to stop the privilege being transferred to John Barber and Benjamin Tooke (*Correspondence of Sir Richard Steele*, p. 48). Kathleen Lynch is probably right in saying that 'most, if not all, of these lucrative but heavy assignments of government publications to the Tonson press were delegated by Tonson the Elder to his junior partner' (*Jacob Tonson*, p. 111); but it was doubtless Watts who did the work.

With the departure of Watts from the Bow Street house at the end of 1717 and the transfer of business from the elder to the younger Tonson, Watts gained a further independence; but the partnership between him and the elder Tonson was replaced by a new partnership with the younger. This partnership must have carried on the government printing if nothing else, and it is possible they had a separate printing house in Covent Garden for this sort of work, as William Strahan did later in the century.[41] But by 1719 Watts had begun to own copyrights,[42] particularly of plays, and appears in their imprints 'at the Printing-Office in Wild-Court near Lincoln's-Inn-Fields'; this is where Benjamin Franklin worked for him in 1725, describing it as a 'still greater Printing House' than

[39] BL Add. MS 28275, fos 40–1. I owe my knowledge of this to a transcript by Sarah L. C. Clapp, made available to me by John Chalmers.

[40] For transference of copyrights, see Lynch, *Jacob Tonson*, p. 17. Bodl. MS Eng. misc. b. 45 records payment of rents on the Bow Street house by Tonson the younger to the Duke of Bedford on 17 Apr. 1726 (fo. 3ᵛ) and 22 June 1737 (fo. 13ʳ) and his refitting of the house in 1736 and 1737 (fos. 8–9 and later).

[41] Negus's list of London printers compiled in 1724 records 'Watts and Tonson, Covent Garden' among those 'well affected to King George' (Nichols, *Anecdotes*, i. 292). J. A. Cochrane discusses Strahan's separate printing house in *Dr. Johnson's printer* (1964), p. 128.

[42] The first entry to Watts in the Register is of a half-sheet essay paper, *The rhapsody*, entered by John Morphew on 1 Jan. 1712; it was followed over the next few years by 3 more works, but the extensive entries start on 17 Mar. 1719.

Samuel Palmer's, and it remained his address until it was taken over by Richard Hett junior.[43] He sold a large batch of copyrights to Thomas Lownds in 1758 and died on 26 September 1763, aged eighty-five.[44] But despite the independence Watts gained, there seems no doubt he continued to print most if not all of the Tonsons' books throughout his long life, and he is therefore of great importance as a printer for English literature as well as for his typography.

And so at last to Pope, though we shall be back to Watts's typography very soon.

POPE'S 'PASTORALS' AND TONSON'S *MISCELLANIES*

Pope first appeared in print in the sixth and last volume of the series of poetical miscellanies produced by Jacob Tonson in 1684 and 1685, 1693 and 1694, 1704 and 1709. Though the later editions announce them as 'Publish'd by Mr. Dryden', Dryden does not seem to have had full responsibility for the four volumes published in his lifetime (he died in 1700), though he wrote the preface to the second volume and the dedication to the third.[45] The most obvious feature of these first four miscellanies is that they all open with about a hundred pages of Dryden's verse, the first volume reprinting *MacFlecknoe*, *Absalom and Achitophel*, and *The medall*, while the others print his translations; there is a parallel here to the Pope–Lintot miscellany, which becomes in later editions a collection of Pope's verse with the addition of work by other authors. I suspect Dryden's primary function in the first four volumes was to provide a core of his own work to which could be added verse by lesser authors. The later miscellanies are likewise built around the work of favoured authors, of whom Pope was one; it may be that the long gaps in the publication of the six Tonson miscellanies were due to the shortage of such materials.

Two other common features of these volumes strike the bibliographer. The first is the use of title-pages within the volume; while there is no real consistency in their use, they do seem to be associated with the key figures in each volume. The second phenomenon is that there is often a gap in the collation between a first and second part of the volume, which suggests that the printing of the two

[43] Franklin says there were near 50 other workmen ('great Guzzlers of Beer') in Watts's shop (*The autobiography of Benjamin Franklin*, ed. L. W. Labaree *et al.* (New Haven and London, 1964), p. 99). For the sale of the Wild-Court office, see Nichols, *Anecdotes*, iii. 607 n.

[44] The sale, on 30 June 1758, was for £67; see Bodl. MS Eng. misc. c. 297, fo. 50ʳ. Watts seems to have turned over the exploitation of his play copyrights to William Feales *c.*1731; see Feales's trade sale of 17 Nov. 1737 (Ward trade sales catalogues, Bodl. John Johnson) and T. Belanger, 'Booksellers' sales of copyright' (unpub. Ph.D. diss., University of Columbia, 1970) and 'Booksellers' trade sales, 1718–1768', *Library*, 5th ser. xxx (1975), 281–302. The details of Watts's death are from Nichols, *Anecdotes*, iii. 739.

[45] H. Macdonald, *John Dryden: A bibliography* (Oxford, 1939), p. 67.

parts proceeded separately (possibly in different shops); the second part may also have its own title-page.[46]

The sixth and last volume of *Poetical miscellanies*, published on 2 May 1709, is dominated by the 'Pastorals' of Pope and Ambrose Philips. Tonson had written to Pope asking to publish his 'Pastorals' on 20 April 1706, and Philips's 'Pastorals', we learn from Addison, had arrived just too late to be included in the fifth volume of 1704.[47] Their importance is recognized by the provision of separate titles for Philips's 'Pastorals' at the beginning of the volume, dated 1708, and for Pope's at the end, dated 1709 (Figure 14). What is apparent from the volume itself, though not from a bibliographical collation, is that parts of it are printed on a much whiter paper, and these correspond to the first sections by Philips and by Pope. They are printed on what the trade called 'Genoa paper' after the port from which it was shipped, as opposed to the other quality paper, which was Dutch (or 'Holland'). Genoa paper is readily distinguished by the presence of a small countermark in the corner of the sheet; common forms are initials, a pair of spectacles, or a heart. It has usually kept a brilliant white colour, whereas Dutch paper is creamy, if not *café au lait*; it may be that the distinction is greater after two centuries of ageing.[48]

Table 2 may help an understanding of the volume. The sections printed on fine paper are sheets B–D[8] and N–P[8]. The first contains Philips's 'Pastorals'; it opens with the special title-page and a preface, which occupy two leaves, and because these leaves have no page numbers, pagination for the rest of the volume is four pages short of what I have labelled the 'ideal pagination', which would start with p. 1 on B1 recto. Philips's 'Pastorals' run over on to the first four pages of sheet E, but the pagination still runs on smoothly until we reach Pope's first contribution. His adaptation of the 'Merchant's tale' from Chaucer starts another fine-paper section on sheet N, but here no allowance has been made for the four unnumbered pages at the beginning (Figure 15); it looks in fact as though this section was set up on the assumption that pagination would begin on B1 recto, and that it was printed off before the decision had been made to provide the unnumbered title and preface to Philips's 'Pastorals'. Pope's contribution also runs four pages over the fine paper into signature Q, but at the beginning of Q the

---

[46] Thus in vol. I (1684) the closing section has a separate title, 'Virgil's ecologues. Translated by several hands. Printed in the year, 1684', and begins a new alphabet in the collation (basically A–E[8] F[4], which follows B–X[8]). The other vols. follow a similar pattern. Vol. II has two alphabets (basically B–L[8]M[4] Aa–Ii[8]), the pagination leaping from 168 to 353 between them; vol. III has a separate section devoted to Tate's translation of Fracastoro's *Syphilis*; vols. IV and V are simpler.

[47] Pope, *Correspondence*, i. 17, and Addison, *Letters*, p. 49. According to the *NCBEL* ii. 780, Rowe edited vols. V and VI of the *Miscellanies*. Pope may later have had a hand in drawing him away from Tonson to Lintot; see M. Boddy, 'Tonson's "loss of Rowe"', *NQ* ccxi (1966), 213–14, and A. W. Hesse, 'Pope's role in Tonson's "loss of Rowe"', *NQ* ccxxii (1977), 234–5.

[48] Some of the other paper in the vol. may be Italian, but it lacks the corner countermark and is now much yellower. The watermarks show five sorts of paper: (1) the Genoa with spectacles or boot countermark (B–D[8] and N–P[8]); (2) poor quality brown paper with no marks (Dd–Kk[8]); (3) medium quality with horse marks (R[8] and Bb–Cc[8]); (4) medium quality with 'M' initials (Ll–Oo[8]); (5) medium quality with 'H' initials (the rest of the vol., including Pope's 'Pastorals').

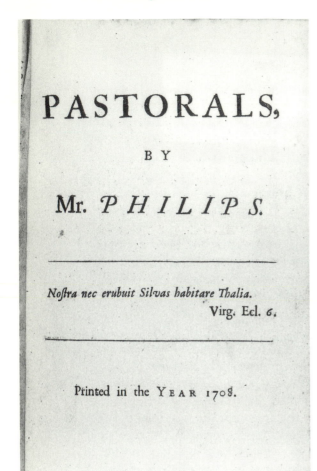

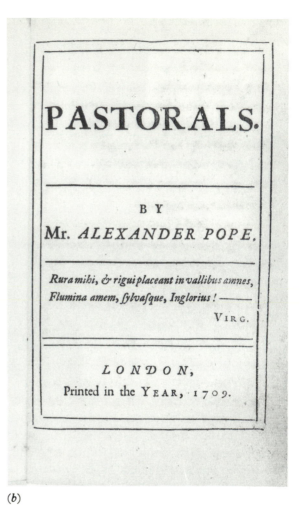

(a)   (b)

FIG. 14. Section-titles for Ambrose Philips and Pope: *Poetical miscellanies* (1709), pp. Bl[r] and 721 (Bodl. 85 b. 20; 189 × 120)

volume reverts to the earlier system of pagination, making allowance for the four unnumbered pages, and so pages 221–4 are repeated.

There are two other interruptions, this time in the normal sequence of signatures, connected with Pope's contributions. The most obvious is the final sequence Aaa–Bbb[8] for Pope's 'Pastorals' with their own title-leaf, which has the pagination that would be appropriate if the second alphabet had been completed (instead of stopping at Ss[6])[49] and if there had been no four-page deficiency in the

---

[49] Ss[6] is 2 leaves short because 2 leaves were used to print the cancels for Hh8 (a half-title in which 'A translation of the first satire of the first book of Horace' is changed to 'An imitation of the first . . .') and Mm3 where the verso suffered from frisket bite. In the copy at the Clark Library the cancels have not been removed and the sheet has been folded normally.

TABLE 2. Miscellanies: The sixth volume, 1709

Collation	Contents (Pope and Philips)	Actual pagination	Ideal pagination
A⁴	Title and preface	none	
B–D⁸ (fine paper)	Philips, 'Pastorals' (to E3)	[i–iv] 1–44	1–48
E–M⁸		45–172	49–176
N–P⁸ (fine paper)	Pope, 'January and May' (to Q2, 224) (recd. 4 Mar 1708)	177–224	177–224
Q–T⁸ U⁸ (–U8)		221–98 [299–300]	225–302 [303–4]
X–Rr⁸ Ss⁶	Pope, 'Episode of Sarpedon' (to Y4, 323) (recd. 13 Jan 1709)	301–632	305–636
Aaa–Bbb⁸	Pope, 'Pastorals' (recd. 4 Mar 1708)	[721–2] 723–51 [752 blank]	721–52

pagination. Apparently this section was printed before the rest of the volume was completed, and it may have been expected that the second alphabet would be completed, though in fact the convention of starting a new alphabet for an appended work is well established (compare Lintot's *Miscellany* below). The other break is the absence of U8, which suggests that a blank leaf (or the cancel for Y6) was removed to link the sequence of poems on pp. 225–98 with Pope's translation from Homer starting on X1, p. 301; this in its turn presupposes that Pope's translation had been printed before the Q–U sequence was complete, though since the pagination of both Q–U and the Homer translation allows for the four blank pages at the beginning of the Philips 'Pastorals', they must have been printed after the 'Pastorals'.[50]

What is clear is that the sequence of the volume is interrupted at three points, and each of these marks the beginning of a Pope contribution on a new signature, printed before the preceding sheets had been set or printed. Whatever the precise order of printing of the Pope and Philips sections, we have to ask why Pope's contributions were put into type before those of the other authors. Three possibilities suggest themselves. The first presupposes a shortage of fine paper and a desire to use what was available on the most favoured contributors; this does not explain the inferior paper used for Pope's 'Pastorals' on Aaa–Bbb⁸, but it seems probable that one reason for supplying a title on Aaa1 was that offprints of the 'Pastorals' were to be printed, and a small reserve of fine paper would suffice for this. The second hypothesis turns more directly on the concept of offprints—that the works of Pope and Philips were printed first so that the authors could circu-

---

[50] A. E. Case, *A bibliography of English poetical miscellanies, 1521–1750* (Oxford, 1935), p. 120, suggests an order consonant with the evidence of the pagination and Tonson's receipt (reproduced by G. Sherburn, *The early career of Alexander Pope* (Oxford, 1934), facing p. 85): Pope's Chaucer and 'Pastorals' (Pope recd. 10 guineas on 4 Mar. 1708); Philips's 'Pastorals'; Pope's Homer (recd. 3 guineas on 13 Jan. 1709). But this overlooks the fact that Philips's title is dated 1708 and Pope's 1709. Pope's 'Pastorals' must have been put to one side, perhaps with their pagination and alphabet already allocated.

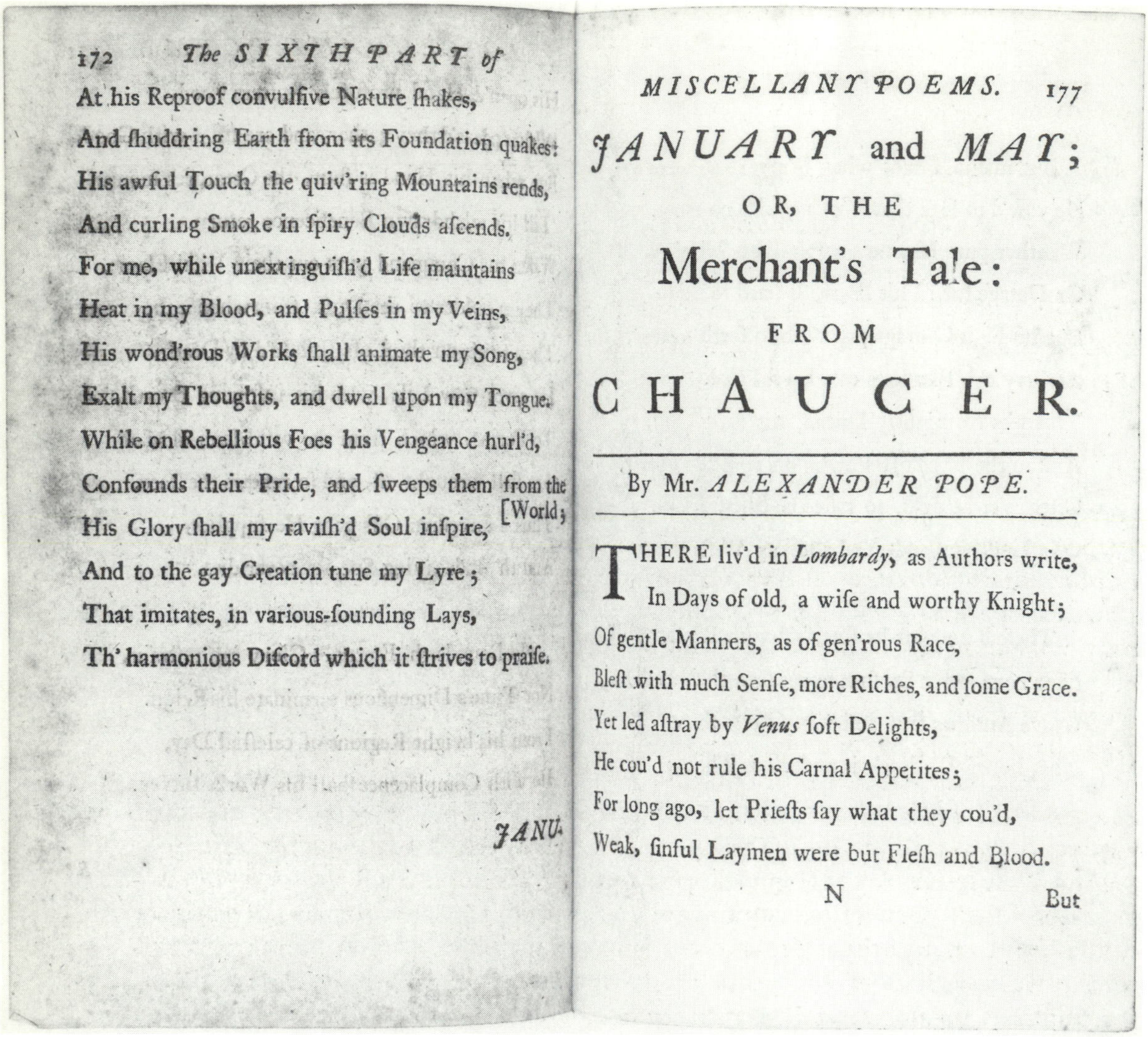

172 *The SIXTH PART of*

At his Reproof convulfive Nature fhakes,
And fhuddring Earth from its Foundation quakes:
His awful Touch the quiv'ring Mountains rends,
And curling Smoke in fpiry Clouds afcends.
For me, while unextinguifh'd Life maintains
Heat in my Blood, and Pulfes in my Veins,
His wond'rous Works fhall animate my Song,
Exalt my Thoughts, and dwell upon my Tongue.
While on Rebellious Foes his Vengeance hurl'd,
Confounds their Pride, and fweeps them from the [World;
His Glory fhall my ravifh'd Soul infpire,
And to the gay Creation tune my Lyre;
That imitates, in various-founding Lays,
Th' harmonious Difcord which it ftrives to praife.

JANU.

MISCELLANY POEMS. 177

JANUARY and *MAY;*
OR, THE
Merchant's Tale:
FROM
CHAUCER.

By Mr. *ALEXANDER POPE.*

THERE liv'd in *Lombardy,* as Authors write,
In Days of old, a wife and worthy Knight;
Of gentle Manners, as of gen'rous Race,
Bleft with much Senfe, more Riches, and fome Grace.
Yet led aftray by *Venus* foft Delights,
He cou'd not rule his Carnal Appetites;
For long ago, let Priefts fay what they cou'd,
Weak, finful Laymen were but Flefh and Blood.

N   But

FIG. 15. Beginning of Pope's adaptation of the 'The merchant's tale', showing a gap in pagination: *Poetical miscellanies* (1709), pp. 172, 177 (Bodl. 85 b. 20; 189 × 209)

late advance copies of their contributions to their friends and patrons. An off-print of Philips's 'Pastorals' does survive in Princeton University Library, with the word 'FINIS' added at the foot of p. 48.[51] But looking at the problem with some knowledge of how heavily Pope revised his proofs, I am tempted by the third explanation, that these sections were set in type in advance of the other contributors' so that Pope could revise the proofs without delaying or disturbing the

---

[51] Shelfmark Ex. 3889.2.371.11; I am very grateful to Marjorie Boulton for this information. There were offprints of the *Rape of the locke* from Lintot's *Miscellany* in 1712 (*Correspondence*, i. 144).

printing of less important authors. All these hypotheses assume Pope was a peculiarly favoured author, but since Tonson had sought to print his 'Pastorals' three years earlier, it is not surprising if Pope could hold out for publication on his own conditions. Certainly this sets a pattern we shall find recurring.

An essay on criticism AND THE Elzeviers

Pope's first separately published work was his *Essay on criticism*, which was published on 15 May 1711. Fortunately Pope's manuscript from which it was printed survives—apparently the only one of his surviving manuscripts (except for parts of the *Odyssey*) which served as printer's copy—and in Chapter 4 I shall consider in detail Pope's relationship with his printer, who again was John Watts. Here I want to comment only on its copyright and the format of the early editions.

I doubt if Pope sold the copyright of *An essay on criticism* to William Lewis outright. Lintot paid £15 for the copyright five years later (17 July 1716) and Nichols lists that payment as being to Pope (*Anecdotes*, viii. 300). When he printed the list of Lintot's copyright purchases, Nichols did rearrange them in alphabetical order, so that, though the entry appears under Pope, the purchase of copyright could have been from Lewis, to whom it had been entered in the Stationers' Register on 11 May 1711. On the other hand, Pope may well have limited his assignment of copyright to Lewis to five years, or in some other way; in the light of his later practice he may effectively have kept it to himself.

The quarto format of the first edition was normal enough for a critical pamphlet: the pamphlets of John Dennis, for example, are pretty evenly divided between quarto and octavo. The quarto format allowed more scope for typographical elegance. According to Pope's letter to Caryll of 19 July 1711 (*Correspondence*, i. 128), a thousand copies were printed. A second edition was printed in octavo eighteen months later, priced sixpence instead of a shilling, but the first edition was still in print five years later.[52] It is the third edition of 1713 in duodecimo that is most interesting, for it seems likely it was produced to follow a new typographical fashion, even though there was no need for a new edition. Figure 16 shows a fourth edition, but copies called third or fourth edition are from the same press-run with the title changed in the course of printing. In this it follows Addison's *Cato* (seventh or eighth edition), as it does typographically. These seem to be early examples of this form of deceit, but before we condemn it too heartily we should remember that Oxford University Press did exactly the same for a fellow of All Souls earlier in the same year: Edward Young's *A poem on the last day* (1713) is an exact parallel.[53] Lewis as bookseller must take the formal re-

---

[52] Advertised in the *Rape of the lock* (1718) at 1s. (unless unsold copies of the octavo had been raised in price above the duodecimo).

[53] Deceit is perhaps too strong a word when the edn. no. is changed for every 1,000 copies, as may happen in the 19th cent., with some continental precedent. At this point I guess the nos. are usually

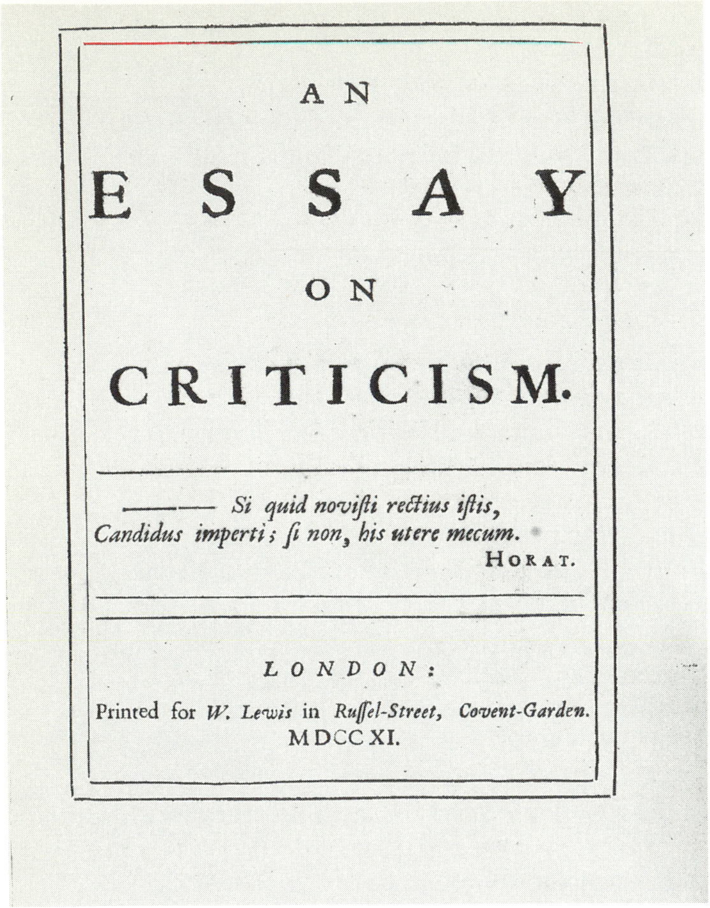

AN

ESSAY

ON

CRITICISM.

——— *Si quid novisti rectius istis,*
Candidus imperti; si non, his utere mecum.
HORAT.

LONDON:
Printed for *W. Lewis* in *Russel-Street, Covent-Garden.*
MDCCXI.

(*a*)

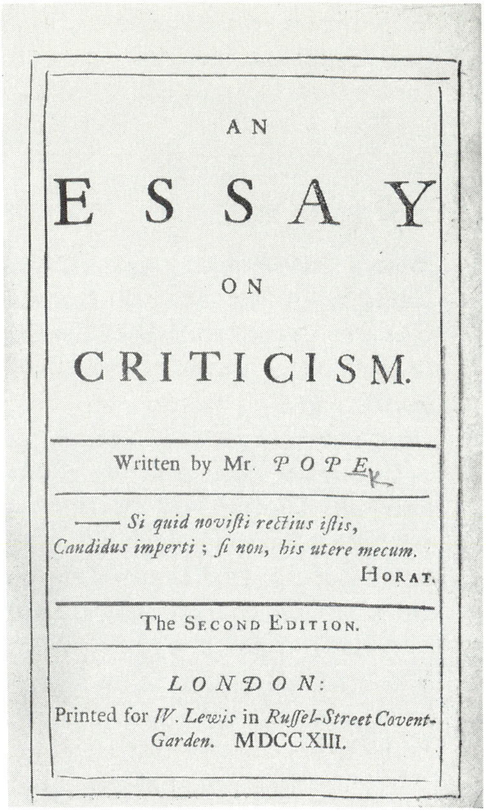

AN

ESSAY

ON

CRITICISM.

Written by Mr. *POPE.*

—— *Si quid novisti rectius istis,*
Candidus imperti ; si non, his utere mecum.
HORAT.

The SECOND EDITION.

LONDON:
Printed for *W. Lewis* in *Russel-Street Covent-Garden.* MDCCXIII.

(*b*)

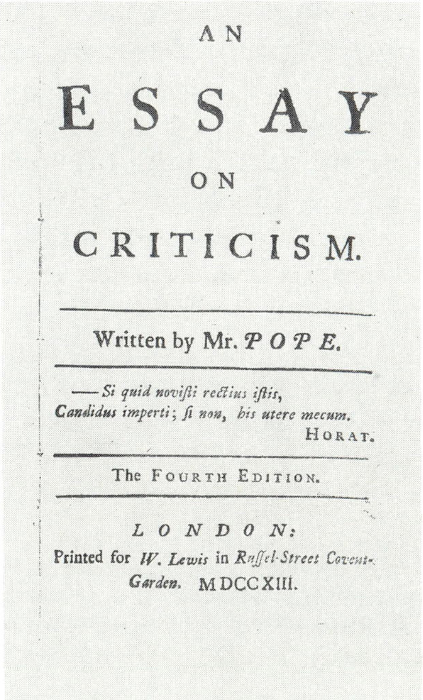

AN

ESSAY

ON

CRITICISM.

Written by Mr. *POPE.*

—— *Si quid novisti rectius istis,*
Candidus imperti; si non, his utere mecum.
HORAT.

The FOURTH EDITION.

LONDON:
Printed for *W. Lewis* in *Russel-Street Covent-Garden.* MDCCXIII.

(*c*)

FIG. 16. Title-pages of first three editions of *Essay on criticism*: (*a*) quarto of 1711 (Beinecke; approx. 219 × 169); (*b*) octavo of 1713 (BL 11631. bbb. 45; 190 × 114); (*c*) duodecimo of 1713 ['fourth edition'] (Bodl. 12 θ 797; 151 × 95) sponsibility for the action but Pope was (as I shall show) closely involved in the production of this edition. Perhaps he compensated for this misdemeanour by printing two third editions (actually impressions with corrections) of the *Dunciad* in 1728; later he rarely put edition numbers on his title-pages. The original title is printed in red and black; note the abandonment of the double rule around the title, part of the new typography.

In England duodecimos had never achieved the respectability they had on the Continent; they were used for devotional books and novels and for cheap re-prints, whereas in France they had for decades been used for elegant first edi-tions.[54] To take a strictly relevant parallel, Mme Dacier's translation of Homer's *Iliad* in 1711 was first published in duodecimo. To an awareness of French prac-tice, which must have been strong in Tonson's circle, we must add the tradition of classical editions from Aldus through Stephanus (Robert Estienne, or 'Stephens' in English) and Colinaeus (Simon de Colines) to Elzevier. As Pope wrote in his 'Verses design't to be prefix'd to Mr. Lintott's Miscellany' of 1712—with his tongue in his cheek:

> Some *Colinaeus* praise, some *Bleau*
> Others account them but so, so;
> Some *Stephens* to the rest prefer
> And some esteem old *Elzevir*:
> Others with *Aldus* would besot us;
> I, for my part, admire *Lintottus*.[55]

Elzevier was the nearest both in time and place, and his types were available in Holland. James Talbot, who had edited the Cambridge quarto of Horace which Tonson published, produced at his own expense a duodecimo edition at the University Press dated 1701, and wrote to Prior: 'I wish you would order my friend and your humble servant, old Elzevir, to recommend this impression in his namesake's types to Leers of Roterdam.'[56]

By 1711 the name of Elzevier had begun to find its way into advertisements for pocket editions; Tonson advertised the ninth edition of *Paradise lost* in *Spectator* xxix, 3 April 1711, 'Just Publish'd, and Printed very Correctly, with a neat Elzever Letter, in 12mo. for the Pocket' (*Paradise regained* followed in 1713, as did Dryden's *Juvenal*). In the same year Lintot followed suit, advertising 'A very

---

smaller. Other probable examples of the practice are John Millan's 2nd, 3rd, and 4th edns. of James Thomson's *Winter* (1726) and the 2nd to 5th edns. of his *A poem sacred to the memory of Sir Isaac Newton* (1727). In 1732 Bowyer printed Swift's *A soldier and a scholar*, of which he owned a part (at least) of the copyright, and the 1st edn. of 500 copies was divided into 300 first and 200 'second edition' copies.

[54] For a stimulating and informed survey of developments in printing throughout the 18th cent., see B. Bronson, 'Printing as an index of taste' in his *Facets of the Enlightenment* (Berkeley and Los Angeles, 1968), pp. 326–65.

[55] *Miscellaneous poems and translations* (1712), pp. 174–5.

[56] Quoted by McKenzie, *The Cambridge University Press*, i. 197. He added, 'I am my own bookseller, and without Jacob Tonson's assistance have already six hundred copies bespoken by the schoolmasters and tutors.'

neat Pocket Edition of the Lying Lover; a comedy, Printed on a new Elziver Letter' in *Spectator* clxxxvii, 4 October 1711. But the real landmark is the advertisement in *Spectator* dlv, 6 December 1712: 'Next Week will be Publish'd, Christus Patiens Carmen Heroicum. Londini, ex Officina J. Tonson & J. Watts. Price 6d. N.B. The Edition of this Poem is Printed much finer than any of the old Elzivers.'[57]

It is perhaps no coincidence that the last of the Elzevier dynasty, Abraham II, had died in the preceding summer. His grandfather Abraham had built up the reputation of the firm, and particularly that of its pocket classics, in Leiden from 1622 to 1652. The Leiden business in its last years under Abraham II was notable mainly for disputes with the University, and the printing declined sadly; a German traveller, Dr Lämmermann, wrote in 1710, 'nowhere in Europe is printing done in a more vicious manner than here'.[58] The old reputation, however, remained; and it is not inconceivable that Tonson saw himself as taking over the Elzevier tradition, for *Christus patiens* (Figure 17) was the forerunner of Maittaire's editions of the classical authors, printed by Tonson and Watts and protected by a royal privilege granted to Maittaire on 4 April 1713. The typography of these editions differs only in the use of red and black for their titles.

Figure 18 shows a large-paper copy of the Lucretius; the Bodleian copy happens to have the signature 'J Gay', and if I am right in thinking it is that of the poet, it encourages my belief that Gay like Pope was interested in typography. The unusual feature of these classics is that they all have very extensive indexes but usually no other apparatus. The volumes dated 1713 were Lucretius, Phaedrus, Sallust, Terence, Trogus Pompeius, and Velleius Paterculus. There is some doubt whether they were actually published until the following year, since Henry Clements subscribed for these titles on 15 April 1714, and such subscriptions normally precede publication. These titles were in fact listed in the *Monthly catalogue* of June 1714; I have found no newspaper advertisements for them. Maittaire's Greek New Testament, dated 1714 and subscribed for by Clements on 24 May, was advertised in the *Daily courant* of 16 June 1714.

A possible reason for delay was that the Stationers' Company regarded the Tonsons' publication as infringing the rights of the English Stock. On 7 June 'the Master Reported that Mr Tonson had attended the Stock keepers concerning his printing severall Copyes belonging to the Company' and it was 'Ordered that the Master Wardens and Stock keepers do looke into the severall Grants and see by what meanes the Company have a Right to the printing of the severall Coppyes of

---

[57] *Tatler*, cii (3 Dec. 1709) advertised 'a very neat Editioin, fitted for the Pocket, on extraordinary good Paper, a new *Brevier* Letter, like the *Elzevir* Editions' in reply to Hills's piracy of its first hundred numbers (see R. P. Bond, 'The pirate and the *Tatler*', *Library*, 5th ser. xviii (1963), 266). One should also note a very elegant edn. of Thomas Hill's *Nundinae Sturbrigienses* printed by John Watts for Jacob Tonson in Feb. 1709.

[58] D. W. Davies, *The world of the Elseviers, 1580–1712* (The Hague, 1954), p. 94. I have taken all my other facts from this source.

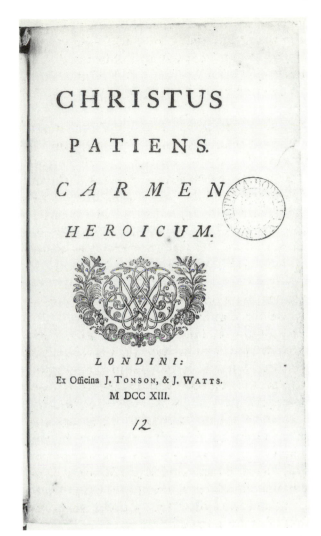

FIG. 17. René Rapin, *Christus patiens* (1713), title-page (Bodl. G. Pamph. 1607(12); 193 × 114)

FIG. 18. Lucretius, *De rerum natura* (1713), title-page (Bodl. 2979 e. 34; 190 × 110)

Schoole Bookes and to Report the same to the Court'.[59] There is no doubt that copies of the Maittaire classics were stitched and issued in the sheep bindings typical of school-books; in 1773 they were advertised as 'a genteel and useful present to young gentlemen at school, or at the University'.[60] Whether all the titles formed part of the school-book patent seems doubtful; I suspect most were

[59] Court Book G, 219ʳ (7 June 1714), referred to by J. H. Brodowski, 'Literary piracy in England from the Restoration to the early eighteenth century' (unpub. doctoral diss., Columbia University, 1973), p. 163.
[60] C. J. Longman, *The house of Longman 1724–1800* (1936), p. 243.

covered by the patent for printing thirty-seven Latin and Greek works granted to Captain Seymour in 1669 for forty-one years. The English stock bought out that patent in 1678 and it would have expired in 1710.[61] Whether the Stationers decided to base their claim on the twenty-one years granted to existing copyright holders by the Copyright Act of 1710 or on the tradition of perpetuity one cannot tell, but on 5 December 1715 they ordered that the dividends of both Tonsons 'be stopt untill such time they give Satisfaction to the Company for their printing the Companys Copyes'.[62] The royal licence of the exclusive right of printing to Maittaire seems to have been ignored.

Not long after Maittaire's classics had started printing, Tonson began to reprint some of his favourite English authors in a similar format, and I find it hard to escape the implication that, along with Milton and Shakespeare, these were the English classics. This was certainly the feeling of the continental authors at being published by the Elzeviers. Jean Guez de Balzac, for example, wrote to the firm: 'I have been made part of the immortal republic. I have been received in the society of demi-gods. In effect, we all live together at Leyden under the same roof. Thanks to you, sometimes I am a neighbour of Pliny; sometimes I find myself beside Seneca, sometimes above Tacitus and Livy.'[63] The contemporary authors honoured by Tonson were: Addison, *Cato* (seventh edition, 1713)[64] and *The Campaign* (fifth edition) with *Rosamund* (third edition, 1713); Tickell, *A poem . . . on the prospect of peace* (sixth edition, 1714); and Garth, *Dispensary* (seventh edition, 1714). Rowe's Shakespeare was also reprinted in 1714 in this same size and type (but unleaded), and so were three of Rowe's own plays, including *The fair penitent* (1714).[65] It is at the beginning of this sequence, immediately after Addison, that Pope's third edition of the *Essay on criticism*, printed by Watts in the same style but with wider margins, fits. The *Evening post* of 28 November 1713 advertises it as '*Just Published*, (in a neat *Elziver Letter 12mo.*) . . . for W. Lewis . . . and sold by John Morphew'. Even though Pope had deserted Tonson for Lewis, he clearly wanted to be part of the new typographic movement.

[61] Blagden, *Stationers' Company*, pp. 193–6; see also H. Carter, *A history of the Oxford University Press* (Oxford, 1975), pp. 70, 96, 147, 219, 226, for Oxford's situation in regard to school-books and pocket classics. On 8 June 1678 the Court of the Stationers' Company instructed its clerk to write to Mr Elzevier to inform him that Cicero's *Opera* were to be printed in folio in London 'by their direction And that if he shall send any that he hath printed or shall print over hither This Company will Seize the Same as forfeited' (Brodowski, 'Literary piracy', p. 296, quoting Court Book E, 320ᵛ–321ʳ).

[62] Court Book G, 235ʳ⁻ᵛ.

[63] Quoted by Davies, *World of the Elseviers*, p. 150.

[64] Addison presented a copy of this edn. of *Cato* to Pope; see M. Mack's finding list from Pope's library in his *Collected in himself* (Newark, London, and Toronto, 1982), p. 395. Some copies have a press-variant title reading 'The eighth edition', like Pope's *Essay on criticism*. I have not traced an advertisement, but that for *The campaign* in the *Daily courant* of 20 Nov. describes it as 'very proper to be bound up with Cato'. For other Elzeviers in Pope's library, see Mack's nos. 19, 90, 170.

[65] *The ambitious step-mother* (3rd edn.) and *Tammerlane* (3rd edn.) were the companions to *The fair penitent* (some copies of which have a press-variant title 'The second edition'). Rowe's *Ulysses* and *The royal convert* were also repr. in 1714 in the same size of duodecimo, but in a larger type.

If there are any doubts that this was an important new departure and a turning-point in eighteenth-century typography, they can be set at rest by noting the speed with which Curll adopted the new fashion—he may have been a rogue, but he was no fool where trends in the trade were concerned. In July 1714 he published translations of Boileau's *Lutrin* and *Art of poetry*, the third edition of Rowe's *Poems on several occasions*, and the miscellany *Original poems and translations* (Case 275), which on its final page advertises four works as 'lately publish'd on an Elzevir Letter, in a neat Pocket-Volume'. Moreover, by 2 September 1714 he reissued the last two together as *The Elzivir miscellany* (dated 1715 on the title). If I am right in thinking that Pope approved of this fashion he must have been galled when Lady Mary Wortley Montagu's *Court poems* and their accretions were reprinted in this format by Curll as *Pope's miscellany* in two parts in 1717.[66] Tonson meanwhile continued to use the duodecimo for English works and within ten years a catalogue of his publications contained something like the whole of English literature in duodecimo.

When we look at four 'Elzivirs' printed by Watts (Figure 19), the first thing to notice is that the Milton, the Maittaire classics (represented here by the Phaedrus), and Addison's *Campaign* have no catchwords.[67] These are the earliest English examples of this abandonment I know and it does not seem to have become a regular feature of Watts's printing, though it is also found in the duodecimo edition of Tonson's *Miscellany*, 1716, which is like the Maittaire classics in format. The Foulises printed many books without catchwords from 1747 and perhaps both they and Watts were influenced by the practice of the Estienne family in the sixteenth century,[68]—or by the return to a purer classical style. What is of particular interest to us is that when Pope controlled the printing of his own books, as in the *Dunciad variorum* (1729), the collected *Essay on man* (1734), and the second volume of his *Works* (1735), he abandoned catchwords too.

The Milton and Maittaire classics are both printed in a smaller type than that in the books shown below them; it measures 60 mm. for twenty lines when set solid as in the classics (it is leaded for Milton), and the size would seem to be bourgeois.[69] Lightly cut copies of the Milton and parallel works measure about

[66] Curll's Elzeviers were often bound up with Tonson's: e.g., Rowe's *Poems on several occasions* are found bound with Tonson's 1714 duodecimos of Rowe's plays, and were perhaps printed for that purpose; Bodl. Don. f. 497 contains Tonson's *Odes and satyrs of Horace* (Case 287), Garth's *Dispensary*, Addison's *Campaign*, Tickell's *Poem*, and Pope's *Essay on criticism*; and from Curll, Young's *Poem on the last day* and both parts of *Pope's miscellany*.

[67] *Paradise lost* and the Maittaire classics have catchwords on the last page of each gathering; some of the classics at least have catchwords throughout the indexes, like Pope's *Dunciad variorum* (1729). *The campaign* is a hybrid example; it omits catchwords only when (because of white lines between paragraphs or for other reasons) there are more than the normal 28 lines to a page. The exception is p. 3 (A4ʳ) where there are 29 lines and a catchword.

[68] See P. Gaskell, *A bibliography of the Foulis Press* (1964), p. 49, and R. A. Sayce, 'Compositorial practices and the localization of printed books, 1530–1800', *Library*, 5th ser. xxi (1966), 33.

[69] In the large-paper copies of the first two works, Rapin's *Christus patiens* and the Lucretius, 20 lines measure 63 mm. I am tempted to believe that the type is very lightly 'leaded'—probably in fact spaced with thin card. I have not been able to compare fine-paper and ordinary copies side by side.

FIG. 19. Four books printed by Watts in Elzevier style: (*a*) Milton, *Paradise lost* (1711), p. 192 (Bodl. 280 m. 88; 126 × 72); (*b*) Phaedrus, *Fabularum Aesopiarum* (1713), p. 43 (Bodl. Douce P 601; 140 × 84); (*c*) Addison, *The campaign*, 5th edn. (1713), p. 13 (Bodl. Malone K 237(3); 159 × 97); (*d*) Pope, *Essay on criticism*, 'fourth' edn. (1713), p. 8 (Bodl. 12 θ 797; 151 × 95)

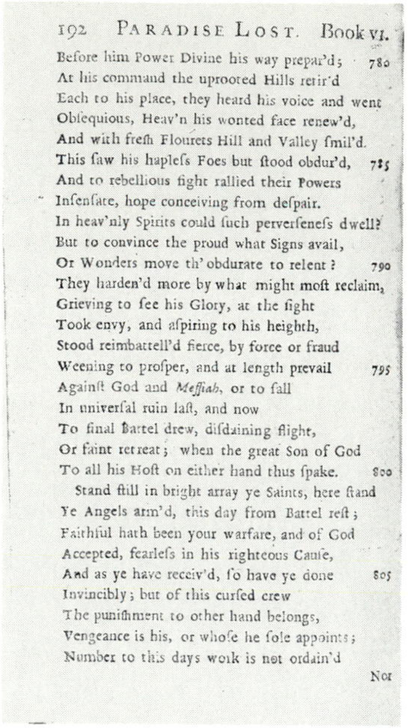

192 PARADISE LOST. Book VI.

Before him Power Divine his way prepar'd;
At his command the uprooted Hills retir'd
Each to his place, they heard his voice and went
Obſequious, Heav'n his wonted face renew'd,
And with freſh Flourets Hill and Valley ſmil'd.
This ſaw his hapleſs Foes but ſtood obdur'd,
And to rebellious fight rallied their Powers
Inſenſate, hope conceiving from deſpair.
In heav'nly Spirits could ſuch perverſeneſs dwell?
But to convince the proud what Signs avail,
Or Wonders move th' obdurate to relent?
They harden'd more by what might moſt reclaim,
Grieving to ſee his Glory, at the ſight
Took envy, and aſpiring to his heighth,
Stood reimbattell'd fierce, by force or fraud
Weening to proſper, and at length prevail
Againſt God and *Meſſiah*, or to fall
In univerſal ruin laſt, and now
To final Battel drew, diſdaining flight,
Or faint retreat; when the great Son of God
To all his Hoſt on either hand thus ſpake.
 Stand ſtill in bright array ye Saints, here ſtand
Ye Angels arm'd, this day from Battel reſt;
Faithful hath been your warfare, and of God
Accepted, fearleſs in his righteous Cauſe,
And as ye have receiv'd, ſo have ye done
Invincibly; but of this curſed crew
The puniſhment to other hand belongs,
Vengeance is his, or whoſe he ſole appoints;
Number to this days work is not ordain'd

Nor

(*a*)

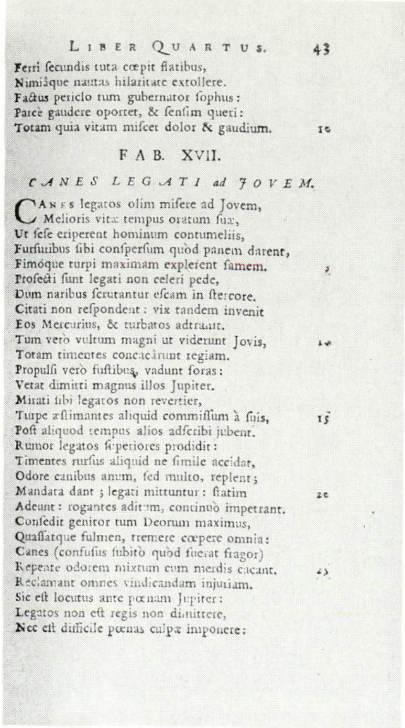

LIBER QUARTUS. 43

Ferri ſecundis tuta cœpit flatibus,
Nimiâque nautas hilaritate extollere.
Factus periclo tum gubernator ſophus:
Parcè gaudere oportet, & ſenſim queri:
Totam quia vitam miſcet dolor & gaudium.

FAB. XVII.

CANES LEGATI ad JOVEM.

CANES legatos olim miſere ad Jovem,
Melioris vitæ tempus oratum ſuæ,
Ut ſeſe eriperent hominum contumeliis,
Furfuribus ſibi conſperſum quòd panem darent,
Fimóque turpi maximam explerent famem.
Profecti ſunt legati non celeri pede,
Dum naribus ſcrutantur eſcam in ſtercore.
Citati non reſpondent: vix tandem invenit
Eos Mercurius, & turbatos adtrahit.
Tum verò vultum magni ut viderunt Jovis,
Totam timentes concacârunt regiam.
Propulſi verò fuſtibus, vadunt foras:
Vetat dimitti magnus illos Jupiter.
Mirati ſibi legatos non revertier,
Turpe æſtimantes aliquid commiſſum à ſuis,
Poſt aliquod tempus alios adſcribi jubent.
Rumor legatos ſuperiores prodidit:
Timentes rurſus aliquid ne ſimile accidat,
Odore canibus anum, ſed multo, replent;
Mandata dant; legati mittuntur: ſtatim
Adeunt: rogantes aditum, continuò impetrant.
Conſedit genitor tum Deorum maximus,
Quaſſatque fulmen, tremere cœpere omnia:
Canes (confuſus ſubitò quòd fuerat fragor)
Repente odorem mixtum cum merdis cacant.
Reclamant omnes vindicandam injuriam.
Sic eſt locutus ante pœnam Jupiter:
Legatos non eſt regis non dimittere,
Nec eſt difficile pœnas culpæ imponere:

(*b*)

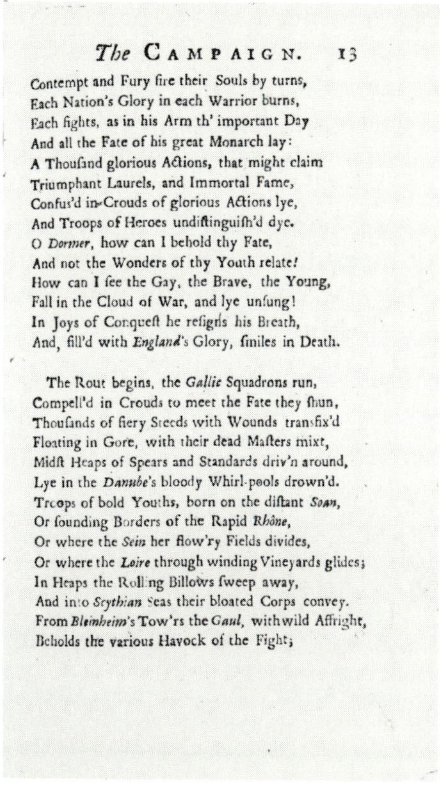

The CAMPAIGN. 13

Contempt and Fury fire their Souls by turns,
Each Nation's Glory in each Warrior burns,
Each fights, as in his Arm th' important Day
And all the Fate of his great Monarch lay:
A Thouſand glorious Actions, that might claim
Triumphant Laurels, and Immortal Fame,
Confus'd in Crouds of glorious Actions lye,
And Troops of Heroes undiſtinguiſh'd dye.
O *Dormer*, how can I behold thy Fate,
And not the Wonders of thy Youth relate!
How can I ſee the Gay, the Brave, the Young,
Fall in the Cloud of War, and lye unſung!
In Joys of Conqueſt he reſigns his Breath,
And, fill'd with *England*'s Glory, ſmiles in Death.

The Rout begins, the *Gallic* Squadrons run,
Compell'd in Crouds to meet the Fate they ſhun,
Thouſands of fiery Steeds with Wounds transfix'd
Floating in Gore, with their dead Maſters mixt,
Midſt Heaps of Spears and Standards driv'n around,
Lye in the *Danube*'s bloody Whirl-pools drown'd.
Troops of bold Youths, born on the diſtant *Soan*,
Or ſounding Borders of the Rapid *Rhône*,
Or where the *Sein* her flow'ry Fields divides,
Or where the *Loire* through winding Vineyards glides;
In Heaps the Rolling Billows ſweep away,
And into *Scythian* Seas their bloated Corps convey.
From *Blenheim*'s Tow'rs the *Gaul*, with wild Affright,
Beholds the various Havock of the Fight;

(*c*)

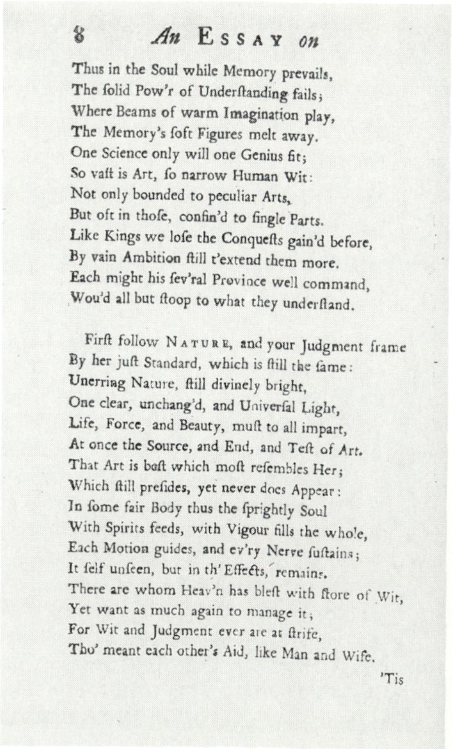

8 *An* ESSAY *on*

Thus in the Soul while Memory prevails,
The ſolid Pow'r of Underſtanding fails;
Where Beams of warm Imagination play,
The Memory's ſoft Figures melt away.
One Science only will one Genius fit;
So vaſt is Art, ſo narrow Human Wit:
Not only bounded to peculiar Arts,
But oft in thoſe, confin'd to ſingle Parts.
Like Kings we loſe the Conqueſts gain'd before,
By vain Ambition ſtill t'extend them more.
Each might his ſev'ral Province well command,
Wou'd all but ſtoop to what they underſtand.

First follow NATURE, and your Judgment frame
By her juſt Standard, which is ſtill the ſame:
Unerring Nature, ſtill divinely bright,
One clear, unchang'd, and Univerſal Light,
Life, Force, and Beauty, muſt to all impart,
At once the Source, and End, and Teſt of Art.
That Art is beſt which moſt reſembles Her;
Which ſtill preſides, yet never does Appear:
In ſome fair Body thus the ſprightly Soul
With Spirits feeds, with Vigour fills the whole,
Each Motion guides, and ev'ry Nerve ſuſtains;
It ſelf unſeen, but in th' Effects, remains.
There are whom Heav'n has bleſt with ſtore of Wit,
Yet want as much again to manage it;
For Wit and Judgment ever are at ſtrife,
Tho' meant each other's Aid, like Man and Wife.

'Tis

(*d*)

130 × 75 mm., while the Maittaire classics are about 145 × 85 mm.; they might be printed on foolscap and crown papers respectively. The measure of the Milton is 58 mm. and these duodecimos are usually set 30 lines to the page; the amount of leading varies. The Maittaire classics are normally set 36 lines to a page, with measurements 112 × 62 mm.

The English classics (as represented by Addison's *Campaign* on the left) are printed in a larger type, variously leaded. It seems to correspond with the type used in Pope's later octavo works, which measures about 66 mm. for twenty lines when set solid: Woodfall in his ledgers described it as long primer.[70] The *Essay on criticism* is printed on the same size of paper, cut copies being about 160 × 95 mm.; the paper is demy or medium. Both the *Essay* and the *Campaign* have larger margins than the other English classics, the type being set to a 68 mm. measure instead of 73 mm., while the page depth is about 10 mm. shorter.[71]

The *Essay* corresponds to the Maittaire classics in having the title printed in red and black, whereas Tonson's English classics are in black (with the exception of Addison's *Cato*, which perhaps prefigures later duodecimo plays, which normally have red and black titles). It is worth noting that all these editions have abandoned the frame of rules around the title-page.

Pope's *Essay on criticism* thus fits into the development of Watts and Tonson's new style, with an added elegance of wide margins and red and black title. Gay's lines to Lintot in his *Miscellany* of 1712 predate the full bloom of the Elzevier revival; they have been often quoted, but with insufficient attention to their positive content:

> While neat old *Elzevir* is reckon'd better
> Than *Pirate Hill*'s brown Sheets, and scurvy Letter . . .
> So long shall live thy Praise in Books of Fame . . .

Pope, I suggest, associated the Elzevier style with books of fame, and the first book he had printed when he separated himself from normal relations with the trade, the *Dunciad* of 1728 (Figure 20), corresponds almost precisely with the *Essay on criticism* fifteen years earlier.[72] The type is the same size, though more heavily leaded; the measure is again 68 mm., though the additional 3 ens for the line numbers brings it up to the size of the other classics (73 mm.). The page is shorter still than the normal *Essay* page, giving a larger margin. When reimposed in octavo on the same size of sheet, the effect is even more impressive.

In 1735, when the second volume of Pope's *Works* was published in quarto and

[70] P.T.P., 'Pope and Woodfall', *NQ* xi (1855), 377–8.

[71] It looks as though 68 mm. was the original measure, enlarged to 73 mm. to allow for speech prefixes in Addison's *Cato*, Shakespeare, and Rowe's plays; it would also accommodate line nos., as in Pope's later edns. in this format.

[72] The duodecimo *Iliad*s from 1720 on ('printed on an Elzevir Letter', *Daily courant*, 27 June 1720) seem to have the same pattern, with the text leaded and the notes set solid. Cut copies measure 159 × 95 mm., and Woodfall describes his 1736 reprint as 'demy, L. Primer and Brevier'. Pope may have had a say in this format too; he translated much of the *Iliad* from a duodecimo (Spence, *Anecdotes*, no. 107), identified by Mack (no. 87 in his library listing, in *Collected in himself*).

folio, the advertisements said, 'This present Volume will with all convenient Speed be published in Twelves' and 'on a neat Elzevir letter'.[73] It actually appeared in two volumes small octavo, and Lintot joined in with two more volumes reprinting the first volume of *Works*; other volumes containing Pope's letters and new poems were added later. The leaf in these octavos (an example can be seen on the right-hand side of Figure 20) is naturally longer than in the earlier duodecimos (170 × 100 mm.), but with the same size of type. Different printers used different amounts of leading for the text (the notes are set solid), but a consistent feature is the large margin at the head.[74] Woodfall in his ledgers refers to these editions as crown octavo, long primer. The fine-paper copies are on writing demy like the large-paper *Dunciad* and are even more handsome. I suspect Pope changed from duodecimo to octavo for the sake of the margins, though he still spoke of printing 'in twelves'. The titles are always in red and black.

There is good evidence that Pope preferred these neat little octavos to his grand quartos and folios. He wrote to Ralph Allen: 'I have done with expensive Editions for ever, which are only a Complement to a few curious people at the expence of the Publisher, & to the displeasure of the Many . . . for the time to come, the World shall not pay, nor make Me pay, more for my Works than they are worth'.[75] If the quartos and folios with their variant readings and notes explanatory and critical follow Boileau's example in producing an edition of a modern author with the apparatus previously reserved for the classics,[76] then these octavos with their Elzevier associations fit the classical pattern as well if not better. It might seem that the so-called death-bed editions which were produced with Warburton's help in the last year of Pope's life showed a change of heart, since they are printed in quarto; but we now know that they were originally printed in this same Elzevier style.[77] It may be that we owe the quartos to Warburton's influence. To sum up, Pope adopted the Elzevier style the moment it began to establish itself in England, and remained faithful to it all his life.

[73] *Grub Street journal*, 31 July 1735, 'In Duodecimo, and on a neat Elzevir Letter'.

[74] The measure is 67 mm. in some early edns. (e.g. books 431–2 in R. H. Griffith, *Alexander Pope: A bibliography* (Austin, 1922, 1927)) and in 2 vols. of 1743 (Griffith 583–4); the narrower measure causes problems with line nos. That the change from duodecimo to octavo was probably due to Pope is suggested by the fact that one result of his reorganizing the Swift *Miscellanies* for Bathurst in 1742 was a similar change to octavo.

[75] *Correspondence*, iv. 350 (?14 July 1741); cf. Spence, *Anecdotes*, no. 210: 'I was first forced to print in little by other printers' beginning to do so from my folios.' At the time of the letter to Allen, Pope was nearly £200 in debt to his printer.

[76] Jonathan Richardson proposed such a project to Pope. For a possible direct debt to the Geneva edition of Boileau's works, see J. McLaverty, 'The mode of existence of literary works of art: The case of the *Dunciad variorum*', *SB* xxxvii (1984), 95–105.

[77] My information comes from unpub. research by my former pupil, Richard Noble; the point will be taken up in Ch. 3 below.

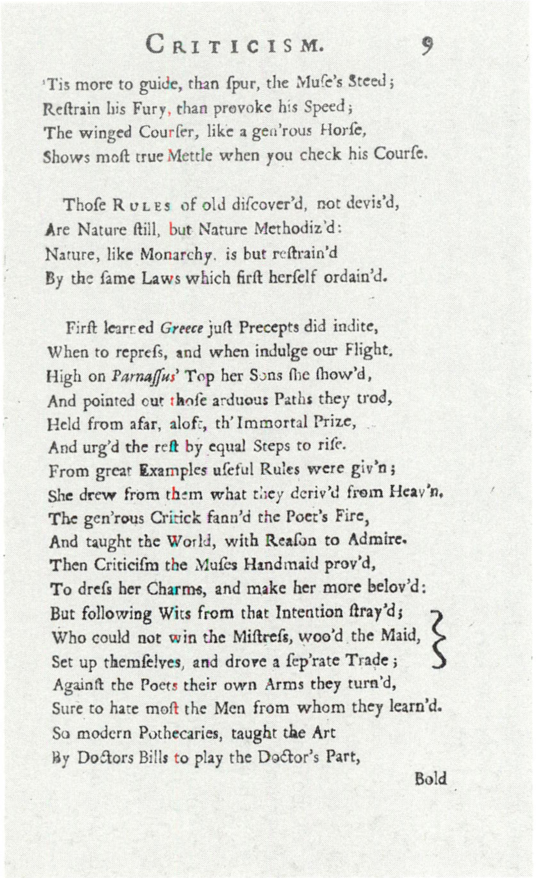

'Tis more to guide, than spur the Muse's Steed;
Restrain his Fury, than provoke his Speed;
The winged Courser, like a gen'rous Horse,
Shows most true Mettle when you check his Course.

Those RULES of old discover'd, not devis'd,
Are Nature still, but Nature Methodiz'd:
Nature, like Monarchy, is but restrain'd
By the same Laws which first herself ordain'd.

First learn'd *Greece* just Precepts did indite,
When to repress, and when indulge our Flight,
High on *Parnassus'* Top her Sons she show'd,
And pointed out those arduous Paths they trod,
Held from afar, aloft, th'Immortal Prize,
And urg'd the rest by equal Steps to rise.
From great Examples useful Rules were giv'n;
She drew from them what they deriv'd from Heav'n,
The gen'rous Critick fann'd the Poet's Fire,
And taught the World, with Reason to Admire.
Then Criticism the Muse's Handmaid prov'd,
To dress her Charms, and make her more belov'd:
But following Wits from that Intention stray'd;
Who could not win the Mistress, woo'd the Maid;
Set up themselves, and drove a sep'rate Trade;
Against the Poets their own Arms they turn'd,
Sure to hate most the Men from whom they learn'd.
So modern Pothecaries, taught the Art
By Doctors Bills to play the Doctor's Part,

Bold

(a)

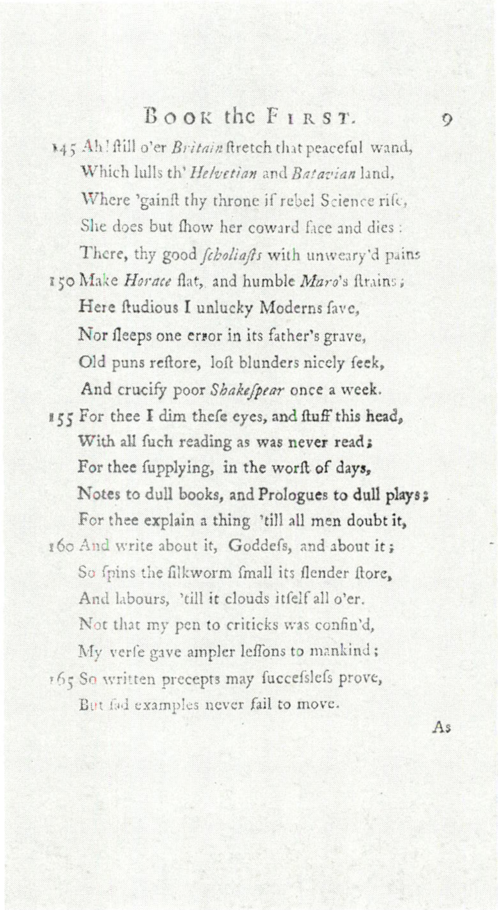

BOOK the FIRST.

145 Ah! still o'er *Britain* stretch that peaceful wand,
Which lulls th' *Helvetian* and *Batavian* land.
Where 'gainst thy throne if rebel Science rise,
She does but show her coward face and dies:
There, thy good *Scholiasts* with unweary'd pains
150 Make *Horace* flat, and humble *Maro's* strains;
Here studious I unlucky Moderns save,
Nor sleeps one error in its father's grave,
Old puns restore, lost blunders nicely seek,
And crucify poor *Shakespear* once a week.
155 For thee I dim these eyes, and stuff this head,
With all such reading as was never read;
For thee supplying, in the worst of days,
Notes to dull books, and Prologues to dull plays;
For the explain a thing 'till all men doubt it,
160 And write about it, Goddess, and about it:
So spins the silkworm small its slender store,
And labours, 'till it clouds itself all o'er.
Not that my pen to criticks was confin'd,
My verse gave ampler lessons to mankind;
165 So written precepts may successless prove,
But sed examples never fail to move.

As

(b)

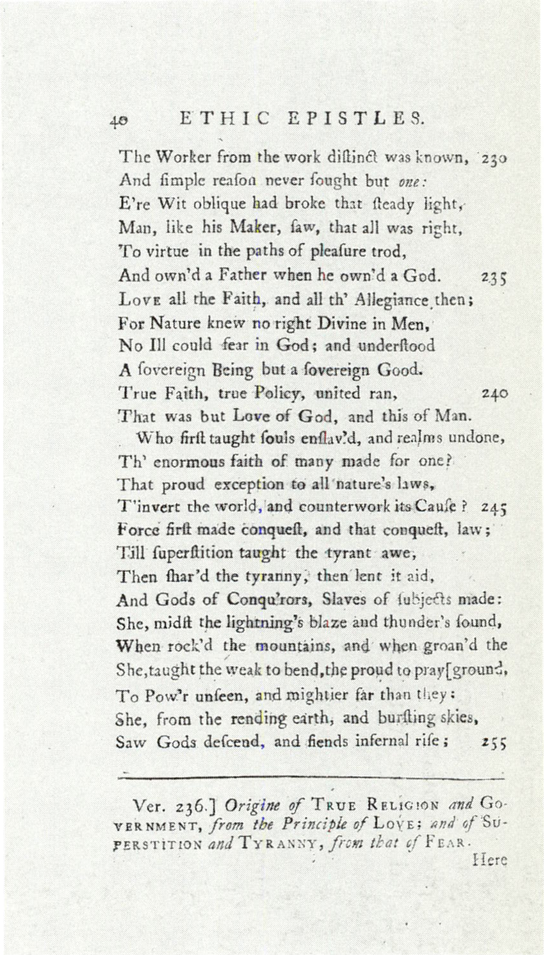

The Worker from the work distinct was known, 230
And simple reason never sought but *one*:
E're Wit oblique had broke that steady light,
Man, like his Maker, saw, that all was right,
To virtue in the paths of pleasure trod,
And own'd a Father when he own'd a God. 235
Love all the Faith, and all th' Allegiance then;
For Nature knew no right Divine in Men,
No Ill could fear in God; and underflood
A sovereign Being but a sovereign Good.
True Faith, true Policy, united ran, 240
That was but Love of God, and this of Man.
Who first taught souls enslav'd, and realms undone,
Th' enormous faith of many made for one?
That proud exception to all nature's laws,
T'invert the world, and counterwork its Cause? 245
Force first made conquest, and that conquest, law;
Till superstition taught the tyrant awe,
Then shar'd the tyranny, then lent it aid,
And Gods of Conqu'rors, Slaves of subjects made:
She, midst the lightning's blaze and thunder's sound,
When rock'd the mountains, and when groan'd the [ground,]
She, taught the weak to bend, the proud to pray[ground,]
To Pow'r unseen, and mightier far than they:
She, from the rending earth, and bursting skies,
Saw Gods descend, and fiends infernal rise; 255

Ver. 236.] *Origine of* TRUE RELIGION *and* GO-
VERNMENT, *from the Principle of* LOVE; *and of* SU-
PERSTITION *and* TYRANNY, *from that of* FEAR.

Here

(c)

FIG. 20. Three Pope books in Elzevier style: (*a*) *Essay on criticism*, 'fourth' edn. (1713), p. 9 (Bodl. 12 θ 797; 151 × 95); (*b*) duodecimo *Dunciad* (1728), p. 9 (BL C. 59. ff. 13(4); 159 × 90); (*c*) octavo *Works* II (1735), p. 40 (Bodl. Radcliffe. f. 244; 169 × 102).

MISCELLANEOUS POEMS AND TRANSLATIONS, 1712

Normal Ault argued very persuasively in his *New light on Pope* (1949), pp. 27–38, that Pope was the editor of Lintot's miscellany from its first edition of 1712 to the fifth of 1726–7, and I need not repeat his arguments and evidence here. It is of interest to note in the list of Lintot's copyrights that only two authors were paid for their contributions, Pope to a total of £26 19s. 0d. and 'Betterton' for 'The Miller's Tale, with some characters from Chaucer' £5 7s. 6d.[78] Betterton had, in fact, died two years earlier on 28 April 1710; Pope claimed his acquaintance from a boy and, to quote Sherburn, 'it is probable he aided in revising and publishing some modernizations of Chaucer that Betterton had made'.[79] The payment of 7 April 1712 was presumably made to Mrs Betterton a day or two before her death; she was buried in Westminster Abbey on 13 April 1712. Caryll, writing to Pope on 23 May 1712 (and out of touch with the latest news) says: 'I am very glad for the sake of the Widow and for the credit of the deceas'd, that *Betterton*'s remains are fallen into such hands as may render 'em reputable to the one and beneficial to the other' (*Correspondence*, i. 142).

Ault was, however, mistaken in adopting with variations the hypothesis put forward by W. F. Prideaux and accepted by T. J. Wise and Robert K. Root[80] that this miscellany originally contained *Windsor Forest* and the *Ode for musick*, which were withdrawn for separate publication before the volume was issued. The hypothesis rested on a gap in the collation between signatures X and Aa which was in fact occupied by these poems when the collection was reissued in 1714. Unfortunately, however, the catchword 'The' on X8ᵛ (Figure 21) would fit neither *Windsor Forest* nor the *Ode*, while it perfectly fits *The rape of the locke*.

Once again a study of the paper gives us a clue. As Table 3 shows, most of the book is printed on paper with the watermark IKW in script, which now seems inferior to the rest. It is used for the opening part of the book and for the three final signatures which contain the *Rape of the locke* and the end of Betterton's modernization of 'The Reeve's tale of the Miller of Trompington'; it ran out in the course of printing signature P, since some copies are printed on the old, some on the new stock. As a group of Pope's poems ('Vertumnus and Pomona' from Ovid, and three minor poems) starts on K1, one might suspect that (as in Tonson's miscellany of 1709) printing began at three separate points so that Pope had an opportunity to rewrite in proof. John Watts was again the printer, and doubtless accustomed to Pope's ways.

It is interesting to note the dates on which Lintot paid for the copyrights. On 19 February 1712 he paid for the translation of Statius (which occupies B1 to E4ᵛ with its own running title) and of Ovid, 'The fable of Vertumnus and Pomona'

[78] Nichols, *Anecdotes*, viii. 294, 299. On p. 304n. Nichols says: 'The Collection is commonly ascribed to Mr. Pope; but was formed by Lintot from contributions of various Friends.'

[79] *Early career*, p. 50. Harte told Warton they were really by Pope.

[80] Wise, *A Pope library* (1931, repr. 1973), p. 8; Root, 'Pope's contributions to the Lintot miscellanies of 1712 and 1714', *ELH* vii (1940), 265–71.

320 *Miſcellaneous* POEMS *and*

O falſe abuſive Knave! (the Wife reply'd)
In ev'ry Word the Villain ſpake he ly'd.
I wak'd, and heard our harmleſs Child complain;
And roſe, to know the Cauſe, and eaſe her Pain.
I found her torn with Gripes, a Dram I brought,
And made her take a comfortable Draught.
Then lay down by her, chaff'd her ſwelling Breaſt,
And lull'd her in theſe very Arms to Reſt.
All was Contrivance, Malice all and Spight,
I have not parted from her all this Night.
Then is ſhe Innocent ? Ay by my Life,
As pure and ſpotleſs------as thy Boſom Wife.
I'm ſatisfied, ſays *Sim.* O that damn'd *Hall!*
I'll do the beſt I can to ſtarve 'em All.
    And thus the Miller of his Fear is eas'd,
    The Mother and the Daughter both well pleas'd.

                                        THE

THE

RAPE of the *LOCKE.*

AN

HEROI-COMICAL

POEM.

Nolueram, Belinda, *tuos violare capillos,*
Sed juvat hoc precibus me tribuiſſe tuis.
                            MART. Lib. 12. Ep. 86.

Printed for BERNARD LINTOTT. 1712.

FIG. 21. Catchword linking 'The miller of Trompington' to *The rape of the locke* in *Miscellaneous poems and translations* (1712), pp. 320–1 (Bodl. 12 θ 794; 190 × 227)

(which occupies only K1 to K4ᵛ). The remaining three short poems in this group, 'To a young lady with the works of Voiture', 'On silence', and 'To the author of a poem, intitled *Successio*', were not paid for until 9 April, and still only reach as far as L2ᵛ, so it scarcely seems necessary to assume that a separate start was made on K1 for a mere twenty pages of minor poems. It would have made sense to start here, though, if Pope had originally planned to print a larger group of poems, and the parallel with the procedure of Tonson's miscellany of 1709 (where the second group started on N) tempts speculation. It could indeed be that it was here that Pope originally intended to print *Windsor Forest* and the *Ode*, and that the hypothesis of their removal is right in essence if wrong in detail. What is clear is that

TABLE 3. Miscellaneous poems and translations, 1712

Collation	Pages	Paper	Contents (Pope and Betterton)
*A*⁴			
B–I⁸	[1–5], 6–128	IKW	Statius 1–56 (19 Feb 1712)
K–O⁸	129–208	IKW	Ovid 129–36 (19 Feb 1712); 'To a young lady', 'On silence', 'To the author', 137–48 (9 Apr 1712)
P⁸	209–24	[IKW]	
Q–V⁸	225–304		Betterton 245–82 (7 Apr 1712)
X⁸	305–20	IKW	Betterton 301–20 (7 Apr 1712)
Aa–Bb⁸	[353]–76	IKW	'Rape of the locke' (21 Mar 1712)

the *Rape of the locke* was paid for on 21 March, and that it was almost certainly set in type before sheet K was completed with the poems paid for on 9 April; separate copies, with their own title-page printed on Aa1, were sent to friends. On 11 May Pope wrote to Edward Bedingfield asking him to send or deliver advance copies, and on 16 May Bedingfield replied that he had sent copies to Lord Petre and Belle Fermor. This was four days before publication. Later Pope wrote to Caryll (28 May 1712): 'I hope Lewis has conveyed to you by this time the *Rape of the Lock*, with what other things of mine are in Lintot's collection; the whole book I will put into your hands when I have the satisfaction to meet you at Reading' (*Correspondence*, i. 144). None of these offprints seems to have survived.[81]

If Lintot's payments correspond with the submission of copy, then we can see that the Betterton modernizations of Chaucer arrived on 7 April, two days before the poems that were to complete sheet K. Since work could not go forward until K was completed, it is understandable that the printer should have used the shorter Betterton piece of Chaucer to work backwards from X. This does indeed raise the question, how did the printer decide that Y and Z would not be needed? To which one must reply that it is easy to cast off verse, and the contributions must by now have been assembled with fair certainty. Within six weeks of Pope's final contribution, the remaining eleven sheets, K–V⁸ were printed off and the book on sale by 20 May.

Use of Lintot's copyright dates in this way gives a not impossible story; even if one rejects the dates, the evidence of the paper shows that X was printed before signatures P to V. If asked how one could ever hope to make the poems fit the space, I would answer that between Betterton's versification of Chaucer's 'Prologue', with its own title-leaf (R3ʳ–T5ᵛ, 245–82)[82] and his 'Reeve's tale', there are eight short poems on eighteen pages (T6ʳ–V6ᵛ, 283–300) which look as though they have been fitted together as padding.

---

[81] They would probably be discarded as imperfect fragments if they had passed through the book trade (or, indeed, most libraries); so with Allan Ramsay's publications with pagination beginning and ending nowhere in particular.

[82] Betterton was the only poet besides Pope to have a title-leaf to his poem, and in the *Spectator* of 22 May his modernizations were the only works named by author and title.

In 1714 the collection was reissued with the addition of *Windsor Forest* and the *Ode for musick* printed on sheets Y–Z, and somewhat crowded to get them in the space—they are printed 21 lines to a page as opposed to the 18 lines to a page elsewhere. The *Essay on criticism* is reprinted on Cc–Ee⁸, and no doubt for this reason the cancel title to the volume adds William Lewis, the owner of its copyright, to Lintot's name in the imprint. The cancel title also lists nine poems in the volume which are by Pope; clearly the volume had sold badly, and the reissue was an attempt to move the stock by trading on the prestige of Pope's name, rather than that of Betterton.[83]

The addition of these poems would have caused little difficulty except for the fact that the last four leaves of Bb were occupied by a catalogue of Lintot's books, which had to be moved or removed so that the *Essay on criticism* could follow *The rape of the locke*. But in addition to the cancel title and half-title to the volume, it proved necessary to cancel Y8 (presumably because of a printer's error, for the text of the cancellans is identical with the folio edition of *Windsor Forest*) and to print an errata leaf. At this moment someone—and I guess it was Pope—decided that it would be agreeable to fill a whole sheet by printing titles for the three added poems, and to add two short poems on a two-leaf section Y² of which the second leaf was the half-title to *Windsor Forest*. Y1 bore the poems 'Upon a girl of seven years old' and 'Epigram upon two or three' and duplicated the pagination of 321–2. Whether because the second poem was slightly naughty, or because the duplicate pagination confused the binders, this leaf is sometimes missing. It can, in fact, be said that no two copies I have seen are identical; the most frequent error is the omission of the errata.

When the miscellany was reprinted in 1720 it was expanded to two volumes duodecimo (in part by incorporating some poems from 'Pope's Own Miscellany', *Poems on several occasions*, 1717), with Pope's poems occupying the first two-thirds of volume I and his name alone on that title; the second volume lists ten other names on its title. This was the form in which Lintot chose to reprint Pope's poems from the *Works* of 1717 until he joined in the octavo edition of the *Works* in 1736 after a typically disingenuous advertisement by Pope (*Grub Street journal*, 31 July) complained Lintot 'would never be induced to publish them compleat, but only a part of them, to which he tack'd and impos'd on the Buyer a whole additional Volume of other Men's Poems'. The further changes are discussed by Ault, who considers that Pope had a hand in all editions except the sixth of 1732, by which time relations between Pope and Lintot had been broken

---

[83] That the miscellany sold badly is confirmed by the fact that the wholesaling conger bought copies from Lintot on 22 Aug. 1712, three months after publication, at 2s. 8d. each (*Notebook of Thomas Bennet and Henry Clements*, p. 158). Henry Clements (and presumably his 15 partners) took 10 copies each. Normally these transactions were arranged before publication, and assured the bookseller who sold copies to the conger a certain return of money in exchange for sale at a reduced price. Lintot had in fact subscribed the book to the trade on 21 May 1712, the day before it was published, and Clements had taken 9 copies at 2s. 6d. (ibid., p. 119). To sell more books to the conger after publication is an admission that normal sales through the trade had been disappointing.

off.[84] Pope's association with the fifth edition is clear from Bowyer's ledger entry of 14 December 1726 for 'Mr Pope's Miscellany in 2 Vols'; Bowyer charged a guinea extra 'For alterations overrunnings & pages cancell'd &c' which clearly shows Pope at work. The existence of copies of the first volume with a title dated 1726 (instead of 1727), and an advertisement (recorded under Griffith 164) in Parnell's *Poems* issued by 12 June 1726 is no doubt to be explained by the delay that Pope caused at the printers. Bowyer's paper ledger shows that 55 reams of paper were delivered as early as 22 November 1725, the next 35 not until 6 August 1726, 6 more on 7 November, and a final 4 on 7 December 1726.[85]

POPE AND THE BOOK TRADE 1709–1717: AN OVERVIEW

It is time to take a general look at Pope's relations with the trade in this early period. (I am concerning myself with his major poems and ignoring his squibs.) Table 4 shows that his first poems were printed in Tonson's *Poetical miscellanies* VI, and after that he had no further dealings with Tonson except for his contributions to *Ovid's epistles*, 1712, and Steele's *Poetical miscellanies*, 1714. His copyright in an *Essay on criticism* may have been assigned for a limited period to Lewis, but from then on all his dealings are with Lintot, who was trying to break Tonson's supremacy in literary publishing and would therefore pay more for copyrights.

All his early works were printed by John Watts; there is a shift to Bowyer with the third edition of the *Rape of the lock* in the summer of 1714, by which time Bowyer was probably committed to printing the *Iliad*, but Watts was still responsible for reprinting the *Essay on criticism* for Lintot in 1716. What is important for our subsequent purposes is that these two printers (probably the best in London) were responsible for printing all Pope's poetical works up to the completion of the *Odyssey* in 1726, with the exception of the first edition of the *Temple of fame*, 1715, and some reprints of 1718–19. As a result we can compare the way in which Pope's poems were printed with the other products of the same printers; and when we find changes in the presentation and accidentals of Pope's poems between editions produced by the same printer, we may be able to dis-

---

[84] Pope to Buckley (16 June 1732): 'he has (upon Rumours, for I never converse with Him) lately been importuning me, and receivd no other answer than a very true one, that I would never imploy him more' (*Correspondence*, iii. 294).

[85] The Bowyer printing ledgers, the property of the Grolier Club, New York, were described by J. D. Fleeman, '18th-century printing ledgers', *TLS*, 19 Dec. 1963, p. 1056; they are now deposited in Bodl. (MSS Dep. b. 243–4 and Dep. c. 718–23). [I have read them in that Library's photocopy and used that foliation. JMcL] The Bowyer paper stock ledger is also in Bodl. (MS Don. b. 4) and was described by H. Davis, 'Bowyer's paper stock ledger', *Library*, 5th ser. vi (1951), 73–87. The accounts cited here are in printing ledger i, fo. 88ʳ and paper ledger, fo. 75ʳ. Lintot may have subscribed the vol. On 14 Dec. 100 copies went 'To Mr How [possibly Lintot's binder] by Lintot's order', and on 18 Dec., 150 to Lintot, 250 to Osborn, 264 to Brotherton, and 260 to Knapton.

count any suggestion that they are merely the result of a different 'house style'—a concept which in any case is inappropriate to this period.

The general impression one has of the first five years of Pope's career is that his fame was in no way matched by the sale of the books he contributed to. An *Essay on criticism* took at least eighteen months to sell a thousand copies, and the so-called 'third' and 'fourth' editions suggest an attempt to encourage poor sales. Lintot's miscellany needed a cancel title in 1714, with the words 'second edition', and the addition of Pope's other poems; and still did not have to be reprinted until 1720. *Windsor Forest* with its topical allusion to the peace of Utrecht did at least achieve a second impression, called the 'second edition', but so did the topical poems of much less favoured authors; it was still in print in 1718.[86] It is only with the *Rape of the lock* that there is any sign of popular success, and yet the subscription list for the *Iliad* had been started among his friends and patrons some five months earlier.[87] Pope actually signed his agreement with Lintot for the *Iliad* three weeks after the *Rape of the lock* was published (on 23 March 1714), and he may well have got better terms as a result of its success. All the same, the contrast between the 30 guineas or so a year that Pope received in those early years and the 800 a year or more that the *Iliad* produced is indeed remarkable. It seems to me as much an example of collective patronage as a sign of the rise of the man of letters. But before we consider the *Iliad* in detail, I want to comment briefly on the production of the other works.

Windsor Forest and the *Ode for musick* were printed in the large folio traditional for such occasional pieces: or to put it more practically, in a format which would justify a price sufficient to make a profit—1s. for *Windsor Forest* in five sheets, 6d. for the *Ode* in three. It is interesting that the advertisements for *Windsor Forest* in the *Daily courant* for 6 and 7 March add that it is sold by 'John Parker at the Blue-Ball in Pall-Mall by Pall-Mall Court'; this bookseller is otherwise unknown until years later, but Lintot obviously wanted to supply the court end of town.

I shall for a moment leave on one side the *Rape of the lock*. The first edition of the *Temple of fame* is remarkably ill-printed, as can be seen when it is compared

---

[86] *Twickenham edition of the poems of Alexander Pope*, ed. John Butt *et al.* (London and New Haven, 1939–69), i. 130 contrasts the 6 edns. of Tickell's *A poem . . . on the prospect of peace*. For the additions which Pope made to *Windsor Forest* to make it topical, see A. J. Varney, 'The composition of Pope's *Windsor Forest*', *DUJ* NS xxxvi (1974), 57–67. *Windsor Forest* was still being advertised in the 5th edn. of the *Rape of the lock* (1718).

[87] Lord Lansdowne's letter to Pope of 21 Oct. 1713 (*Correspondence*, i. 194–5) seems the earliest reference. Proposals were not printed at this stage; a letter of 19 Nov. 1713 says, 'Proposals in writing have been delivered into all the chief hands' (Sherburn, 'Letters of Alexander Pope, chiefly to Sir William Trumbull', *RES* NS ix (1958), 398). Pope's letters to Caryll of 9 Jan., 25 Feb., and 12 Mar. 1714 all ask for a list of the subscribers he has collected (38 according to Elwin) 'in order to insert them in the printed catalogue now just about to be published' (*Correspondence*, i. 207, 210, 214). This catalogue no doubt formed part of the proposals printed after he had signed his contract with Lintot, and which he sent to John Hughes on 19 Apr. (*Correspondence*, i. 218). Griffith gives the date of publication as 1 May 1714 (ii. 539), which seems plausible, since Pope usually circulated printed proposals privately before he advertised them, though I cannot trace his source; they were advertised 'with a List of those who have already subscribed' in Lintot's *Monthly catalogue* for May 1714 (probably published early in June).

TABLE 4. Pope's Publications, 1709–20

Publication	title-page date	pub date	imprint	copyright	printer
Tonson's *Miscellany* VI	1709 8°	2 May	for J. Tonson	Tonson (10 gns. 4 Mar 1708; 3 gns. 13 Jan 1709)	Watts
	1716 12°				Watts
Essay on criticism	1711 4°	15 May	for W. Lewis	Lewis?	Watts 1000
2nd edn.	1713 8°	27 Nov 1712	for W. Lewis		Watts
3rd/4th edn.	1713 12°	28 Nov	for W. Lewis		Watts
5th edn.	1716 12°		for B. Lintot	Lintot (£15 17 Jul)	Watts
6th edn.	1719 8°	2 May	for B. Lintot		Watts
Ovid's epistles	1712 8°	18 May	for J. Tonson	Tonson (Sapho to Phaon)	Watts
Miscellaneous poems	1712 8°	20 May	for B. Lintott	Lintot (15 gns. 19 Feb; £7 21 Mar; £3.16.6. 9 Apr)	Watts
'2nd edn.' + *Essay*	1714 8°	4 Dec 1713	Lintott & Lewis		Watts
3rd edn.	1720 12°	15 Aug	for B. Lintot		Bowyer ?2000
Windsor Forest	1713 2°	7 Mar	for B. Lintott	Lintot (30 gns. 23 Feb)	Watts
'2nd edn.'	1713 2°	9 Apr	for B. Lintott		Watts
4th edn.	1720 8°	2 Dec 1719	for B. Lintot		Bowyer 1000
Ode for musick	1713 2°	16 Jul	for B. Lintott	Lintot (£15 23 Jul)	Watts
3rd edn.	1719 8°	15 Oct	for B. Lintot		Watts
Steele's Miscellany	1714 8°	29 Dec 1713	for J. Tonson	Tonson (15 gns 5 Oct 1713)	Watts

Rape of the lock	1714	8°	4 Mar	for B. Lintott	Lintot (£15 20 Feb)	Watts ?3000
2nd edn.	1714	8°	11 Mar	for B. Lintott		Watts ?2000
3rd edn.	1714	8°	27 Jul	for B. Lintott		Bowyer 1000
4th edn.	1715	8°	Sept	for B. Lintott		Bowyer 1000
5th edn.	1718	8°		for B. Lintott		
Temple of fame	1715	8°	1 Feb	for B. Lintott	Lintot (30 gns. 1 Feb)	J. Darby
2nd edn.	1715	8°	8 Oct	for B. Lintott		Bowyer 1000
Key to the lock	1715	8°	25 Apr	for J. Roberts	Lintot (10 gns. 31 Apr!)	
(2nd 1715) (3rd 1718)						
Works	1717	4°	3 Jun	for B. Lintot & J. Tonson	Lintot $\frac{3}{4}$ Tonson $\frac{1}{4}$	Bowyer 500 + 250 1000 + 250
		2°				
Pope's Miscellany	1717	8°	13 Jul	for B. Lintot		Bowyer 1000
Eloisa to Abelard 2nd edn. +	1720	8°	13 Oct 1719	for B. Lintot	(from Works)	Bowyer 2500

with the second edition, and Pope can have had no say in its production, though he doubtless read proof; Gay's letter to Ford of 30 December 1714 reports that Pope 'will publish his Temple of Fame as soon as he comes to Town'. What may be significant is that its last four pages contain Lintot's proposals for Urry's edition of Chaucer's works, dated 19 January 1715, and I am tempted to think Lintot may have hoped to arouse interest in Chaucer by the use of Pope's poem.[88] The second edition, published on 8 October, after Urry's death, substitutes four pages of advertisements for Lintot's books.

I have included the pseudonymous *Key to the lock* in my table primarily to illustrate the use of a publisher's name in the imprint as soon as anonymity comes into question, although Lintot as usual paid for the copyright.

The *Rape of the lock* in its expanded and illustrated form was much the most successful of Pope's early poems.[89] Pope wrote to Caryll on 12 March 1714 that the poem 'has in four days time sold to the number [of] three thousand, and is already reprinted tho' not in so fair a manner as the first Impression' (*Correspondence*, i. 214). Since the second edition was not advertised until 11 March, this could mean that the first edition was of three thousand copies. The *Key to the lock*, which was published a year later, on 25 April 1715, speaks of 'the uncommon sale of this Book (for above 6000 of 'em have been already vended)'; if this is a precisely accurate statement, it means that the first three editions totalled more than 6,000 copies. We know the third edition was of 1,000, and if the first was 3,000, the second would have to be at least 2,000 (allowing for the fine-paper copies to provide the overplus of 6,000). These are surprisingly large figures for verse: the nearest I know is the 2,000 and 250 fine copies of Gay's *Trivia* printed in January 1716.

Moreover, this was a luxury edition. Not only were there six copper plates engraved by Claude Du Bosc after Louis Du Guernier (Figure 22), but there were also fine-paper copies.[90] In Pope's previous works, such fine-paper copies had not been on sale to the public, but kept for presentation to friends; this time they were advertised in the *Post boy* of 28 January 1714: 'There will be a small Number . . . printed on fine Paper; those who are willing to have These, are desired to send in their Names to Bernard Lintott . . . No more being to be thus printed than are bespoke.' It promised publication 'in a few days'; the book was actually advertised as published in the *Post boy* for 4 March, five weeks later,

---

[88] See the *Letters of John Gay*, ed. C. F. Burgess (Oxford, 1966), p. 16. Urry's royal privilege is dated 20 July 1714; Lintot made an agreement with Urry on 17 Dec. 1714 by which Lintot would pay for paper, print, copper plates, and all incidental expenses and the subscription money would be divided into thirds between Lintot, Urry, and Christ Church, Oxford (Nichols, *Anecdotes*, viii. 304). But Urry died 18 Mar. 1715, and the vol. was not completed until 1721.

[89] G. Tillotson in *Twickenham*, ii. 103–5, gives an admirable account on which I have drawn freely.

[90] For a discussion of these plates, including the suggestion that Pope may have had a hand in their design, see R. Halsband, '*The rape of the lock*' and its illustrations, *1714–1896* (Oxford, 1980), pp. 1–23. Du Guernier and Du Bosc had undertaken in Jan. 1714 to produce a series of engravings commemorating the battles of the Duke of Marlborough and Prince Eugene; it appeared in 1717. They also produced the plates for Gay's *The shepherds week* (1714).

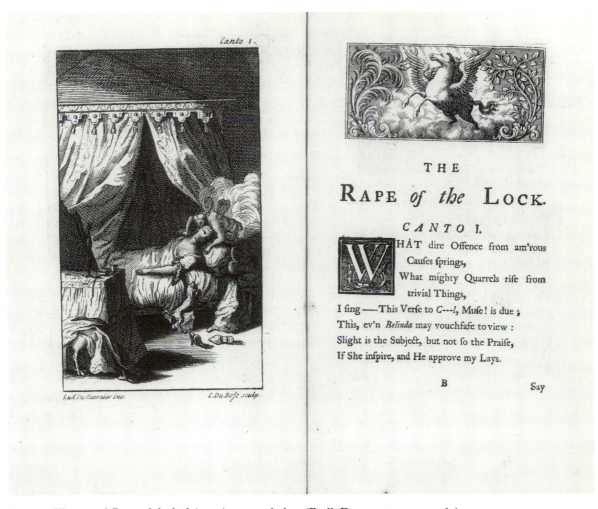

FIG. 22. Illustrated *Rape of the lock* (1714), p. 1 and plate (Bodl. Don. e. 115; 214 × 260)

though Pope was able to send two copies to Caryll a week earlier (*Correspondence*, i. 210). We cannot tell whether the delay was due to Pope's being dilatory in reading proof or to the engravings; these would certainly have taken some time to design (doubtless under Pope's supervision), engrave, and print, if parallel cases are anything to go by.

It may, however, have been an added touch of luxury, thought of at the last minute, that caused the delay. For the first time to my knowledge, an English book of verse in octavo (and a slim one at that) was given the engraved head-pieces, tailpiece, and initial letter that had previously been reserved for the pompous folio (Figures 23 and 24); though the French, of course, had pioneered this style, and would have felt that the initial letter of Figure 23 was too big for the page. However, its size allows its symbolism to be clear; it represents Pegasus, the horse of the Muses, with his attribute of Zeus's thunderbolt, striking Mount Helicon with his hoof and causing the stream of Hippocrene to flow. (The same initial was used for Pope's preface to the *Works* of 1717, and elsewhere.)

I had hoped to be able to attribute this new departure entirely to Pope, but these engravings, by Simon Gribelin, had been used before for the inaugural lecture of Oxford's first Professor of Poetry, Joseph Trapp (Figure 25). The first volume of his *Praelectiones poeticae* was printed at the Sheldonian Theatre for Bernard Lintot in 1711, and in his preface he speaks of his bookseller's care for the appearance of the book (I translate from the Latin):

The references for our quotations from the poets are not set out in the margin of the page; since all who are not strangers to their writings will sufficiently recognize the passages; and we thought that the elegance of the page would be diminished by notes added in this way. The bookseller was very concerned about this, and for the same reason arranged to have the book embellished with copper plates; perhaps with the idea that at all events there might be *some elegance* in this work; and the art of the engraver might accomplish what the wit of the author could not perhaps surpass.

Lintot had paid Trapp £20 for the copyright on 3 January 1710/11 (Nichols, *Anecdotes*, viii. 304), and no doubt wished to do his best for the new professor and for his own reputation, but the venture does not seem to have been a financial success; the second and third volumes of 1715 and 1719 were printed at Oxford for Henry Clements.

The ornaments by Simon Gribelin for the first volume must have been returned from Oxford to Lintot; they were used by John Watts for printing the first edition of the *Rape of the lock*, though not for the second, which, as Pope wrote, was not reprinted in 'so fair a manner as the first impression'. The engravings appear again in Pope's *Iliad* and the *Works* of 1717, later printed by Bowyer for Lintot, and these include other engraved initials that had been used in Trapp's *Praelectiones* as well as more in the same style. It may well be that Gribelin had produced more engraved decorations for Lintot for some earlier work which I have not identified.

Whether the use of these ornaments in the *Rape of the lock* was due to Pope or Lintot, the fashion did not rapidly spread beyond Pope's circle. In 1716 the fine-paper copies of Gay's *Trivia*, printed by Bowyer for Lintot, had three engraved headpieces substituted (on pp. 1, 25, and 53); only the first of these was new, the others being from Trapp's *Praelectiones*. Similarly, the fine-paper copies of Parnell's translation of *Homer's battle of the frogs and mice*, which was printed for Lintot in 1717 and seen through the press by Pope,[91] had engraved headpieces and initials substituted on the rectos of A3, B1, C6, and E5; the last three leaves were cancelled to make the substitution, which suggests that the change was an afterthought; the type had to be reset to make room for the engraved initials. The headpieces are again from Trapp, while the initials are those used in Pope's quartos.

There was no rush to follow this fashion of decorating poetic octavos; I know

[91] John Chalmers suggests Richard Harbin may have been the printer.

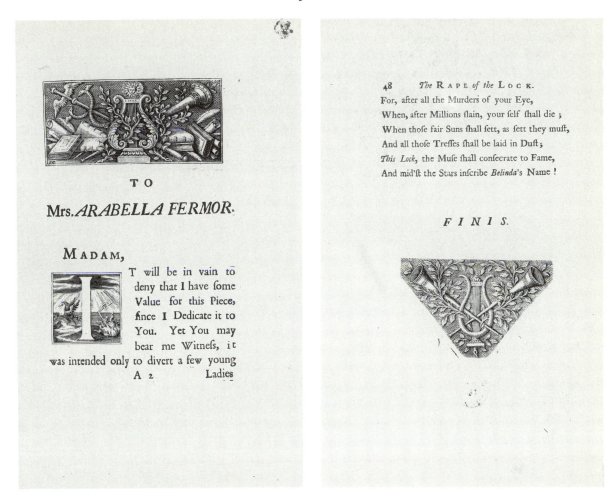

FIG. 23. Engraved headpiece and initial: *Rape of the lock* (1714), p. A2ʳ (Bodl. Don. e. 115; 214 × 138)

FIG. 24. Engraved tailpiece: *Rape of the lock* (1714), p. 48 (Bodl. Don. e. 115; 214 × 138)

of only two further examples. It was necessarily an expensive and difficult opera-tion, because the sheets had to have the text printed first, and then go through the engraver's rolling press with the plates placed in precisely the correct position in relation to the text. In 1719 Francis Peck's *Sighs upon the never enough lamented death of Queen Anne* was published by Henry Clements with engraved head-pieces and initials; this time they were drawn (with the exception of the vignette on the title-page) from the second volume of Trapp's *Praelectiones* (1715)—a nice parallel with Pope.[92] William Benson's translations of the first two *Georgics*, *Vir-*

---

[92] As with Pope's *Works* there is an engraved initial in this vol. not found, and not needed, in Trapp. Perhaps both Lintot and Clements ordered a complete set of initials from their respective engravers. Peck's *Proposals* (Bodl. fo. θ 663 (11)) say that if they 'permit such farther Expence as a good Frontis-piece, an Head and Tail-Piece, and an initial Letter to adorn the Design, they shall be done by a good Artist from the Designs of the Author'.

46 *De Natura & Origine* &c.

cos, Sapphicos, Anapæsticos, aliosque pluri-
mos quibus abundat Poeseos fœcunditas, ex-
pendant. Non, quod contemnenda Nobis
ista videantur; sed quod Pueris, non Acade-
micis sint explicanda. Neque nostri est insti-
tuti in arena tam sterili atque arida versari,
vel circa exteriorem Poeseos corticem ludere;
sed ipsius penitissimos recessus rimari, & non
tam de verbis, quam de rebus differere.

P R Æ-

[47]

PRÆLECTIO

QUARTA.

De Stylo Poetico.

UM nobis propositum sit,
istam, quæ à Poetis usur-
patur, scribendi atque co-
gitandi rationem discute-
re; oportet ut in primo
limine hoc præmoneamus,
plurima esse ad Argumen-
tum quod præ manibus ha-
bemus spectantia, quæ verbis vix sunt asse-
quenda, sed multo commodius mente pos-
sunt concipi, quam dictis proferri. Ve-
rum esse hoc constabit, cuivis exploratum
habenti Poesis quam sit delicatæ atque ele-
gantis naturæ; quam varia sint illi rerum,
vocum·

FIG. 25. Engraved headpiece, initial, and tailpiece: Joseph Trapp, *Praelectiones poeticae* (1711),
pp. 46–7 (Bodl. 8° G 177 Art; 190 × 240)

gil's husbandry, of 1724–5 have a few agreeable engravings by John Pine and Ger-
ard Van der Gucht, but that seems to exhaust my notes for the first half of the
century. I know nothing of fashions in prose, and for engravings in larger formats
we must wait until we consider Pope's translations of Homer in the next chapter;
but once again Pope must be seen as a pioneer, if one with few followers on this
occasion.

WORKS AND POEMS ON SEVERAL OCCASIONS, 1717

We can complete our survey of the first phase of Pope's career by looking at the practical arrangements for Pope's collected *Works* of 1717; its aesthetic aspects I shall consider in relation to the *Iliad*, with which it is closely related. The two are in fact closely connected, if only because the *Works* were advertised for publication at the same time as the subscribers' copies of the third volume of the *Iliad* on 3 June 1717 (*Post boy*, 28 May); no prices are quoted, but presumably volumes sold at the same price as volumes of the *Iliad*. Lintot already owned the copyright of most of Pope's poems, and was therefore free to reprint them in folio and quarto to match the volumes of the *Iliad*. Pope clearly co-operated closely in adding new material, which makes up about one-seventh of the volume (excluding the preliminary leaves). In addition, on 5 October 1713 when Pope was paid by Tonson for his contribution to Steele's miscellany, they had made an agreement that Pope could reprint the works to which Tonson held the copyright in a collected edition, the bookseller 'allowing me books in proportion to the number of sheets the said poems amount to'. Conversely, if Tonson were to print Pope's collected works, he would include poems previously printed by other booksellers.[93] Tonson actually provided about a third of the volume; the number of sheets in the quarto is roughly Lintot, 28; Tonson, 18; Pope, 8. Tonson's share was reckoned as one-quarter, and so entered in the Stationers' Register; Bowyer's accounts, after listing the quarto copies, 500 royal and 250 writing royal, records '125/63 Titles red & black with Mr. Tonson's Name'.[94] These were cancel titles for the quarto; it looks as though the folio copies had Tonson's name added to his share by stop-press correction ('The Titles red & black for the fine Paper with Alterations'). No doubt the 'Agreement about Pope's Works' with Tonson, recorded in Lintot's memorandum book under 13 June 1717 (Nichols, *Anecdotes*, viii. 303) relates to arrangements about Tonson's share of the edition.

Pope's agreement with Lintot is recorded seven months after the publication, by an indenture dated 28 December 1717, which goes into great detail to ensure that Lintot receives every right possible under the Copyright Act of 1710 (BL MS Egerton Charter 129). It recites that the author or his assignee has the sole rights of printing for fourteen years from the first day of publishing, and that the right then returns to the author. Pope then grants the sole right of printing (other than the listed works formerly sold to the two Jacob Tonsons) to Lintot as long as he has any authority by the Act of Parliament. At the end of fourteen years Pope

---

[93] BL MS Egerton 1951, fo. 1, repr. in *Correspondence*, i. 191–2.

[94] The writing royal copies, like those of the *Iliad*, can be distinguished by their having 3 lines in the bend of the Strasburg bend where the seconds have 4 lines (see *Twickenham*, x. 589–90). These accounts were printed by Fleeman, '18th-century printing ledgers', p. 1056. One entry in them I do not understand is the '5 Forms 4to wrought Headpieces at 5s. each'. There are 10 headpieces in addition to many initial letters and tailpieces, and more like 40 formes involved. It may be that most of the engravings were printed in another shop, and are excluded from Bowyer's accounts.

shall not sell the rights to any other person, and if he be then living shall upon request and at Lintot's charge make a new or further grant for another term of fourteen years, and during all and every other term as he may have a right to grant. In return for this, Pope is to receive 'One Hundred and Twenty of the said bookes printed on Royall paper in quarto in Quires'. This is parallel to the arrangements made by Prior and Gay for the collected editions of their poems.[95]

In addition, Lintot binds himself to pay £20 to Pope over and above any penalties to which he is liable by Act of Parliament, if he shall print any poems or works as Pope's other than those contained in the published volume or for which he may hereafter have Pope's licence under his hand and seal. Norman Ault was no doubt right in thinking that this referred to those of his juvenilia which Pope included in the miscellany *Poems on several occasions* which (as he showed) was edited by Pope and had been published by Lintot on 13 July 1717, six weeks after the *Works*.[96] It could also have been a precaution against Lintot's daring to print such Curll productions as *To the ingenious Mr. Moore* and *A Roman Catholic version of the First Psalm*. There is no payment to Pope for the miscellany in Lintot's accounts, and one could suggest that Pope's editing of this collection was a favour to recompense Lintot for his disappointment at the lack of profit from the *Iliad*. There is at the same time some force in Ault's contention that Pope, having produced the definitive collection of poems he wished to acknowledge, was anxious to save some of his juvenilia from oblivion.

Almost all the works included in the miscellany are by friends of Pope, but apart from the poems of Broome and James Ward (a friend of Parnell) which are found in groups through the volume, the most conspicuous contributors are the Duke of Buckingham and Lady Winchilsea. Ault says that 'all except four of the Duke's seventeen poems were first printed in this miscellany';[97] and I think there is bibliographical evidence to show that Pope gave special attention to his work (he was to edit the Duke's works for the Duchess in 1723), and also made space for an additional contribution from Lady Winchilsea.

As in the Tonson miscellany of 1709 and the Lintot of 1712, we find two batches of paper and evidence that the volume was not printed normally from beginning to end. As Table 5 shows, the first part of the book, A–K⁸, is printed on paper watermarked GI, and so is O⁸; the intervening and final sections are watermarked I[N]S (the middle letter is obscure). This clearly suggests that O was printed off during the production of A to K, and since O was already printed when additional material came to be inserted, an extra gathering *N⁸ had to be inserted. We can, I think, explain how this came about.

---

[95] The contracts are printed in S. Rivington, *The publishing family of Rivington* (1919), pp. 58–60. They were subscription edns.: Gay received 250 copies free and was to pay 11*s*. 6*d*. each for the next 500 subscribers' copies; Prior received 1,100 free copies and paid Tonson £1 each for 900 extra copies (as well as £80 for recomposing 73 sheets, which had to be reprinted because of the flow of subscribers).

[96] Ault (ed.), *Pope's own miscellany* (1935), p. xxvii.

[97] Ibid., p. xciii; the early poems are titled as 'by the Earl of Mulgrave'.

TABLE 5. *Poems on several occasions*, 1717

Collation	Paper		Contents
A⁴			
B–K⁸	GI	G1^r–G6^r	Pope
L–N⁸ *N⁸	INS	M6^v–N4^v	Finch
		N5^r–*N4^r	Winchilsea
		*N4^v–*N8^v	Buckeridge
O⁸	GI	O1^r–O6^v	Buckingham
		O7^r–P4^r	Pope
P⁸	INS	P4^v–Q2^r	Buckingham
Q²	INS	Q2^v	Winchilsea[a]

[a] Q² was actually printed as Q⁴, with Q2 as Lintot's booklist, Q3 as cancel O6. There are two states of sheet A/Q, the earlier at Bodley and Clark; Ault says they are from two settings of type; I have only been able to compare the titles, and here the authors' names are reset but the rest is from standing type.

O1 begins the group of six choruses by the Duke of Buckingham. We have seen on earlier occasions that printing started early on sections of particular importance in miscellanies, either in order that they could be revised in proof or printed off early for presentation to friends.[98] This seems to have been a similar situation, though whatever care was taken, second thoughts called for a cancel for O6, the last leaf of this group, so that in stanza 4 of the second chorus for *Marcus Brutus* the first two lines read

> Our senators great *Jove* restrain
> From angry piques they prudence call;

instead of 'From cautious votes'.[99] When we look at the intervening section for an addition that might have called for the additional signature *N⁸ the one homogeneous section of any length is the group of four poems by Lady Winchilsea which occupies N5^r–*N4^r (pp. 185–'183'): fifteen pages out of the sixteen we need to explain the added signature. Moreover, Bainbrigg Buckeridge's poem 'On Buckingham-House' which intervenes between Lady Winchilsea's poems and the Duke of Buckingham's choruses on O1 has obviously been spaced out with short pages to fill one more page than it would normally have done. And if we ask why Lady Winchilsea's poems had to be inserted here rather than printed at the end of the book, the answer is not only that more of Buckingham's poems

[98] It may be that G1 was another early start. This is the beginning of the group of poems which have been generally accepted as Pope's; they occupy only 11 pages, as Buckingham's choruses occupy 12.

[99] The cancelland survives in the Yale copy; in the *Works* of 1723 ed. Pope the reading is 'private piques'. Sherburn (*Correspondence*, i. 386–7) quotes two undated notes and a fragment from Buckingham to Pope, the first two of which seem to relate to the publication of the choruses in this miscellany. In the first, having made a change to remove the word 'Lust', he goes on 'The other changes I referr intirely to you, not only as my freind but the best iudg I know'.

were to go there (P4^v–Q2^r), keeping one of Pope's major contributors in a conspicuous place, but more importantly that Lady Winchilsea's group of four poems are enabled to follow immediately on two by 'the honourable Mrs Finch'—earlier works by the same author.[100]

With this third miscellany, we can leave Pope's early career.

[100] There is one further possibility. If this second group of Buckingham's poems had followed directly after O6, the 6 leaves would have neatly completed the vol. with a P^4 to be printed with the preliminaries A^4, and have finished the vol. with a solid group of Buckinghham's poems in the traditional fashion. I wonder whether the Duke hesitated about their inclusion, and in the mean time 11 pages containing 4 translations now accepted as by Pope were interposed. These finish on P4^r, and if the Duke had decided to withhold his second group, the vol. would have ended there. In this case, as in the vol. as it exists, the last page could have had the 6-line poem by 'Mrs Finch'.

2

HOMER: BUSINESS AND AESTHETICS

BUSINESS: THE *ILIAD*

The production of Pope's translation of the *Iliad* is revolutionary both commercially and aesthetically, and in some respects commerce and aesthetics interact. It appeared at a time when the publication of works in weekly or monthly parts was becoming popular, and the idea of publishing it in six annual volumes rather than in one or two large volumes, as had previously been usual, meant a similar reduction in the working capital employed.[1] In principle, Lintot had only to pay for the production of one volume, the sales of which would in turn provide the money for the second volume, and so on to the end. As a result, Pope could negotiate better terms with Lintot than if Lintot had had to provide capital for printing the whole work at once; that would have been about £1,800 in addition to Pope's 1,200 guineas for the copyright, or £3,000 in all—a truly formidable figure. In order to make part publication possible, and extract the maximum profit from it, the work had to be swelled out by the use of large formats and large type, even for the annotations to each book, which occupy almost as many pages as the text. What Mme Dacier printed in three volumes duodecimo, Pope and Lintot swelled to six volumes in quarto or folio.

Publishing in six volumes on the instalment system also meant that subscribers could be asked for more money because it was spread over six years. The terms are given in the *Rape of the lock*, third edition, in July 1714:

It is proposed at the rate of one Guinea for each Volume: The first Volume to be deliver'd in Quires within the space of a Year from the Date of this Proposal, and the rest in like manner annually: Only the Subscribers are to pay two Guineas in hand, advancing one in regard of the Expence the Undertaker must be at in collecting the several Editions, Criticks and Commentators, which are very numerous upon this Author.

A third Guinea to be given upon delivery of the second Volume; and so on to the sixth,

---

[1] The *Iliad* falls outside the scope of R. Wiles's *Serial publication in England before 1750* (Cambridge, 1957), which is limited to books published in fascicules at least once a month. Wiles shows Pope himself was a subscriber to Herman Moll, *Atlas geographus*, starting in May 1708 (pp. 84 and 271–2).

for which nothing will be required, on consideration of the Guinea advanced at first. Subscriptions are taken by *Bernard Lintott*.[2]

The subscribers were only asked for 2 guineas in the first place, though the total would be 6; and this money went directly to Pope. (The 'Expence the Undertaker must be at in collecting the several Editions, Criticks and Commentators' must have been amply covered by the extra 650 guineas or so included in the first instalment.) By contrast, only those 101 noble subscribers to Dryden's Virgil whose names were printed below the plates subscribed 3 guineas in advance; the 250 second subscribers only paid a guinea.[3]

A further inducement to subscribers was that only their copies would contain 'Ornaments and initial Letters engraven in Copper'. So far as I am aware, in earlier projects the advantage offered to subscribers was either a reduced price or (as in Dryden's Virgil) fine-paper copies not available to those who bought the trade edition—in addition, of course, to having the honour of their names printed at the beginning of the volume. In the case of the *Iliad* the subscribers' copies were additionally distinctive in being printed in quarto rather than in folio, which I shall suggest shows a change in taste. Both folios and quartos were printed from the same basic setting of type, differently imposed for the two formats. The fine-paper folios cost the public a guinea in sheets like the subscribers' quartos, but were without the engravings.

Before we turn to the aesthetics of the *Iliad*, we must take the opportunity presented to us by the records to attempt an estimate of the financial consequences of Pope's contract with Lintot. In addition to the contract—a massive parchment appropriate to the size of the deal, and similar to those for the *Works* of 1717 and the *Odyssey*—and a subsidiary agreement on the delivery of the subscribers' copies, we have a record of Lintot's payments and the entries in Bowyer's ledgers for all the printing and for the paper of the last three volumes: the one unknown is the cost of the paper, and this we can estimate to within about 20 per cent.[4] The contract provides that Lintot shall pay 200 guineas for each volume to have 'the sole and absolute property' for all such times as Pope 'may can might or could' have it, and print 750 copies of each volume 'on a Royall Paper of a Quarto size of the same sort whereof a Specimen is hereunto annexed'.[5] These copies for the

---

[2] Quoted in *Twickenham*, vii, p. xxxvi n.

[3] The first subscribers paid 5 guineas in total, the second subscribers 2; see J. Barnard, 'Dryden, Tonson, and subscriptions for the 1697 Virgil', *PBSA* lvii (1963), 129–51.

[4] The copies of the contract for the *Iliad* are in Bodl. (MS Don. a. 6) and BL (MS Egerton Charter 128); Lintot's payments are recorded in Nichols, *Anecdotes*, viii. 299–300 and in unidentified extracts (perhaps from the *Literary gazette*) in Bodl. John Johnson; the Bowyer ledgers are also in Bodl. (see Ch. 1, n. 85 above).

[5] The first 100 guineas to be paid on the signing of the contract and the second on delivery of copy for vol. I. Thereafter a further 100 guineas was payable within three months of publication of vol. I and the second on delivery of copy for vol. II, and so on. If Pope did not deliver the copy for any volume within a year of the advance, he was to pay 6 per cent on the 100 guineas from the expiry of the year. He was also to repay any advance with interest if he were to die 'or through sickness or otherwise desist from proceeding' in it. Any part he had completed would then become Lintot's without any payment. For differences between the Bodleian and British Library copies, see Appendix A below.

subscribers are to be delivered to Pope or to such persons as he appoints at Lintot's shop, and no copy beside these 750 shall be printed 'on the same or any other Royall Paper or with the Same head pieces tail pieces or initial letters or any other engraven on Copper'. The last condition was broken by Henry Lintot, who apparently allowed 75 copies of the first volume to be reprinted in 1738 to make up sets when Gilliver took over Lintot's stock of the quartos.[6]

The 750 copies were, I suppose, a maximum figure for free subscribers' copies; the contract for Gay's *Poems on several occasions* (1720) specifically says this.[7] In the event there were only 575 subscribers for 654 copies (Griffith, i. 41), and Bowyer's ledgers show that 660 copies were printed, 200 on writing royal and the rest on second or printing royal paper.[8] When Pope referred back to the contract in 1740 he could not explain this discrepancy—'Bowyer printed all along of the Iliad but 660, 4⁰ in stead of 750 articled to be mine' (*Correspondence*, iv. 224)—but I think we can do so by supposing that Pope and Lintot agreed to apply the difference in cost between 660 and 750 copies to provide fine-paper copies. Bowyer's first printing ledger gives plenty of examples of the price of paper for the years 1719–22, and Dutch printing royal is priced between 28s. and 32s. 6d.; writing royal is 54s. (for 'insides') or £3 a ream.[9] If, as seems likely by the inclusion of a sample, the paper specified by the contract was best Holland royal at 32s. 6d. a ream, the cost of paper per gathering would have been £2. 8s. 9d. for 750 copies. If, when the total was reduced to 660 copies, an inferior printing royal was used costing 28s., it would be possible to print 200 copies on writing royal without increasing Lintot's overall expenditure on paper.[10] While the figures I give in Table 6 can in no way be verified, the relationship between different qualities and prices should remain constant. By establishing these prices we can get near to calculating what Pope's subscription copies cost Lintot; we cannot calculate Lintot's profits without knowing what he paid for the paper for the folio

[6] See *Correspondence*, iv. 223–4, and Appendix A below.

[7] Rivington, *The publishing family of Rivington*, pp. 59–60. Gay was to have 250 copies for his new copyrights and 500 more at 11s. 6d. each; no more were to be printed.

[8] Printing ledger i, fos. 8–10, and paper stock ledger, fo. 41ᵛ. Lintot records 650 copies (Nichols, *Anecdotes*, viii. 300) but I think Bowyer's the more reliable figures. The writing royal paper can be distinguished from the seconds by the presence of 3 lines in the bend of the Strasburg bend watermark; the seconds have 4 lines. For a discussion of the *Iliad* paper, see *Twickenham*, x. 589–90.

[9] See ledger i, fos. 100, 102,, 104, 110; cf. Spence, *Anecdotes*, no. 200, where Pope gave the price of paper for his later vols. as 'The quarto paper, Genoa, 27s. or 28. Royal paper 55s. and some 60s.' I have taken a ream to contain 500 sheets rather than the traditional 480 on the evidence of many entries in Bowyer's and Ackers's ledgers. Most routine calculations are made on this basis; for an exception see *A ledger of Charles Ackers*, ed. D. F. McKenzie and J. C. Ross (Oxford, 1968), p. 79, where there are details of allowance for making perfect, amounting to some 8 per cent against the minimum 4 per cent necessary for 480 sheet reams. Bowyer ledger i, fo. 47 has a calculation (rounded up) allowing something over 2 per cent. An undated auction list of Genoa papers 'in partnership between Janssen and Roberts, at their late Dwelling-house in *Dean's Court* the South Side of St. *Pauls*' lists most lots as either 'perfect' or '25 sheets' (copy in the Halliwell-Phillipps collection, Chetham's Library, Manchester). I suspect paper merchants found the trade preferred to have their making perfect done for them.

[10] I have taken 56s. a ream as a figure for the writing royal to allow some perfecting on the lower price of 54s.

TABLE 6. Cost of Fine Paper for *Iliad*

Original estimate of paper cost per gathering:

Best Holland royal at 32*s*. 6*d*. per ream
Cost for 750 copies: $\dfrac{750 \times 32s.\ 6d.}{500}$ £2. 8. 9.

Alternative:

Second Holland royal at 28*s*. per ream
Cost for 460 copies: $\dfrac{460 \times 28s.}{500}$ £1. 5. 9.

Holland writing royal at 56*s*. per ream
Cost for 200 copies: $\dfrac{200 \times 56s.}{500}$ £1. 2. 5.

£2. 8. 2.

copies he could sell to the trade. In an advertisement in the *Post man* of 14 July 1720 the large-paper folios are said to be 'printed on fine Engl. writing Demy' and the small folios 'on fine Genoa Pot'. After a good deal of hesitation I have decided on round figures of £1 a ream for the writing demy and 10*s*. for the pott.[11] With the help of Bowyer's accounts we can begin to calculate Lintot's profits and loss: and the crucial feature is that having printed 1,750 copies in small folio of the first volume, he reduced it to 1,000 for the later volumes.

I have set out Lintot's costs in Table 7. The left-hand column shows Lintot's payments to Pope for his copyrights—the change from £215 to £210 after the first three volumes is due to the devaluation of the guinea from £1. 1*s*. 6*d*. to £1 1*s*. 0*d*. in December 1717. The dates are those of the receipts on the Bodleian copy of the contract; their relationship to the two payments per volume laid down in the contract is obscure. The second column shows the price Lintot records for the '650 Royal Paper' for each volume. The rest of the table is given over to a calculation of the cost of each volume, based on Bowyer's accounts (ledger i, fos. 8–10). It gives the number of sheets, the cost for print, and the estimated cost for paper; the dates are those in the accounts. Quartos, to the left, and folios, to the right, are totalled separately. The first thing to consider is the relationship between the cost Lintot noted for the subscribers' books in the second column and the calculated cost in the third. Volume I was by far the biggest because of

---

[11] Bowyer's prices for pott range from 7*s*. 6*d*. to as high as 13*s*. for 'fine Dutch pot'. The appearance and watermarks of vol. I correspond to good Genoa paper; the later vols. are browner and without corner countermarks, though the paper could still be Italian. In the circumstances a middle figure seems reasonable. It is surprising to find English writing demy specified when in 1712 the papermakers were petitioning Parliament with the claim that as yet English paper was only good enough for 'newspapers and pamphlets'. In fact the official handbook, *Instructions to be observed by the officers employ'd in the duties on paper*, 1713 (copy at Yale) includes writing demy in its table of dimensions of 'papers, that are usually made in England'. Woodfall charges fine writing demy at 25*s*. in 1735 to 1738 (P.T.P., 'Pope and Woodfall', pp. 377–8), while Wright charged Pope 18*s*. for Dutch paper of this size according to Spence, *Anecdotes*, no. 200.

TABLE 7. Lintot's Costs for *Iliad*

Copyright payment	Lintot figure for subscriber copies	Estimated printing costs							
		Quartos for subscribers				Folios for trade			
		Total	Paper	Print	Shts	Shts	Print	Paper	Total
Vol I £215 23 Mar 1714	£176	£171 =	£132 +	£39	55 (20 May 1715)	99	£79 +	£50 + £173 =	£302 (1750 pott)
Vol II £215 9 Feb 7 May 1716	£150	£134 =	£103 +	£31	43 (19 May 1716)	78	£49 +	£39 + £78 =	£166 (1000 pott)
Vol III £215 9 Aug 1717 6 Jan 1718	£150	£115 =	£89 +	£26	37 (22 May 1717)	67	£42 +	£33 + £67 =	£142 (1000 pott)
Vol IV £210 3 Mar 1718	£150	£139 =	£107 +	£32	44½ (3 June 1718)	79½	£50 +	£40 + £79 =	£169 (1000 pott)
Vol V £210 17 Oct 1718 6 Apr 1719	£150	£128 =	£99 +	£29	41 (9 May 1720)	76	£48 +	£38 + £76 =	£162 (1000 pott)
Vol VI £210 26 Feb 7 May 1720	£150	£113 =	£86 +	£27	36 (9 May 1720)	64½	£42 +	£32 + £65 =	£139 (1000 pott)

the preface and the essay on Homer, and the discrepancy between £176 and £171 is slight. Moreover, Lintot had to pay for the engraving of headpieces, tailpieces, and initial letters as well as for their printing; this is not covered by Bowyer's accounts—no doubt the printing of the engravings was done by another firm.[12] Perhaps the discrepancy of £16 for volume II would allow for the cost of engraving, but Lintot has clearly chosen to give a round figure of £150 per volume as the cost of the later volumes, regardless of the bill for print and paper: since some of these volumes were quite thin, he seems to be overestimating his costs. His figures for copyright and production give an average cost per volume of £367 for the quartos and I shall use that figure in subsequent calculations as his payment to Pope.

We must now try to see how Lintot could match this outlay by his profit on the folios he could sell to the trade (Table 8). We know from the *Monthly catalogue*

[12] According to Carter's *History of Oxford University Press*, the prices at the rolling press were 'absurdly low'; on p. 130 he cites 3s. for 1,000 impressions of a title-page.

TABLE 8. Lintot's Potential Profits on *Iliad*

			£	s. d.
1. *On sale of 250 demy and 1750 pott folio*				
Demy:	trade price			15 9
	unit cost			4
	profit per vol.			11 9
	profit per edn. of 250		145	
Pott:	trade price			9
	unit cost			2 4
	profit per vol.			6 8
	profit per edn. of 1750		583	
Less payment to Pope			367	
Profit on edn. of 1 vol.			361	
Total profit on edn. of 6 vols.			2166	
2. *On sale of 250 demy but only 1000 pott folio*				
Demy			145	
Pott			333	
Less payment to Pope			367	
Profit on edn. of 1 vol.			111	
Total profit on edn. of 6 vols.			666	
3. *On sale of duodecimos*				
Maximum profit per set of 6 vols.				6
Edn. of 2500, 27 June 1720			750	
Edn. of 5000, 8 December 1720			1500	
Total profit on 2 edns.			2250	

for June 1715 that the ordinary paper or pott copies sold at 12s. stitched or 14s. bound; the large paper or demy at one guinea stitched or 25s. bound.[13] We know from Henry Clements's notebook that ordinary copies sold to the trade in bulk ('by subscription' in the booksellers' as opposed to common usage) cost 8s. and that the regular trade price was 10s.[14] We can take a selling figure of 9s. as the average for our present purposes. For the large paper we can take a similar figure of three-quarters retail price or 15s. 9d. According to my calculations the average cost of pott copies was 2s. 4d., about a fifth of the retail price, and the large paper about 4s. On this basis the potential profit was about £145 for 250 of the large

[13] The 5th edn. of the *Rape of the lock* (1718) lists the first 3 vols. at £3 15s. 0d. and £2 2s. 0d., which corresponds to the bound price above. The advertisement for the duodecimo edn. in the *Daily courant* for 27 June 1720 lists Pope's *Works* of 1717 and the *Iliad* sets for 10 guineas for fine-paper quarto, 8 guineas for ordinary quarto, 7 guineas for large-paper folio, and 4 guineas for small folio; again there is no indication of binding, but the prices correspond to those for sets bound and gilt in 1727 (Griffith, i. 70) and work out at £1 10s. and £1 4s. for the quartos, and a guinea and 12s. for the folios. Vol. IV of the 2nd duodecimo *Iliad*, dated 1721, lists the large-paper vols. at one guinea, the ordinary at 14s., with no indication of binding, and in June 1726 (Griffith, i. 130) the prices are the same. In the advertisement of 1727 referred to above, the small folio sets are reduced to £4 or 11s. 6d. a vol.

[14] On 24 Jan. 1717 the wholesaling conger took copies of vols. I and II; the trade price is here listed as 20s. the two vols. In contrast to this clearing of stock after publication, Henry Clements subscribed for 25 copies of vol. IV on 13 June 1718 (just before publication) at 8s. a copy (*Notebook of Thomas Bennet and Henry Clements*, p. 172).

paper and £583 for 1,750 of the small, giving Lintot a potential return of £728 per volume against his payment to Pope of £367, a profit of £361. That looks fairly reasonable and divides the returns 50:50.

If Lintot had succeeded in selling all these copies, Pope would not have driven a hard bargain; but as we can see, the printing order for the ordinary copies dropped from 1,750 to 1,000 for volumes II to VI. I suppose that it might be sensible to print more copies of a first volume than of later ones in the expectation that some customers who buy the first will fail to come back for the later volumes, but it is common to find a first volume reprinted because it sold better than the others. (The first volume of the subscribers' quarto was in fact reprinted in 1738 to complete sets.) There seems to me little doubt that the reduction in the printing order was the result of poor sales, and as a consequence Lintot's possible profit after his payment to Pope dropped to £111 a volume, or about £666 overall; a very different story.

This, of course, assumes that Lintot could at least sell all the thousand copies he printed, but we know that this was far from the case. Some twenty years later, in May 1739, Thomas Osborne of Gray's Inn advertised that Pope's *Iliad* and *Odyssey* in folio would be sold at 8s. a volume in the large paper and 6s. a volume the small paper, in sheets, until 1 August.[15] Henry Lintot told Pope that he had 'sold a large Number of the small folios' (*Correspondence*, iv. 223), and Osborne was able to sell them at half price. Only 250 small-paper copies would need to remain unsold in order to wipe out any profit for Lintot (quite apart from interest on his capital), so we can begin to understand his frustration.

The traditional explanation for his poor sales is that given by Johnson, no doubt working on information given him by John Nichols:

It is unpleasant to relate that the bookseller, after all his hopes and all his liberality, was, by a very unjust and illegal action, defrauded of his profit. An edition of the English *Iliad* was printed in Holland in duodecimo, and imported clandestinely for the gratification of those who were impatient to read what they could not yet afford to buy. This fraud could only be counteracted by an edition equally cheap and more commodious; and Lintot was compelled to contract his folio at once into a duodecimo, and lose the advantage of an intermediate gradation. The notes, which in the Dutch copies were placed at the end of each book, as they had been in the large volumes, were now subjoined to the text in the same page, and are therefore more easily consulted. Of this edition two thousand five hundred were first printed, and five thousand a few weeks afterwards; but indeed great numbers were necessary to produce considerable profit.[16]

I have become increasingly sceptical about the accuracy of this interpretation. It is certainly true that Thomas Johnson of The Hague reprinted the *Iliad* in a

---

[15] *Daily gazetteer*, 11–29 May; the *Iliad* and *Odyssey* were available separately. A note records 'The above Books will bind to the Quarto size of Mr. Pope's Works'. In Jan. 1740 Osborne printed further advertisements beginning 'There being a very small Number left of Mr. Pope's Homer's Iliads, in six Volumes, Folio, in large and small Paper, as also in Quarto, large Paper . . .'. As Sherburn notes, Pope was curious to know how quarto copies intended for subscribers came to be in Osborne's hands and asked Henry Lintot about it (*Correspondence*, iv. 222 n).

[16] Johnson, *Lives of the English poets*, ed. G. Birkbeck Hill (Oxford, 1905), iii. 111–12.

cheap edition in small octavo, the first three volumes dated 1718 and the fourth 1719. Lintot had received a royal privilege dated 6 May 1715, which as usual forbade reprints or abridgements 'within our Kingdoms and Dominions' and also the importation and selling of copies printed abroad. The first prohibition was a way of preventing the work being reprinted in Ireland, which was not subject to the Copyright Act of 1709/10; the second seems to add nothing to the provisions of the Act. The crucial question is whether smuggled copies of the Dutch edition were a sufficient threat to make Lintot change his plans. There is no doubt whatever that copies of Johnson's piracies were smuggled in, particularly to Scotland; but though they may have circulated in the provinces, it seems unlikely that they would have had much currency in London, which was the heart of the book trade. The respectable trade was accustomed to combine against piracy, and Lintot could take legal action against those who infringed the law; I suspect that the piracies of Pope's later works (from 1728 on) had what success they did just because Pope, and not an influential member of the trade, held the copyright. Pope could only go to law, whereas the trade could first apply pressure. So although the Dutch edition might have had some effect on provincial sales, I doubt whether it would have been catastrophic, any more than the Scotch piracies which subsequently became so widespread.

The other weakness in Johnson's account is his assumption that Lintot lost 'the advantage of an intermediate gradation', by which I suppose is meant an edition in octavo. In fact, in such parallel cases as Prior's and Gay's *Poems*, and Addison's *Works*, or Pope's Shakespeare, the grand edition was succeeded directly by a duodecimo, but only after a delay of three years or more. This period gave the bookseller time to sell off sets of the completed work in the large format without competition from a cheap edition, and Lintot would no doubt have sold more of his folios if he had followed suit. At the same time it is clear that Lintot was having difficulty in recovering his costs on the *Iliad*, and there were therefore arguments in favour of getting a quick return by as large a cheap edition as possible. The real reason for Lintot's difficulties in making a profit on the folios was not the Dutch piracy, but his own over-optimism when he made his contract with Pope. I notice that when he published the *Odyssey* six years later, he produced a duodecimo edition within a month of the folio, even though there was no Dutch piracy to compete with.

Having made his decision to produce a duodecimo, Lintot's strategy was clearly to have it ready for the public very soon after the larger formats. The subscribers' copies of volumes V and VI were published on 12 May and the folios on 19 May 1720, and the cheap edition went on sale on 27 June, the third volume having been printed (with some revisions by Pope) as early as 1 September 1719 (*Twickenham*, x. 588).

The duodecimo sets sold well; the first edition of 2,500 copies was followed within six months by a second edition of 5,000 (advertised in the *Daily courant* of 8 December 1720). This was over-optimistic, since a third edition was not needed for another twelve years; it was then reprinted at about six-year intervals until the Lintot copyrights were sold in 1759. It is less easy to calculate Lintot's profits from the duodecimo because it was advertised at 'half a Crown each Volume bound Sheep' (*Daily courant*, 27 June 1720), and the problem of who paid for the binding and how it affected the trade price is still obscure. I have taken 6*s.* a set as a minimum figure for the profit, and on this basis the profit on the first duodecimo edition was £750 (Table 8), or more than could have been achieved from the first edition once the number of small folios was reduced to 1,000.[17]

Lintot was not only disappointed in his profits, he was already irritated by Pope when he came to publish the first volume. Pope's contract provided that Lintot should sell no copies of his folio until a month after the subscribers' copies had been delivered or notice of their being ready for delivery given in the newspapers. Yet on 10 June 1715, four days after copies of the first volume were announced as ready for subscribers, Lintot wrote to Pope:

All your Books were deliverd pursuant to your directions the middle of the Week after you left Us. . . . Pray detain me not from publishing my Own Book having deliverd the greatest part of the Subscribers allready, upwards of four hundred.
    I designd to publish Monday sevennight pray interrupt me not by an Errata.
    I doubt not the Sale of Homer if you do not dissapoint me by delaying the Publication.
(*Correspondence*, i. 295.)

Lintot had advertised in the *Daily courant* for 6 June that his 'fine Folio edition' would be published 'next week'; in fact it was not advertised as published until the *Post man* of 30 June.[18] It appears Pope stood upon his rights, if not to the very day.

Another bone of contention concerned the delivery of copies to subscribers. Pope wrote to William Fortescue, Gay's Devonshire friend who became Pope's legal adviser, on 18 March [1715]:

Lintot has manifested great Desire, but attended with great Impotence, to play the Scoundrell. I could not but smile when he told me, you took him by the hand at parting, & advisd him, Mr Lintot, whatever you do, don't go to Law! You are to [be] a judge whether by the articles he is obligd to deliver the books at his Shop to the Subscribers

---

[17] There is no problem about the printing costs; Bowyer in 1719 charged 30*s.* for the first 1,000 and 6*s.* for each 500 thereafter as did Woodfall in 1736 (P.T.P., 'Pope and Woodfall', pp. 377–8), though Bowyer charged extra (55*s.* a sheet) for the 2½ sheets of index to vol. VI. If one takes the price of demy paper at 11*s.* 6*d.*, copies in sheets cost 3*s.* a set. To this must be added the cost of binding and the plates from the 1712 translation of Mme Dacier's version, which are apparently a regular part of it. At 5*s.* a set, 6*s.* profit would allow a trade price of 11*s.* against a retail price of 15*s.*

[18] *Twickenham*, vii, p. xxxvii n. Sherburn's suggestion in *Early career*, pp. 132–40, that publication of vol. I was deliberately delayed as part of a waiting game with Tickell is not corroborated by Bowyer's ledgers; there seems no doubt that the difficulties in drying the sheets in a wet winter were a genuine reason for delay.

who send for 'em. The words are—*& to deliver the same books at his House or Shop fr[om ti]me to time to whatever person or persons Mr Pope shall appoint.*[19]

We cannot tell what Fortescue as arbitrator decided, but Lintot certainly did deliver the first volume and receive the subscriptions in June 1715. Whatever the decision, Lintot was certainly under no obligation to collect the subscriptions, and this is sufficient reason for a supplementary document dated 10 February 1716 which records a new agreement about the second volume of the *Iliad*. According to this, Lintot is to receive for his own use all the subscription money for the second volume only, and pay Pope 400 guineas in addition to the copy money for the volume. In return he shall at his own expense furnish subscribers with one or more second volumes according to the number each person has subscribed for. He shall also deliver to Pope on demand 120 second volumes, and 'so many of such said One hundred & Twenty second volumes . . . shall be printed on Royall paper of the first and second sort as will answer the number of so many of the first volumes as have been already delivered to any of the Subscribers printed on such Royal paper'. No doubt the recipients of these 120 copies were identified on the list of subscribers held by Lintot; Pope endorsed the agreement '& to furnish Mr P. with 120 books more for his own use for some particular Subscribers, the double ones, &c.'[20] Because of the problems involved in publication by subscription, which we must soon consider, we cannot put any precise figures on what Lintot might gain; to be highly hypothetical, if the 120 copies were for 120 subscribers, and if Pope could be sure that these 120 of the 575 subscribers would pay him and not Lintot despite the agreement, and if all the others called for their copies and paid Lintot, Lintot would receive 455 guineas in return for paying 400.

The subscribers' copies of the third volume were advertised for publication on the same day as Pope's *Works*, 3 June 1717, and no doubt Lintot agreed to publish this volume of the *Iliad* in return for the chance of selling matching copies of the *Works* to the subscribers. What happened with the fourth volume is a mystery. Lintot advertised the folio copies on large and small paper on 14 June 1718, but I can find no advertisement concerning the subscribers' copies. Bowyer's paper ledger shows that the 200 fine-paper and 448 ordinary copies were delivered to Pope on 3 June and 10 ordinary copies to Lintot on 6 June; there seems little doubt that Pope made his own arrangements for getting them to the subscribers, but we have no idea how. It seems that Pope did not wish to repeat the experiment, for Lintot made another agreement for the fifth volume and again

---

[19] *Correspondence*, ii. 290. Sherburn dates the letter 1724/5, assuming it refers to the *Odyssey*. He notes that the words quoted by Pope are not in the *Odyssey* indenture, but did not know that they approximate to those in that for the *Iliad*. I think 1715 must be the year, for Pope would not have signed the agreement of 10 Feb. 1716, discussed below, until he had cleared up the legal position.

[20] BL MS Egerton 1951, fos. 2–3, from which Sherburn prints extracts in *Early career*, p. 188. Lintot was to pay the first 100 guineas by 30 Apr. and the remaining 300 by Dec. 1716. Pope has a memorandum dated 20 May 1717 that he has received the full 400 guineas.

paid 400 guineas—doubtless on the same basis as for the second volume.[21] There remained the problem of the final volume.

Because subscribers paid two guineas in advance for the first volume, a third guinea on delivery of the second and so on, they would pay their sixth guinea on delivery of the fifth volume and later get the sixth free. So if Lintot were to deliver the fifth volume in return for collecting the subscriptions, there would be no way of getting the free sixth volume delivered—without paying him directly for the service, which was contrary to Pope's attitude to tradesmen. Hence, I think, the decision to publish volumes V and VI together, and Pope's advertisement in the *Evening post* of 19 May 1719:

the 5th Volume of that Translation now lies finished at the Press: But the said Mr. Pope having made greater Progress in the Remainder than he expected, or promised; hereby gives Notice, that he shall deliver the whole to the Subscribers by the Beginning of the next Winter, and that they may then receive the two last Volumes together, paying the Subscription, which is now due, at that time.[22]

But publication was delayed.

As late as 9 November 1719 Pope could write to Dancastle, 'I hope to finish the whole work by Christmass' (*Correspondence*, ii. 19), but he was delayed by illness in the winter, by building work at Twickenham, and perhaps worst of all by the 'Four very laborious and uncommon sorts of Indexes to *Homer*', of which he was 'forc'd, for want of time, to publish two only'.[23] Pope certainly did not deliver the volumes to subscribers at the beginning of the winter—the subscription copies were not available until 12 May and the folios until 19 May 1720.

If Lintot made less profit on the *Iliad* than he hoped, what of Pope? He did not find the 750 subscribers he originally envisaged, only 575 for 654 copies according to the printed list in volume I (Griffith, i. 41). Unfortunately one cannot simply multiply 6 guineas by 654 to find Pope's receipts; subscription lists are full of pitfalls.[24] As we shall see when we come to the *Odyssey*, friends who had helped Pope might be added to the list without their having paid anything, while paying subscribers might be omitted.[25] Benefactors might subscribe for multiple

[21] Nichols, *Anecdotes*, viii. 300; apparently it was at this time that Pope assigned 'the Royal Paper that were then left of his Homer' to Lintot.

[22] Quoted by Sherburn, *Early career*, p. 190. R. H. Griffith in 'A piracy of Pope's *Iliad*', *SP* xxviii (1931), 737–41, argues that vols. V and VI were published together to make it more difficult for Thomas Johnson to compete with either Lintot's folio or his duodecimo edn.; his vols. V and VI are dated 1721. This is an attractive theory and the idea could well have been in Lintot's mind, but I think the proposal is more likely to have come from Pope.

[23] *Correspondence*, ii. 43, which also records two unfinished essays 'one on the *Theology* and *Morality* of *Homer*, and another on the *Oratory* of *Homer* and *Virgil*' which 'must wait for future Editions, or perish'.

[24] See P. J. Wallis, 'Book subscription lists', *Library*, 5th ser. xxix (1974), 255–86, and F. J. G. Robinson and P. J. Wallis, *Book subscription lists: A revised guide* (Newcastle, 1975).

[25] There is at least one receipt by Pope for a subscriber whose name does not appear in the list, though this could be one who subscribed after vol. I was published (see C. Ryskamp, '"Epigrams I more especially delight in": The receipts for Pope's *Iliad*', *PULC* xxiv (1962–3), 36–8; the subscriber is Samuel Lowe). On Pope's subscription lists, see P. Rogers, 'Pope and his subscribers', *PubH* iii (1978), 7–36, and M. Hodgart, 'The subscription list for Pope's *Iliad*, 1715', in *The dress of words*, ed. R. B. White, jun. (Lawrence, Kansas, 1978), pp. 25–34.

copies as a form of patronage, but take only one or two; the best illustration I know is Gay's letter to Addison: 'I have sent you only two Copys of my Poems though by your Subscription you are entitled to ten, whatever Books you want more Tonson or Lintot upon your sending will deliver.'[26] One must not, however, assume that all multiple subscribers would be content with one or two copies if we are to take literally the letter of James Brydges, Earl of Carnarvon and later Duke of Chandos, to Simon Harcourt: 'As I desire to oblidge some friends as early as I can with this great work, I entreat you will subscribe for ten sets for me. I think ye subscription money is 2 Guin. each, & I enclose a Note for ye sum.'[27] The Earl of Carnarvon is down for twelve sets (the largest subscriber) but unless he is being very delicate in giving his patronage, he actually wanted the copies for his friends. Other friends of Pope, like John Caryll or John Elwood, collected lists of subscribers and sometimes forwarded the subscriptions, but were down for only a single copy themselves. It is possible that some benefactors wished to hide their generosity under an entry for only one copy in the list.

In normal cases of subscription, where half the cost was paid in advance and half on delivery of the book, death would take its toll of subscribers. Pope's scheme, calling for four further subscriptions, was open to this danger as well as to defaulters. He wrote to Caryll on 4 February 1718:

I find, upon stating the final account of the last [third] volume of Homer, that not above ten persons of all the living subscribers, have refused to continue and send for their third volumes; (a thing which I'm sure you'll be pleased to hear), of which number Sir Harry Tichbourn is one and Will. Plowden Esqr. another. I beg when you see 'em you would propose to repay 'em the subscription money, and take back their first volume, which may be sent me in one of the hampers. I have taken that course with the rest of my deserters, and may do it with evident profit, having a demand for more entire new sets than I can furnish any other way. (*Correspondence*, i. 463–4.)

It is not easy to reconcile this confident statement with the fact that Lintot was able to advertise sets of the quarto between 1720 and 1727, and later sell 75 sets lacking the first volume to Gilliver. A plausible explanation for the lack of 75 copies of the first volume is that since the difference between the number of copies subscribed for and the number of subscribers is 79, Lintot gave multiple subscribers all the copies of the first volume they were down for, whether they wanted them or not: it would be hard for Pope to ask his benefactors to return the extra copies, though he could easily do it with defaulting subscribers. After Pope made his agreement with Lintot for the second volume, by which Lintot gave him 120 copies and 400 guineas in return for the subscription money, Lintot would be able to reserve unclaimed copies for his own use; Pope might only be

[26] Gay, *Letters*, pp. 5–6, quoted by J. M. Treadwell in *TLS*, 7 July 1972, p. 777.

[27] Dated 9 Jan. 1715 and quoted in *Early career*, p. 139 n. Brydges wrote in response to a present of Pope's first book, presumably an offprint produced for the purpose. There is no entry for such a separate printing in Bowyer's ledgers, however; if part of the regular first vol. was used in this way that could explain a shortage of 75 first vols. later on.

able to supply new subscribers from his 120 copies. But these are speculative solutions, and still leave loose ends. Even more open to speculation is who received the 200 writing-paper quartos, a number far exceeding the multiple subscribers or the 120 copies which Pope received from Lintot 'for some particular Subscribers, the Double ones, &c.' Did the peerage and others receive preferential treatment? or did some subscribers pay extra for them?

It is impossible in view of these uncertainties to be precise about how much money Pope received, but we can produce maximum and minimum figures. The copy-money and full payment of 650 6-guinea subscriptions add up to £5,435; copy-money, full payment of the first subscription of 2 guineas, and four payments of 400 guineas (as stipulated in the later agreement with Lintot) makes the total £4,372. A middle figure of about £5,000 seems reasonable, and a great contrast with Lintot's return. As Lady Mary Wortley Montagu wrote, Pope 'outwitted Lintot in his very trade'.[28]

AESTHETICS: *ILIAD*, *WORKS* (1717), *ODYSSEY*

One of the most influential changes that Pope made in English book production was the introduction of the quarto format for the *Iliad* (as well as for the *Odyssey*, Shakespeare, and his collected works). The traditional format in England for such major works, from Chaucer, through Ben Jonson, Drayton, and Shakespeare, had been folio; and with the larger and grander folios pioneered by Ogilby in the 1650s we are brought to the unmanageable folios in which Dryden's works were finally produced in 1701. Tonson began to break down the tradition by printing Rowe's Shakespeare in octavo in 1709, and Congreve's *Works* of 1710 use the same format in a more elegant way.

In France, where (in company with Don McKenzie) I think we must look for the source of the new styles,[29] folio was used for Ronsard's works in 1623, and there is a grand series of Ovid's *Metamorphoses* of 1619, 1660, and 1667. But quarto seems (on the slender evidence I have accumulated) to have become the fashionable format around the middle of the seventeenth century.[30] I note French translations of the *Aeneid* of 1648 and 1658, an edition of Corneille's works in 1647, and illustrated first editions of the *Fables* of La Fontaine in 1668 and the works of Boileau in 1647. Perhaps most important in fixing the fashion were the series of editions of classical authors 'in usum Delphini', which Tonson

---

[28] 'Pope to Bolingbroke', line 55, in *Essays and poems*, ed. R. Halsband and I. Grundy (Oxford, 1977). Pope estimated Dryden's receipts from his Virgil at *c.*£1,200; see Spence's *Anecdotes*, no. 63. Barnard, 'Dryden, Tonson, and subscriptions for the 1697 *Virgil*', gives an admirable account which suggests this figure could only have been reached by supplementary gifts from dedicatees or patrons.

[29] See McKenzie, 'The London book trade in 1668', *Words*, iv (1974), 90–1; Bronson is also conscious of French influences in 'Printing as an index of taste' in his *Facets of the Enlightenment*, pp. 326–65.

[30] Folio remained the norm for the majestic productions of the Imprimerie Royale, but I am concerned with literary works.

may well have been imitating with his classical editions from the new University Press at Cambridge, and their successor, the Lucretius of 1712.[31] It still remains true that duodecimo editions in several volumes, often adorned with engravings, were the norm in France: the first quarto edition of the works of Racine, believe it or not, was printed in London by Tonson and Watts in 1723 with illustrations after Louis Chéron; the first quarto edition of Molière was that of 1734 with plates after Boucher. But the use of quartos for English works appears to have been another way of giving them the status of classics, in parallel with Tonson's publication of the Elzevier editions discussed in Chapter 1.

In England the use of folio for the luxury book as opposed to the reference work dies out rapidly after publication of the *Iliad* in 1715. If we look at the work of Watts for Tonson we still find folio used for the 1717 edition of Ovid's *Metamorphoses* edited by Garth, for the posthumous 1718 edition of Rowe's Lucan, and for Prior's *Poems on several occasions*, 1718. But with Trapp's translation of the *Aeneid*, 1718–20, Gay's *Poems on several occasions*, 1720, Milton's *Poetical works*, 1720, Addison's *Works*, 1721, and Pope's edition of Shakespeare, 1723–25, the quarto format is fully established. (I must add in passing that the bulky Shakespeare seems a retrograde step from Rowe's octavo, but it probably had to be bulky as a subscription edition for Tonson's benefit.[32]) From this time on, the large folio edition for major literary works became exceptional, though it remained the usual way of presenting pamphlet poems until the 1740s. Probably the real turning-point there was Edward Young's *Complaint: or, night thoughts* of 1742; the first edition of the 'First night' was printed in folio, but all the rest in quarto, and possibly Dodsley persuaded him to change the format. I certainly associate quarto poems from this period with Dodsley in particular, and Dodsley was Pope's protégé.[33]

Pope's contract with Lintot for the *Iliad* put Pope in charge of the design of the volume. A specimen of the royal paper it was at first intended to use is attached to the contract; the work is to be printed 'with a new Letter of such kind and size' as Pope shall choose or direct;[34] but above all, the subscribers' copies 'shall have

---

[31] Tonson also used quarto in printing the *Œuvres meslées* of Saint-Evremond in French, both for the edition by Des Maizeaux in 1705 and the enlarged edition of 1709.

[32] The evidence that the contracts originally provided for an octavo *Iliad* for subscribers, discussed in Appendix A below, suggests Pope may have been contemplating an even more radical departure from the customary folio. If so, his thought may have followed the same path as Tonson's on Shakespeare: first towards octavo but then back to large format for financial reasons.

[33] Johnson says that Pope 'assisted Dodsley with a hundred pounds that he might open a shop', *Lives of the English poets*, iii. 213. Earlier quartos have an association with the Scriblerus club: 3 of Swift's poems printed by Bowyer (*The lady's dressing-room*, 1732; *A soldier and a scholar*, 1732; *A beautiful young nymph going to bed*, 1734), and Arbuthnot's Γνῶθι σεαυτόν: *know your self*, published in 1734. This last is very closely associated with Pope's *Essay on man*, collected in quarto the same year and bearing the same Greek motto in the engraving on the title and as tailpiece on the last page; I have no doubt that Pope was instrumental in getting the work published before Arbuthnot's death in the following year. These are comparatively scarce works since there were no companions with which they could be bound in pamphlet vols.

[34] The *Monthly catalogue* of May 1714, advertising the *Proposals*, says, 'on the finest Paper, and a new Dutch Letter'; the advertisement in the 3rd edn. of the *Rape of the lock* says, 'on a Letter new Cast on purpose'.

THE

ELEVENTH BOOK

OF THE

ILIAD.

THE ARGUMENT.

The third Battel, and the Acts of Agamemnon.

Agamemnon having arm'd himself, leads the Grecians to Battel: Hector prepares the Trojans to receive them; while Jupiter, Juno, and Minerva give the Signals of War. Agamemnon bears all before him; and Hector is commanded by Jupiter (who sends Iris for that purpose) to decline the Engagement, till the King shall be wounded and retire from the Field. He then makes a great Slaughter of the Enemy; Ulysses and Diomed put a stop to him for a while; but the latter being wounded by Paris is obliged to desert his Companion, who is encompass'd by the Trojans, wounded, and in the utmost danger, till Menelaus and Ajax rescue him. Hector comes against Ajax, but that Hero alone opposes Multitudes, and rallies the Greeks. In the mean time Machaon, in the other Wing of the Army, is pierced with an Arrow by Paris, and carried from the Fight in Nestor's Chariot. Achilles (who overlook'd the Action from his Ship) sends Patroclus to enquire which of the Greeks was wounded in that manner? Nestor entertains him in his Tent with an Account of the Accidents of the Day, and a long Recital of some former Wars which he remember'd, tending to put Patroclus upon perswading Achilles to fight for his Countrymen, or at least to permit him to do it, clad in Achilles's Armour. Patroclus in his Return meets Eurypilus also wounded, and assists him in that Distress.

This Book opens with the eight and twentieth Day of the Poem; and the same Day, with its various Actions and Adventures, is extended thro' the twelfth, thirteenth, fourteenth, fifteenth, sixteenth, seventeenth, and part of the eighteenth, Books. The Scene lies in the Field near the Monument of Ilus.

THE

HE Saffron Morn, with early Blushes spread,
Now rose refulgent from *Tithonus'* Bed;
With new-born Day to gladden mortal Sight,
And gild the Courts of Heav'n with sacred Light.
When baleful *Eris*, sent by *Jove's* Command,   5
The Torch of Discord blazing in her Hand,

Thro'

FIG. 26. Engraved headpiece and initial: Pope, quarto *Iliad* (1715–20), iii. 816–17 (Bodl. CC 35 Art; 284 × 436)

OBSERVATIONS

ON THE

Seventeenth Book.

1346   HOMER's ILIAD.   Book XVII.

845 While *Greece* a heavy, thick Retreat maintains,
Wedg'd in one Body like a Flight of Cranes,
That fhriek inceffant, while the Faulcon hung
High on pois'd Pinions, threats their callow Young.
So from the *Trojan* Chiefs the *Grecians* fly,
850 Such the wild Terror, and the mingled Cry.
Within, without the Trench, and all the way,
Strow'd in bright Heaps, their Arms and Armour lay;
Such Horror *Jove* impreft! Yet ftill proceeds
854 The Work of Death, and ftill the Battel bleeds.

OBSER-

FIG. 27. Engraved tailpiece and 'Observations' section-title: Pope, quarto *Iliad* (1715–20) v. 1346–7 (Bodl. CC 37 Art; 280 × 424)

head pieces and tail pieces and initiall letters at the beginning and end of each Book and of the Notes engraven on Copper in such manner and by such Graver as the said Alexander Pope shall direct and appoint'.

Clearly the typography was planned between Pope and Bowyer, and the pattern is clear enough. Each book opens with a pictorical headpiece and engraved initial letter; facing this is a page with 'The Argument' of the book, with a headpiece composed of varying designs of printer's flowers (Figure 26). If there is space at the end of the book, there is an engraved tailpiece (Figure 27). The Observations on each book always start on a recto with a blank page facing; they too have a headpiece of type flowers and an engraved initial—many of these initials come from the set we have seen used in Trapp's *Praelectiones* (Figure 28). Again a tailpiece concludes the Observations if there is room. The result is effective, especially on the writing-paper quartos, but I find the page less impressive than Watts and Tonson's Lucretius of 1712 in a different type (Figure 29). It may well be that the heavier type face which is so effective with the lower-case Latin would look ugly with all the capitals and italic of English verse. I shall argue that Pope led the movement away from their use in English printing, but I suspect that he was more influenced by the desire to appear as a classic author than by considerations of pure typography. The folios have no engravings, nor do they have the headpiece of type flowers to the Argument and Observations of each book; in large paper they have an uncluttered grandeur (Figure 30).

As for the engravings, the contract was signed when Pope had been intensively studying painting with Charles Jervas for a year,[35] and it is not surprising that he should wish to superintend their design. Pope and Jervas were on very intimate terms for many years, and Jervas's house in Cleveland Court was Pope's regular home when he was in town; as late as 1726, subscribers to the *Odyssey* were asked to send to his house for their copies.[36] Among the proofs of the *Iliad* which I shall speak of later, there is one sheet where Lintot suggests the use of a printer's ornament rather than a copperplate at the end of the notes to book V, and Jervas replies: 'Mr Pope is not at home—There are very few Erratas in this sheet. I believe the Common Printers Ornament will not please.' There is even a suggested rewriting of the penultimate line of book VIII by Jervas which Pope adopts into his text.[37]

Jervas was a portrait painter, but he had started his career by copying the Raphael cartoons at Hampton Court and he collected studies of the Old Masters after he went to Rome in 1703. Pope in a letter of 29 November 1716 writes, 'I long to see you a History Painter', and makes it clear that he thinks he is qualified for that role by his studies (*Correspondence*, i. 377). It seems inconceivable that

---

[35] For Pope's relations with Jervas, see M. R. Brownell, *Alexander Pope and the arts of Georgian England* (Oxford, 1978), pp. 10–26, and W. K. Wimsatt, *The portraits of Alexander Pope* (1965), pp. 11–12.

[36] Wimsatt, *Portraits*, p. 13.

[37] Quoted by N. Callan, 'Pope's *Iliad*: A new document', *RES* NS iv (1953), 110–11.

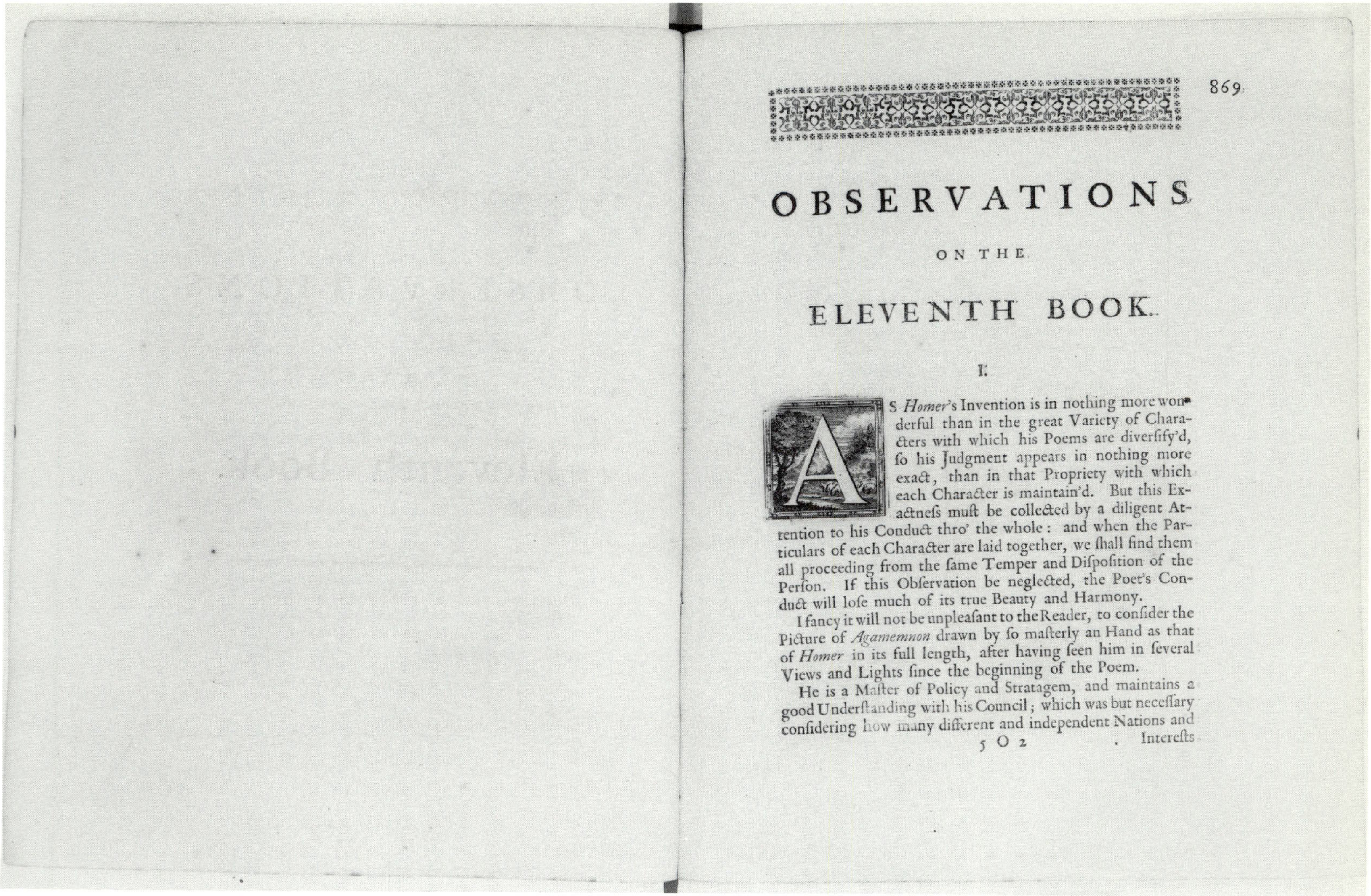

OBSERVATIONS

ON THE

ELEVENTH BOOK.

I.

S *Homer*'s Invention is in nothing more won-
derful than in the great Variety of Chara-
cters with which his Poems are diversify'd,
so his Judgment appears in nothing more
exact, than in that Propriety with which
each Character is maintain'd. But this Ex-
actness must be collected by a diligent Attention to his Conduct thro' the whole: and when the Particulars of each Character are laid together, we shall find them
all proceeding from the same Temper and Disposition of the
Person. If this Observation be neglected, the Poet's Conduct will lose much of its true Beauty and Harmony.

I fancy it will not be unpleasant to the Reader, to consider the
Picture of *Agamemnon* drawn by so masterly an Hand as that
of *Homer* in its full length, after having seen him in several
Views and Lights since the beginning of the Poem.

He is a Master of Policy and Stratagem, and maintains a
good Understanding with his Council; which was but necessary
considering how many different and independent Nations and
Interests

5 O 2

Fig. 28. Beginning of 'Observations' with engraved initial from Trapp's *Praelectiones*: Pope, quarto *Iliad* (1715–20), iii: 868–9 (Bodl. CC 35 Art; 284 × 436)

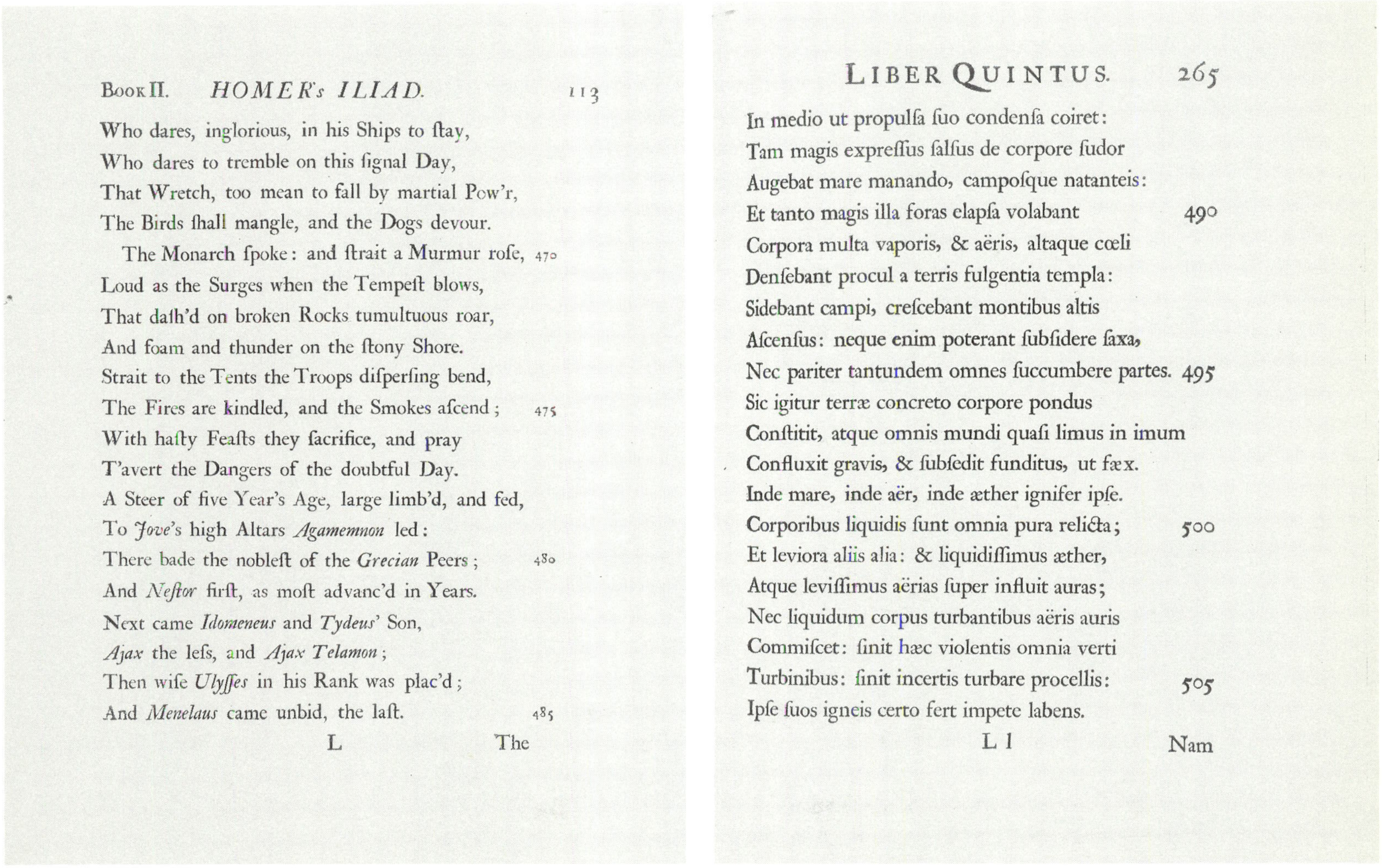

In medio ut propulsa suo condensa coiret:
Tam magis expressius salsus de corpore sudor
Augebat mare manando, camposque natanteis:
Et tanto magis illa foras elapsa volabant   490
Corpora multa vaporis, & aëris, altaque cœli
Densebant procul a terris fulgentia templa:
Sidebant campi, crescebant montibus altis
Ascensus: neque enim poterant subsidere saxa,
Nec pariter tantundem omnes succumbere partes.   495
Sic igitur terre concreto corpore pondus
Constitit, atque omnis mundi quasi limus in imum
Confluxit gravis, & subsedit funditus, ut fæx.
Inde mare, inde aër, inde æther ignifer ipse.
Corporibus liquidis sunt omnia pura relicta;   500
Et leviora aliis alia : & liquidissimus æther,
Atque levissimus aërias super influit auras;
Nec liquidum corpus turbantibus aëris auris
Commiscet: sinit hæc violentis omnia verti
Turbinibus: sinit incertis turbare procellis: 505
Ipse suos igneis certo fert impete labens.

            L l   Nam

113

HOMER's ILIAD.

BOOK II.

Who dares, inglorious, in his Ships to stay,
Who dares to tremble on this signal Day,
That Wretch, too mean to fall by martial Pow'r,
The Birds shall mangle, and the Dogs devour.

  The Monarch spoke: and strait a Murmur rose, 470
Loud as the Surges when the Tempest blows,
That dash'd on broken Rocks tumultuous roar,
And foam and thunder on the stony Shore.
Strait to the Tents the Troops dispersing bend,
The Fires are kindled, and the Smokes ascend ;   475
With hasty Feasts they sacrifice, and pray
T'avert the Dangers of the doubtful Day.
A Steer of five Year's Age, large limb'd, and fed,
To Jove's high Altars Agamemnon led :
There bade the noblest of the Grecian Peers ;   480
And Nestor first, as most advanc'd in Years.
Next came Idomeneus and Tydeus' Son,
Ajax the less, and Ajax Telamon,
Then wise Ulysses in his Rank was plac'd ;
And Menelaus came unbid, the last.   485

            L   The

FIG. 29. Pope, quarto *Iliad* (1715–20), i. 113 (Bodl. CC 33 Art; 288 × 229) and Lucretius, *De rerum natura* (1712), p. 265 (Bodl. AA 6 Jur; 290 × 224)

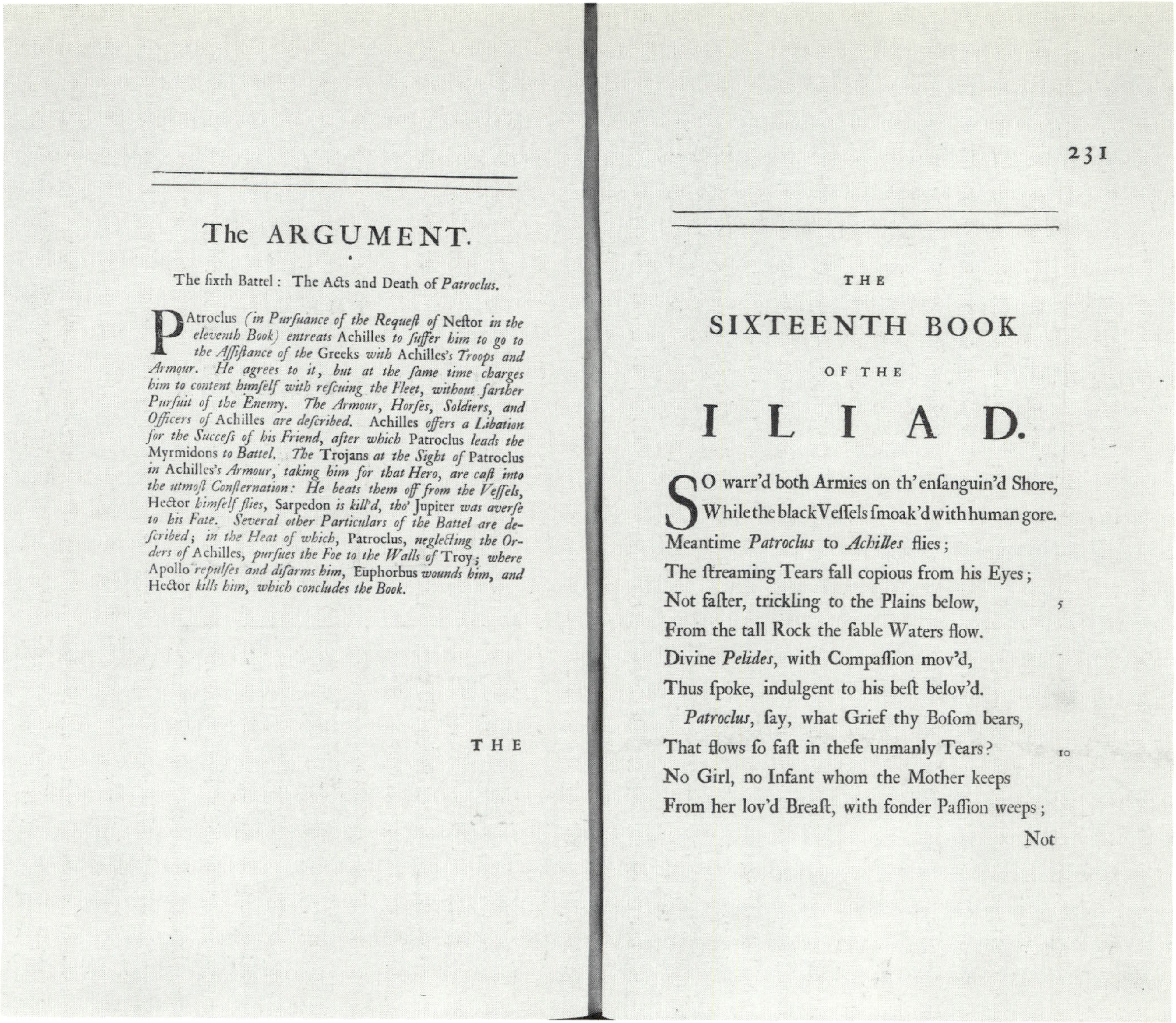

231

The ARGUMENT.

The fixth Battel: The Acts and Death of *Patroclus*.

Patroclus *(in Purfuance of the Requeft of* Neftor *in the eleventh Book) entreats* Achilles *to fuffer him to go to the Affiftance of the Greeks with* Achilles*'s Troops and Armour. He agrees to it, but at the fame time charges him to content himfelf with refufing the Fleet, without farther Purfuit of the Enemy. The Armour, Horfes, Soldiers, and Officers of* Achilles *are defcribed.* Achilles *offers a Libation for the Succefs of his Friend, after which* Patroclus *leads the* Myrmidons *to Battel. The* Trojans *at the Sight of* Patroclus *in* Achilles*'s Armour, taking him for that Hero, are caft into the utmoft Confternation: He beats them off from the Veffels,* Hector *himfelf flies,* Sarpedon *is kill'd, tho'* Jupiter *was averfe to his Fate. Several other Particulars of the Battel are defcribed; in the Heat of which,* Patroclus*, neglecting the Orders of* Achilles*, purfues the Foe to the Walls of* Troy*; where* Apollo *repulfes and difarms him,* Euphorbus *wounds him, and* Hector *kills him, which concludes the Book.*

THE

THE

SIXTEENTH BOOK

OF THE

I L I A D.

So war'd both Armies on th'enfanguin'd Shore,
While the black Veffels fmoak'd with human gore.
Meantime *Patroclus* to *Achilles* flies;
The ftreaming Tears fall copious from his Eyes ;
Not fafter, trickling to the Plains below,   5
From the tall Rock the fable Waters flow.
Divine *Pelides*, with Compaffion mov'd,
Thus fpoke, indulgent to his beft belov'd.
   Patroclus, fay, what Grief thy Bofom bears,
That flows fo faft in thefe unmanly Tears?   10
No Girl, no Infant whom the Mother keeps
From her lov'd Breaft, with fonder Paffion weeps ;

Not

FIG. 30. Pope, large-paper folio *Iliad* (1715–20), iv. 230–1 (Houghton *EC7. P8103. 715 iba; approx. 359 × 229)

Pope should have laid down the manner (or indeed the graver) of the ornaments as the contract permitted him without long consultation with Jervas and his circle, and this is suggested by Pope's letter to Jervas of 16 August 1714: '*Homer* advances so fast, that he begins to look about for the ornaments he is to appear in, like a modish modern author' (*Correspondence*, i. 243). The direct reference of these letters, which we know only from the edited version printed by Pope, is to the head of Homer, designed by Jervas and engraved by Vertue, which served as a frontispiece to the edition; of Jervas's involvement in the other engravings we have no evidence, either way. While I am both unqualified and unwilling to study the designs in detail, I must stress that Pope supervised them himself. It must have been Pope, for example, who laid on their designer the difficult task of producing a long narrow headpiece in place of the traditional upright plate.

This is another revolutionary departure for the English book. Traditionally illustrations were on plates facing each book, while headpieces were formal decorations. Pope turned the headpieces into the illustrations, perhaps trying to bring engraving and letterpress into a closer relationship. There may be French or Dutch precursors of this fashion; what is clear, however, is that just as quarto replaced folio as the standard format for the luxury book about 1718, the pictorial headpiece became fashionable. It is noticeable, however, that when Watts and Tonson used the pictorial headpiece they made it deeper in relation to its length, bringing it nearer to a conventional shape for pictures and lessening the artist's problems while making it less suitable for a quarto page.

We can see the traditional pattern of engraved plate, formal headpiece, and engraved initial in Ogilby's translation of the *Iliad*, 1660 (Figure 31). The first edition of Clarendon's *History of the Rebellion*, printed at Oxford in 1702–4, again uses the formal headpiece and engraved initial (Figure 32), though the dedication to each volume (Figure 33) has a pictorial headpiece which foreshadows later developments.

The pictorial headpiece reaches its most impressive form, though it is still formal rather than illustrative, with the 1712 Caesar on which Tonson had worked for ten years; the headpieces may well have been engraved in Holland like the plates (Figure 34).[38] Though there are many plates, each with its own dedicatee, they are kept apart from the opening of each book, which has its Argument facing it just as Pope's *Iliad* has. It seems to me that this is the chief influence on Pope; though Pope has abandoned the full page plates and has changed the format from folio to quarto, he uses headpieces of very similar proportions, the length being about 2.3 times the height.[39]

---

[38] Tonson's Lucretius of 1712 (ed. Maittaire) has a plate facing the opening of each book, and the first and fifth were certainly engraved in Amsterdam by G. van Gouwen. The first of the symbolic headpieces, however, gives M. Van der Gucht as the engraver; their ratio is about 2:1.

[39] Ogilby's headpieces have a ratio of 4:1; the formal headpieces of Clarendon about 2.7:1; earlier English pictorial headpieces, such as those to Barlow's Aesop, 1666 and 1687, or those to *Miscellaneous poetical novels or tales*, by Tate and Motteux, 1705, have conventional proportions and in no way fit the typography.

Lib.XIII.   279

HOMERS ILIADS.

The Thirteenth Book.

The Argument.

Neptune like Calchas th' Ajaxes first cheers:
To many prime Commanders next appears.
Hector their Camp, through Works deserted, fills.
Idomineus stout Othronius kills.
The rallied Grecians roughly entertain
The enter'd Foe: on both sides many slain.

W-Hen *Jove* had brought the
Trojans to the Fleet,
Where they did rougher
Entertainment meet;
He turning thence his splen-
did Eyes explores,
Renown'd for Chivalry, the
(a)*Thracian* Shores;

And (b) *Mysians* who in drawn up Squadrons fight,
And *Hippomolgs* that so in Milk delight,
A (c)long-liv'd Race for (d) Justice most extold;
Nor longer *Trojan* Bulwarks did behold;
                                                          Presuming
S f

Johanni Lewis de
Ebor: Armi: Fairolam

Rushton in Comitatu
hanc· L M D D D

Fig. 31. Engraved plate, formal headpiece, and engraved initial: Ogilby, *Iliads* (1660), pp. 278–9

FIG. 32. Formal headpiece and engraved initial: Clarendon, *History of the Rebellion* (1702–4), ii. 1 (Bodl. I 3.10 Jur; 430 × 272)

FIG. 33. Pictorial headpiece and engraved initial: Clarendon, *History of the Rebellion* (1702–4), i. 1 (Bodl. I 3.9 Jur; 430 × 272)

FIG. 34. Engraved headpiece and initial: Tonson's Caesar (1712), p. 147 (Bodl. 23643 b. 1; 472 × 280)

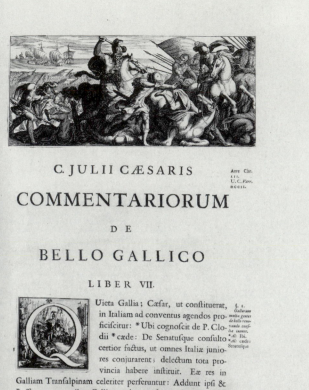

I have said that the pictorial headpiece now becomes fashionable in the publications of Tonson and Watts. Ovid's *Metamorphoses*, 1717, edited by Garth, harks back to the old tradition with full folio plates dedicated to subscribers as Ogilby's were. With the folio subscription editions of Prior's *Poems on several occasions*, 1718 (Figure 35), and Rowe's Lucan, 1718 (Figure 36), we have pictorial headpieces by Louis Chéron which are deeper than those used by Pope, the length being only twice the height. The headpieces to Trapp's translation of the

LUCAN's *PHARSALIA.*

BOOK III.

HRO' the mid Ocean now the Navy fails,
Their yielding Canvaſs ſtretch'd by
Southern Gales.
Each to the vaſt *Ionian* turns his Eye,
Where Seas and Skies the Proſpect
wide ſupply: 5
But *Pompey* backward ever bent his Look,
Nor to the laſt his Native Coaſt forſook.
His wat'ry Eyes the leſs'ning Objects mourn,
And parting Shores that never ſhall return;
Still the lov'd Land attentive they purſue,
'Till the tall Hills are veil'd in cloudy Blue,   10
'Till all is loſt in Air; and vaniſh'd from his View.
At length the weary Chieftain ſunk to Reſt,
And creeping Slumbers footh'd his anxious Breaſt:

When,

THE
FIRST CANTO.

ATTHEW met RICHARD; when
or where
From Story is not mighty clear:
Of many knotty Points They ſpoke;
And *Pro* and *Con* by turns They took:
Ratts half the Manuſcript have eat:
Dire Hunger! which We ſtill regret:
O! may they ne'er again digeſt
The Horrors of ſo ſad a Feaſt.
Yet leſs our Grief, if what remains,
Dear JACOB, by thy Care and Pains
Shall be to future Times convey'd.
It thus begins:
* * * * Here MATTHEW ſaid:

Mmmm2   ALMA

FIG. 35. Pictorial headpiece and engraved initial: Prior, *Poems on several occasions* (1718), p. 319 (Bodl. Vet. A4 b. 36; 454 × 274)

FIG. 36. Pictorial headpiece and engraved initial: Rowe's Lucan, *Pharsalia* (1718), p. 87 (Bodl. H 4.4 Art; 383 × 228)

Aeneid, 1718–20 (Figure 37), are deeper still (length about 1.7 times the height); they seem much less effective, and though this may be partly due both to inferior design (some in the second volume have the name of [Joseph?] Goupy as designer) and to the fact that they usually have a large tailpiece facing them from the end of the preceding book, I think the fundamental weakness is that they are too big, particularly for a quarto page. The headpieces to the subscription edition of Milton's *Poetical works*, 1720 (Figure 38), largely by Chéron, are slightly less deep (length about 1.9 times the height). It seems likely that the plates for these last two works were originally intended for a folio page, where they would be more at home.

While I have convinced myself that Pope was responsible for popularizing both the quarto format and the illustrative headpiece, I don't think they go well together. The narrow headpiece is fine for the formalized and crowded battle scenes of Caesar, but I wonder whether even so brilliant a designer could have made much of the few figures in the *Iliad* illustrations. Making the headpieces deeper certainly made the designer's task easier, and the folios look well; but put the large headpieces on to a quarto page, and all proportion is lost. It is probably no accident that Kent's headpieces to the *Odyssey* of 1725–6 (Figure 39) revert to the formal as opposed to the pictorial and their proportions are almost identical with those of Clarendon's *History* back in 1702. Tonson and Watts seem to have abandoned the grand illustrated book for a time; when they printed Waller's works edited by Fenton in 1729, Vertue's engraved headpieces (some of which abandon the strict rectangular shape rather effectively) are essentially portraits in a formal setting. For Cervantes' *Don Quixote* of 1738 they returned to the traditional use of plates and a narrow engraved headpiece above the text; the same is true for the English translation of *Don Quixote* by our painter Charles Jervas (but spelt Jarvis on the title) which was published posthumously in 1742. The one imitation of Pope's style I know is the translation of Virgil's *Aeneid* which John Theobald planned to publish in parts in 1739 with engraved headpieces by Gravelot, but it came to nothing.[40]

We must return to the *Iliad* and the question of who designed the engravings. The Twickenham editors comment that 'his pilferings may have something to do with his remaining anonymous' and list eighteen parallels between headpieces and initial letters and the designs used by Ogilby and those in the 1712 translation of Mme Dacier's French version by Ozell, Oldisworth, and Broome.[41] In relation to the borrowings from Ogilby we must remember Spence's anecdote:

---

[40] Theobald had published *The second book of Virgils Aeneid* in 1736 with both engraved plates dedicated to individuals in the tradition of Ogilby and Dryden and narrow pictorial headpieces which are unsigned. He followed it with *The fourth book* in 1739, without plates but with pictorial headpieces designed by Gravelot, and a prospectus dated 2 Nov. 1739 offered books II and IV, books I and III to follow in Mar. 1740, and a book every three months thereafter. The ratio of the headpieces is about 3:1.

[41] According to the *Twickenham* notes, the headpieces of books VIII, XX, XXIII and the initial letters to books III, XIV, XVIII follow Ozell; 'three or four more of his headpieces seem obligated in lesser ways' (vii, p. xiv). The headpieces of books X, XIII, XVI, XVII, XXI, XXII, XXIV and the initials to books I, XIV, XXI, XXII, XXIII draw 'to a greater or lesser degree' on Ogilby (viii, p. xiv).

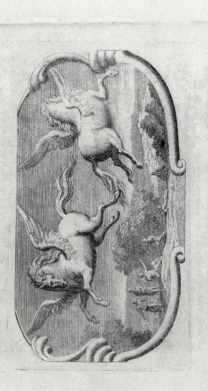

lvi   P R E F A C E.

both as a Conclusion of my Preface, and as a Kind of Poetical Invoca-
tion to my Work:

Hail mighty MARO! May That sacred Name
Kindle my Breast with Thy celestial Flame;
Sublime Ideas, and apt Words infuse:
The Muse instruct my Voice, and THOU inspire the Muse.

V I R G I L's Æ N E I S.

BOOK *the* FIRST.

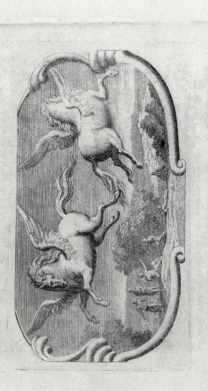

ARMS, and the Man I sing, who first
    from *Troy*
Came to th' *Italian*, and *Lavinian*
    Shores,
    Exil'd by Fate ; Much toss'd on
       Land, and Sea,
By Pow'r Divine, and cruel *Juno's* Rage.
Much too in War he suffer'd; 'till he rear'd

B

5
A

FIG. 37. Pictorial headpiece, engraved initial, and tailpiece: Trapp's *Aeneis* (1718–20), i. lvi, 1
(Bodl. CC 31 Art; 280 × 445)

FIG. 38. Pictorial headpiece and engraved initial: Milton, *Poetical works* (1720), i. 133 (Bodl. Vet. A4 d. 131; 287 × 222)

Ogilby's translation of Homer was one of the first large poems that ever Mr. Pope read, and he still spoke of the pleasure it then gave him, with a sort of rapture only on reflecting on it.

'It was that great edition with pictures. I was then about eight years old.'   (*Anecdotes*, no. 30.)

I find it hard to believe that the initial to book I of the *Iliad*, which echoes Ogilby's plate to that book (Figure 40), is not a deliberate tribute by Pope to his predecessor; some of the other borrowings may also be due to Pope's affectionate recollection of Ogilby. As to the more general question of the indebtedness of the

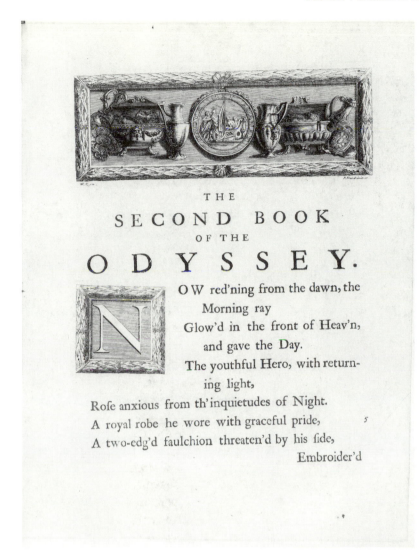

FIG. 39. Formal headpiece and
engraved initial: Pope, quarto
Odyssey (1725–6), i. 69 (Bodl. CC
64 Art; 281 × 212)

illustrations to previous designs, I feel unwilling to pursue the matter far since
the designs by Ogilby and those in the Ozell edition by no means exhaust the possible sources, though they are all that I have seen.[42] Nevertheless, if we tabulate
the headpieces that we know are based on earlier designs, they show an increasing
reliance on copies as the work proceeded, all the last five headpieces being copied
(Table 9). If I am right in thinking that Jervas was Pope's adviser, the fact that he

[42] Though the titles of Ozell's 1712 translation speak of cuts 'by the best Gravers, from the Paris Plates
design'd by Coypel', only the frontispiece bears Coypel's name; plates to books V–VIII read 'B. Picart
delin.' Picart appears to have designed the plates for the Amsterdam 1712 reprint of Mme Dacier's translation, though they are reported to be unsigned. I have not been able to trace the plates designed by
Coypel. The 3rd edn. of La Valterie's translation (Paris, 1708) has 25 plates 'dans le genre de R. de

(a)

(b)

FIG. 40. Plate facing opening of book I of Ogilby's *Iliads* (1660), p. 1 (Bodl. Douce H Subt. 42; 400 × 257) and engraved initial of opening of book I of Pope's *Iliad* (1715–20), p. 1 (Bodl. CC 33 Art; 39 × 38)

spent much of his time in Ireland between 1716 and 1720 might have some bearing on this change; the last time we know he was in England was in the summer of 1717, when he took the subscribers' copies of the third volume of the *Iliad* to Ireland with him (*Correspondence*, i. 410).

I suppose it must be clear that I am tempted by the proposition that Pope himself had a hand in designing the plates. On one level there is nothing surprising about this—an author is the obvious person to decide what should be depicted, and Wytze Hellinga illustrates the range of an author's ability from the increasingly sophisticated sketches of Constantijn Huygens (1644) culminating in a final design for the engraver, to the primitive layout by P. Marchand for the engraved

Hooghe, signées Petrus Balthazar Bouttats'. The Paris edn. of the translation by de La Motte, 1714, has 12 plates by 4 artists, and the Amsterdam reprint has 16 plates 'dont 7 sont signées: Broen sculp.'. See E. Crottet, *Supplément à la 5me édition du Guide de l'amateur de livres à figures du XVIIIe siècle* (Amsterdam, 1890).

TABLE 9. Indebtedness of *Iliad* Headpieces to Ogilby and Ozell

			Ogilby	Ozell	Total
Vol I	books I–IV	(1715)			
Vol II	books V–VIII	(1716)		VIII	1 out of 4
Vol III	books IX–XII	(1717)	X		1 out of 4
Vol IV	books XIII–XVI	(1718)	XIII, XVI		2 out of 4
Vol V	books XVII–XXI	(1720)	XVII, XXI	XX	3 out of 5
Vol VI	books XXII–XXIV	(1720)	XXII, XXIV	XXIII	3 out of 3

headpiece for his *Histoire de l'imprimerie* (1740).[43] In England we know that Dean Aldrich of Christ Church 'was considered to be a master of ornament, the only man for inventing . . . borders for a book'; his designs, preserved at the Press, were used for the first edition of Clarendon's *History of the Rebellion*.[44] Matthew Prior writes in 1718 of his own 'contriving Emblems such as Cupids, Torches and Hearts for great Letters' for his *Poems* of 1718, and Francis Peck's proposals for his *Sighs upon the . . . death of Queen Anne* offer 'a good Frontispiece, an Head and Tail-Piece, and an initial Letter to . . . be done by a good Artist from the Designs of the Author'.[45] If Pope had made designs for the *Iliad* he would not have been doing something very unusual, and Elias Mengel has suggested that later in his career he may have designed the illustrations to the *Dunciad variorum*. What we do not know is how polished a sketch he could have contrived. The only plate we know he designed is the frontispiece to the posthumous octavo of Warburton's edition of the *Essay on man*, 1745 (Figure 41); it is dated 6 February 1744, just before his death.[46]

The design of the first volume of Pope's *Works*, which appeared in 1717, matches that of the *Iliad*, and it was also printed by Bowyer. As I have said, this was Lintot's own venture and one in which Pope co-operated by adding poems not previously published. That he supervised the printing is shown by the note he sent to Broome to be forwarded to Bowyer:

I desire, for fear of mistakes, that you will cause the space for the initial letter to the Dedication to the Rape of the Lock to be made of the size of those in Trapp's Prælectiones [Figure 42]. Only a small ornament at the top of that leaf, not so large as four lines breadth. The rest as I told you before.

I hope they will not neglect to add at the bottom of the page in the Essay on Criticism, where are the lines 'Such was the Muse whose rules,' &c., a note thus: 'Essay on Poetry,

[43] Hellinga, *Copy and print in the Netherlands* (Amsterdam, 1962), plates 113–14, 174–5. For a French example, see L. B. Voet, *The golden compasses* (Amsterdam, 1972), ii. 213 n.

[44] Carter, *History of Oxford University Press*, p. 149.

[45] For Prior, see *The correspondence of Jonathan Swift*, ed. H. Williams (Oxford, 1963–5), ii. 290. A copy of Peck's proposals is in Bodl. (fo. θ 663).

[46] See E. Mengel, 'The *Dunciad* illustrations', *ECS* vii (1973–4), 161. B. Boyce gives a detailed discussion of the *Essay on man* plate, though not of Pope's role, in 'Baroque into satire: Pope's frontispiece for the "Essay on man"', *Criticism*, iv (1962–3), 14–27.

FIG. 41. Plate designed by Pope as frontispiece for *Essay on man* (1745) (Bodl. 12 θ 1006; 185 × 110)

by the present Duke of Buckingham,' and to print the line 'Nature's chief masterpiece' in italic [Figure 43]. Be pleased also to let the second verse of the Rape of the Lock be thus,

What mighty contests rise from trivial things.

(*Correspondence*, i. 394.)

We can assume that Lintot was paying for the new engravings, and these are mainly the nine headpieces by Gribelin (one of which was used both for the *Ode for musick* (Figure 44) and for the preface); many initial letters and tailpieces derive from Trapp's *Praelectiones* and the *Iliad*. The style is quite different from those we have seen before, combining an oval scene with formal decoration which breaks the straight line at top and bottom and impinges on the side of the frame as well (Figure 45). By keeping the pictorial element to the centre oval, problems of composition are avoided. I rather like their grotesques and trophies, but they do not always fit easily with the headings below them. Possibly their symmetry is

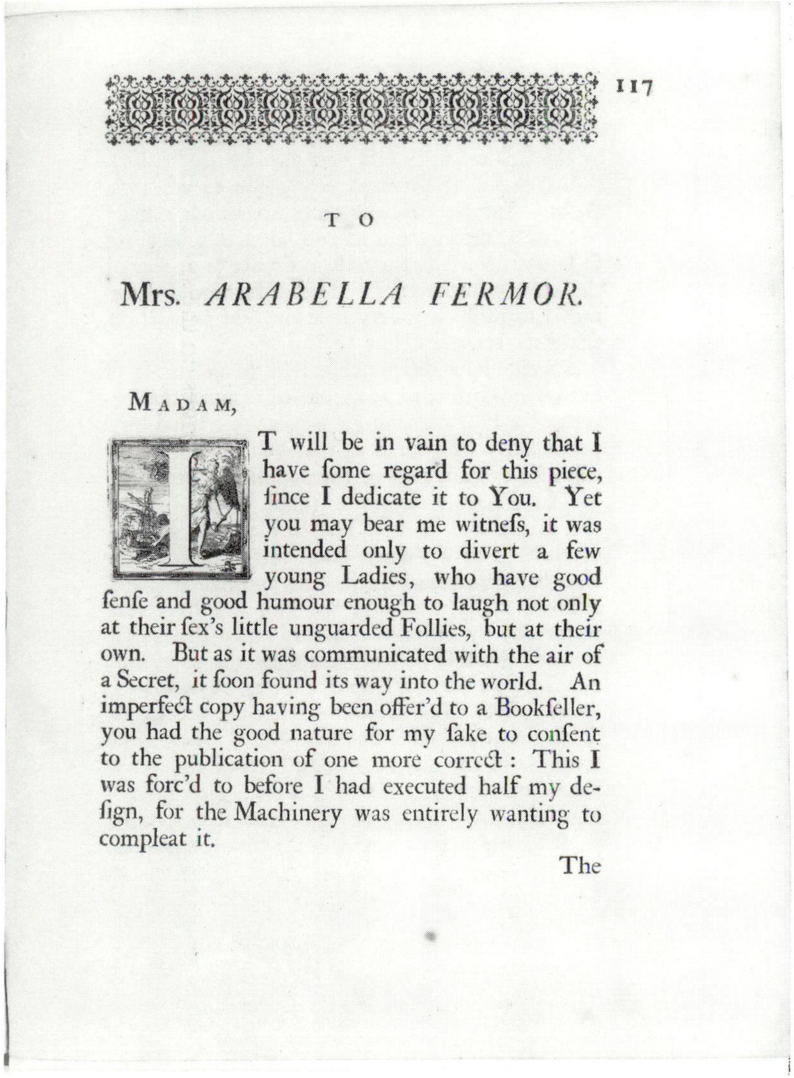

Cremona now shall ever boast thy name,
As next in place to Mantua, next in fame!
But soon by impious arms from Latium chas'd,
Their ancient bounds the banish'd Muses past;
Thence arts o'er all the northern world advance;
But critic learning flourish'd most in France:
The rules, a nation born to serve, obeys;
And Boileau still in right of Horace sways.
But we, brave Britons, foreign laws despis'd,
And kept unconquer'd, and uncivliz'd,
Fierce for the liberties of wit, and bold,
We still defy'd the Romans, as of old.
Yet some there were, among the founder few
Of those who less presum'd, and better knew,
Who durst assert the juster ancient cause,
And here restor'd Wit's fundamental laws
Such was the Muse, whose rules and practice tell,
Nature's * chief master-piece is writing well.

* Essay on Poetry, by the Duke of Buckingham.

Q   Such

FIG. 43. Page of *Essay on criticism* with Pope's note and italic: quarto *Works* (1717), p. 113 (Bodl. Vet. A4 d. 140; 290 × 218)

117

T O

Mrs. *ARABELLA FERMOR.*

MADAM,

I will be in vain to deny that I have some regard for this piece, since I dedicate it to You. Yet you may bear me witness, it was intended only to divert a few young Ladies, who have good sense and good humour enough to laugh not only at their sex's little unguarded Follies, but at their own. But as it was communicated with the air of a Secret, it soon found its way into the world. An imperfect copy having been offer'd to a Bookseller, you had the good nature for my sake to consent to the publication of one more correct: This I was forc'd to before I had executed half my design, for the Machinery was entirely wanting to compleat it.

The

FIG. 42. Dedication to *Rape of the lock* with engraved initial: quarto *Works* (1717), p. 117 (Bodl. Vet. A4 d. 140; 290 × 218)

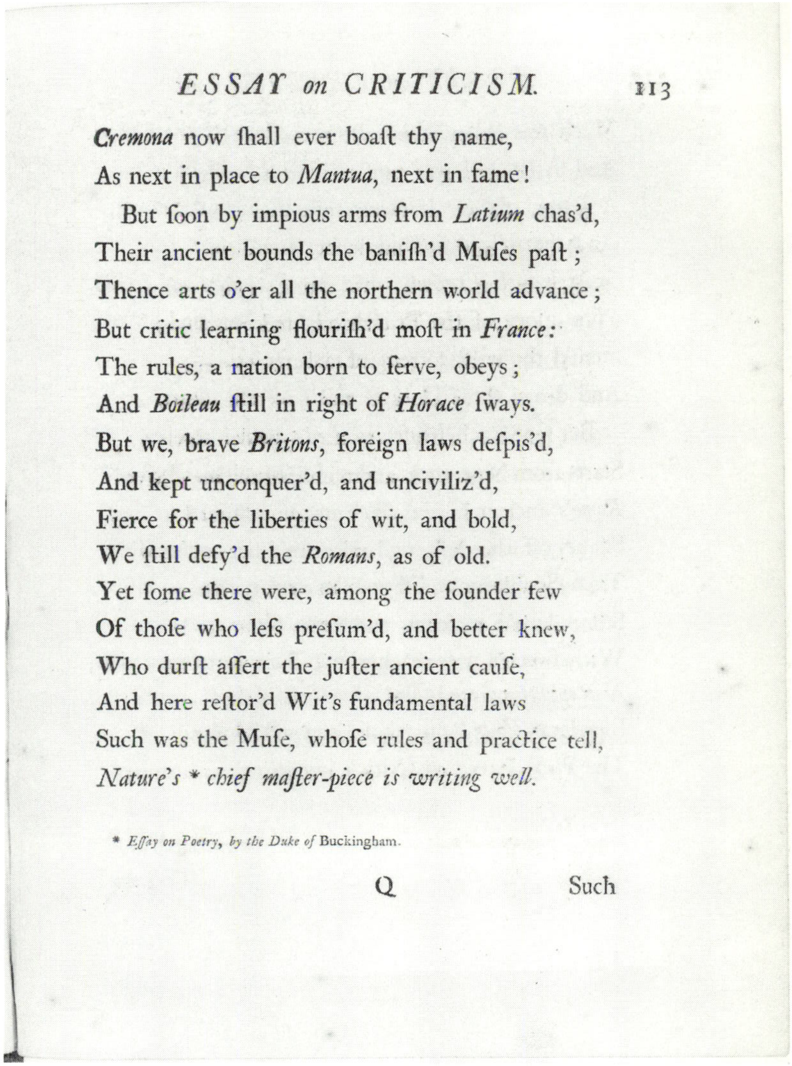

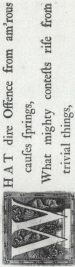

THE
RAPE *of the* LOCK.

CANTO I.

W HAT dire Offence from am'rous
    causes springs,
What mighty contests rise from
    trivial things,
I sing----This verse to C---, Muse! is due:
This, ev'n *Belinda* may vouchsafe to view:
Slight is the subject, but not so the praise,
If She inspire, and He approve my lays.

R

Say

FIG. 45. Engraved headpiece for *Rape of the lock*: quarto *Works* (1717), p. 121 (Bodl. Vet. A4 d. 140; 290 × 218)

ODE for MUSICK
ON
St. *CECILIA*'s Day.

I.

Descend ye nine! descend and sing;
    The breathing instruments inspire,
Wake into voice each silent string,
    And sweep the sounding lyre!
In a sadly-pleasing strain
Let the warbling lute complain:
Let the loud trumpet found,
Till the roofs all around
The shrill echos rebound:

Bbb 2

While

FIG. 44. Engraved headpiece for *Ode for musick*: quarto *Works* (1717), p. 371 (Bodl. Vet. A4 d. 140; 290 × 218)

also desirable; Kent's designs for the *Odyssey*, more formal still, are similarly symmetrical. We cannot tell how much influence Pope had in their design; Gribelin had been Lintot's engraver since Trapp's *Praelectiones* of 1711, and these cuts match those in style. The fine-paper folios have the same headpieces (though the tailpieces vary according to where poems end) and I think Bowyer might have done well to leave a little more space between engraving and heading when he had more height to play with (Figure 46). Nevertheless, to my eye the folio has slightly the better appearance, though this may be due to a personal prejudice against the dumpiness of quartos.

The *Odyssey*, of which the first three volumes were published in April 1725 and the last two in June 1726, was printed by John Watts, as we know from Pope's letter to Tonson, which Sherburn dates in February 1724:

> I must desire a favor of you, in return to this, which is also to redound to the credit of Mr Lintot. I mean in regard to the beauty of the Impression, that you will use your interest with Mr Watts, to cause them to work off the Sheets more carefully than they usually do: & to preserve the blackness of the Letter, by good working, as well as by the best Ink. The sheets I've seen since the first Proof, are not so well in this respect as the first. I beg your Recommendation as to this particular, There's nothing so mu[ch] contributes to the Beauty & credit of a Book, which would be Equally a reputation to Mr Lintot & to me. (*Correspondence*, ii. 217.)

Sherburn quotes a letter from 'Homerides' in the *London journal* of 17 July 1725, hoping that Pope and Lintot 'will not always expect to impose *extravagant Prices* upon us, for *bad Paper*, *old Types*, and *Journey work Poetry*' (*Early career*, p. 264). Compared with the *Iliad*, the *Odyssey* seems as well printed, and the type of the Observations to each book better; the paper of the subscription quartos is from Genoa, and whiter than the Dutch paper used for the writing-paper quartos, which resembles that for the *Iliad*.

To my eye, what cheapens the effect of the printing is the use of printer's ornaments in conjunction with the engravings, both for the Argument facing each book (Figure 47) and for the head of the Observations on each book; in both places Bowyer had used arrangements of printer's flowers which gave a neater effect. (The arguments here are briefer than in the *Iliad*, and result in a less satisfactory balance; it was felt a tailpiece was needed to complete the page.) The other change in appearance is the use of formal, not pictorial, headpieces, initials, and tailpieces designed by William Kent and paid for by Pope. Pope advertised 'Ornaments on Copper . . . fifty in number, designed by Mr. Kent'; this number does not include the initial letters that open each book, for which a complete alphabet was not necessary. The engraved initials for the observations seem to have come from Lintot's stock. I have already mentioned that the headpieces revert to the long narrow shape and the symmetrical, formalized pattern we found in Clarendon's *History of the Rebellion* in 1702. The new initial letters are larger, six lines deep rather than four lines deep as in the *Iliad*: this means the third line of verse has to be broken giving nine lines of print, and to my eye these two changes combine to give a better balance of text and headpiece. In the ordin-ary-paper folios the engraved headpieces are replaced by ordinary ornaments, and these go better with those facing them above and below the Argument; the effect may not be sophisticated, but it gives the buyer a conventionally decorated volume rather than an elegant by-product of the subscribers' edition. Perhaps this change of emphasis is a result of the conflict between Pope and Lintot.

One interesting technical problem arose from Pope's having twenty-four tail-pieces from Kent; whereas in the *Iliad* tailpieces were only used when there was a short page, here the printer had to try to make enough room at the end of each book for a tailpiece to be printed (Figure 48). He succeeded in doing this for every book except XX (in the quarto), though in one or two cases the engraving slightly overlaps text and catchword. The quarto is printed 20 lines to a page (like the *Iliad* and the *Works*), and his only resource was the way he dealt with triplets. It was a rule in printing rhyming couplets that a couplet could not be split between pages, so when there was a triplet on a page, the printer had the choice of printing 19 or 21 lines. The number of triplets could vary from book to book at least from 1 (book XVI) to 18 (book XIX), so the ease with which the printer could adjust his pages varied widely. I cannot help wondering whether on this occasion Pope adjusted the translation to make room for the tailpieces he had paid for.

BUSINESS: SHAKESPEARE AND THE *ODYSSEY*

I shall skirt around the editorial concerns of Pope after the completion of the *Iliad*. In December 1721 he brought out Parnell's *Poems on several occasions* without incident. His editing of Sheffield's *Works*, published on 24 January 1723, caused a major upheaval because of the Jacobite flavour of some of the prose works, to which the government took exception; the books were seized and there was a hue and cry of which Sherburn gives a good account (*Early career*, pp. 219–30). While he seems to have been paid nothing for editing Parnell's *Poems*, the situation in regard to Sheffield's *Works* is more obscure. He certainly wrote to Tonson on 3 September 1721, 'I have resolvd upon further thoughts to . . . have no concern in the profits at all' (*Correspondence*, ii. 81), but this was when he thought he would have little to do; a letter from Tonson to Pope of 1723–4 speaks of a balance of £187. 6s. being ready for Pope when they have cancelled the agreement between them (*Correspondence*, ii. 211–12), and this could be the return from the subscribers' copies. Meanwhile Pope was editing Shakespeare for Tonson, having signed an agreement with him on 22 May 1721 by which Tonson undertook to pay Pope £100, the work to be published within two years of the contract.[47] By comparison with Homer, this was very modest payment for a

---

[47] The contract is in the Houghton Library at Harvard, MS Eng. 233.13 (Fig. 49).

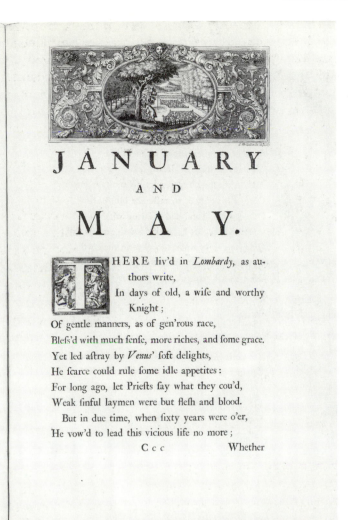

FIG. 46. Engraved headpiece for 'January and May': folio *Works* (1717), p. 189 (Bodl. fo. θ 677; 358 × 230)

great deal of work; clearly Pope did not undertake it for the money. Again Sherburn gives us an admirable account (*Early career*, pp. 232–47), though he lacked the contract and accounts. Pope seems to have finished his task by the end of October 1724, eighteen months late, though the work was not published until 12 March 1725.

The contract, which lays down Pope's duties as 'Correcting & writing a Preface & making Notes & Explaining the obscure passages', bears Pope's receipt for the £100, dated 26 February; perhaps the year is 1725 (Figure 49). The traditional story, based on figures extracted from Tonson's accounts, is that Pope was paid £217. 12*s*., but an earlier extract in the hand of Somerset Draper 'who was

The ARGUMENT.

Telemachus returning to the City, relates to Penelope the sum of his travels. Ulysses is conducted by Eumæus to the Palace, where his old dog Argus acknowledges his Master, after an absence of twenty years, and dies with joy. Eumæus returns into the country, and Ulysses remains among the Suitors, whose behaviour is described.

THE

THE

SEVENTEENTH BOOK

OF THE

ODYSSEY.

OON as *Aurora,* daughter of the
  dawn,
   Sprinkled with roseate light the
    dewy lawn;
    In haste the Prince arose, prepar'd
     to part;
His hand impatient grasps the pointed dart;
Fair on his feet the polish'd sandals shine,
And thus he greets the master of the swine.   s
  Vol. IV.   S   My

FIG. 47. Printer's ornament as tailpiece facing engraved headpiece and initial: quarto *Odyssey* (1725–6), iv. 128–9 (Bodl. CC 67 Art; 277 × 412)

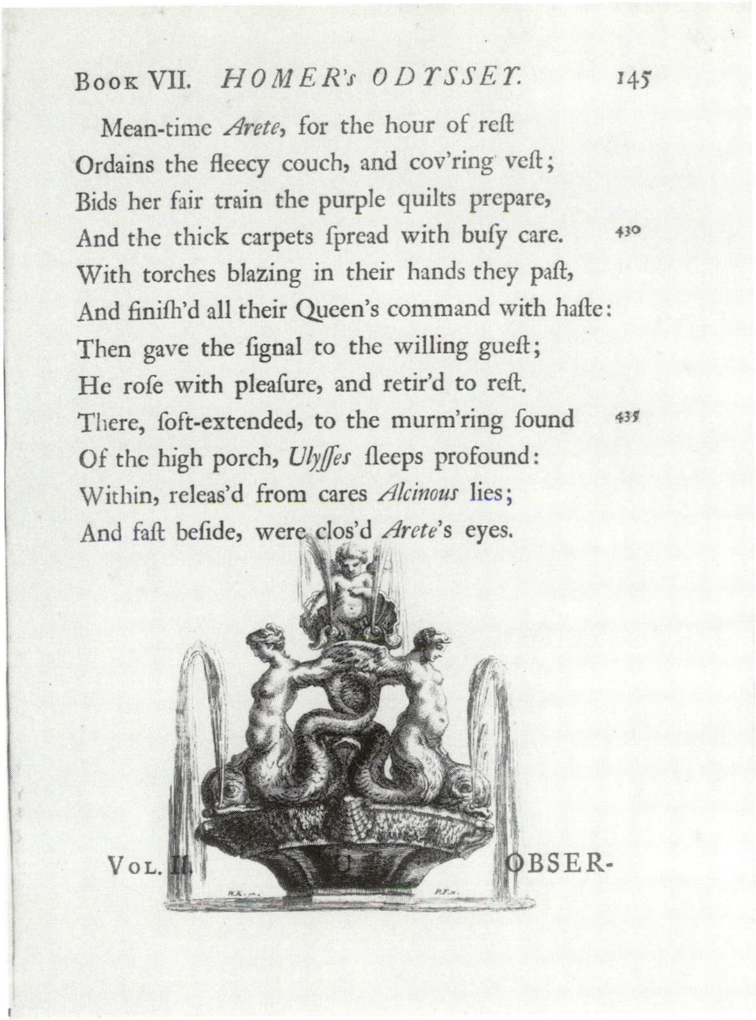

> BOOK VII.   *HOMER's ODYSSEY.*   145
>
>   Mean-time *Arete*, for the hour of reſt
> Ordains the fleecy couch, and cov'ring veſt;
> Bids her fair train the purple quilts prepare,
> And the thick carpets ſpread with buſy care.   430
> With torches blazing in their hands they paſt,
> And finiſh'd all their Queen's command with haſte:
> Then gave the ſignal to the willing gueſt;
> He roſe with pleaſure, and retir'd to reſt.
> There, ſoft-extended, to the murm'ring ſound   435
> Of the high porch, *Ulyſſes* ſleeps profound:
> Within, releas'd from cares *Alcinous* lies;
> And faſt beſide, were clos'd *Arete*'s eyes.
>
> VOL. II.   OBSER-

FIG. 48. Engraved tailpiece overlapping text: quarto *Odyssey* (1725–6), ii. 145 (Bodl. CC 65 Art; 282 × 210)

servant to the late Mr Tonson' enables us to be more precise (Figure 50).[48] Pope wrote to Tonson from Oxford on 3 September 1721 that 'I have got a Man or two here at Oxford to ease me of part of the drudgery of Shakespear, If you'l let me draw upon you (as you told me) by parcells, as far as sixty pounds as they shall have occasion'. Tonson annotated the letter '35*l* paid To Mr Pope's ordr, 25 D° to E: Fentons ordr, both sums paid to Mr popes Waterman. The 2 Notes for these Sums I gave up to Mr pope as being no part of the 1st agreement but which I agreed to allow Mr pope further that he might not want help' (*Correspondence*, ii. 81). So Pope received only the £100 for his own use. The payments of 20 guineas to Fenton and £31. 3*s*. 6*d*. to Gay in Figure 50 may also have been additional to

---

[48] The traditional account is to be found in 'Common-place notes', *Gentleman's magazine*, lvii (1787), 76, and in Johnson's *Lives of the English poets*, iii. 138. Draper's extract is Folger MS S. a. 163.

May 22d 1721. I do agree to pay Mr Alexander Pope one hundred pounds for Correcting & writeing a Preface & makeing Notes & Explaining the obscure passages in the Works of Mr William Shakespear & the said Alexander Pope doth agree to Publish the said Works of Mr William Shakespear in the manner before mentioned within Two years from the date hereof Witness our hands

A. Pope.
Jacob Tonson

Feb. 26. Receivd the said 100 pds.
A. Pope.

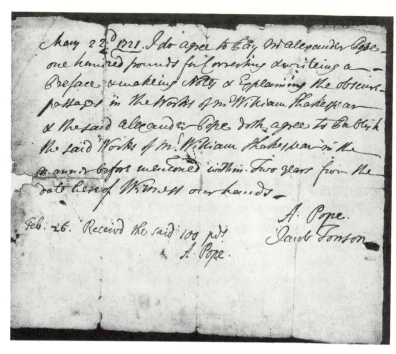

FIG. 49. Agreement between Tonson and Pope for the Shakespeare, 22 May 1721 (Houghton Library MS Eng. 233.13; approx. 165 × 195)

S.a.163

Paid the Editors of Shakespear.

Mr. Rowe	36:10:
Mr. Hughes	28: 7:
Mr. Pope – Money	160:–:
Do — Books	57:12:
Mr. Fenton money	21:
Do — Books	9:12:
Mr. Gay – money	31 : 3 : 6
Do — Books	4:16:
Mr. Whatley	12:
Mr. Theobald 400 on Demy & 110 Royal worth about	600
Do. 2d. Edition	52:10:0
	913:10:6

Bishop Warburton 500
Capells 300
Johnsons 250 Books 250

This is the hand writing of Mr Somerset Draper who was Partner to the late Mr Tonson

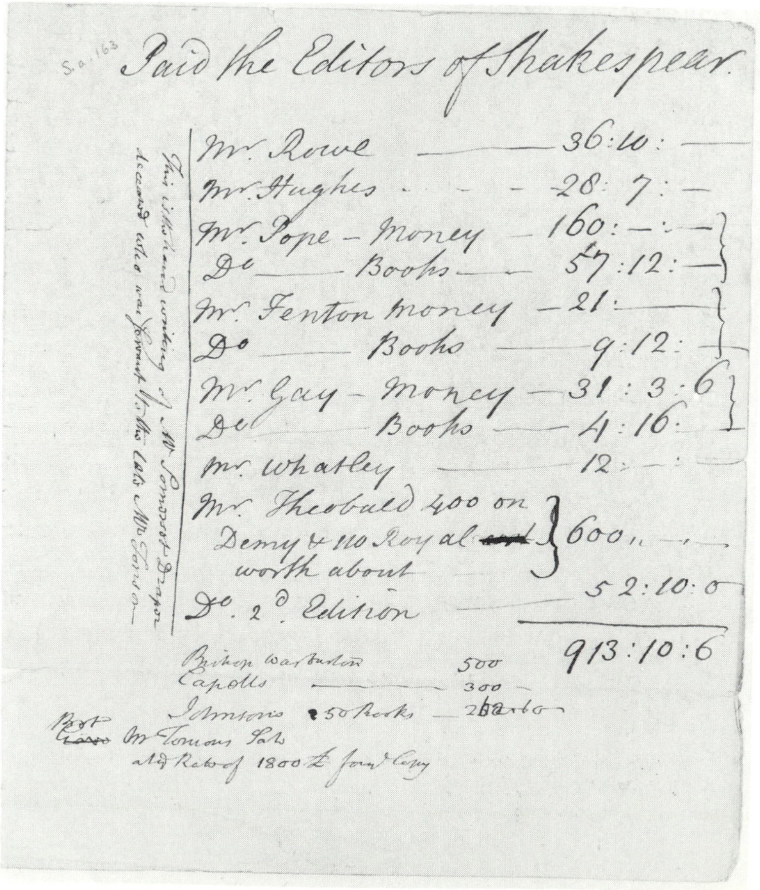

FIG. 50. Somerset Draper's extracts from Tonson's accounts for the Shakespeare (Folger MS S. a. 163; approx. 210 × 180)

Tonson's original intention, but it is to be noted that Pope's agreement does not involve him in making an index. This is mentioned by Pope as early as 3 September 1721; a later letter giving instructions how it should be done refers to 'your first Index-maker' who is clearly to be succeeded by 'whoever you set upon the Index'.[49] Fenton finally did the job, as well as the proof-reading of volume V (*Correspondence*, ii. 244 and 222), and this may explain the payment of 20 guineas; we have no idea what Gay was paid for.

The payments in books are all multiples of £4. 16s. and seem to correspond to one copy for Gay, two for Fenton, and a dozen for Pope, charged at the trade price. Subscribers paid 5 guineas for the six volumes and the price to the public was 6 guineas; £4. 16s. would correspond to a trade price of 16s. a volume.

The translation of the *Odyssey* had started while the edition of Shakespeare was being produced. This time Pope had collaborators; he translated 12 books himself, Fenton translated 4, and Broome was responsible for 8 books and the commentary. There had been some thought of publishing proposals as early as Michaelmas 1722 (*Correspondence*, ii. 111), but clearly work on Shakespeare was taking longer than Pope had expected when he signed the contract with Tonson and agreed to complete it by May 1723. The difficulties over the publication of Sheffield's *Works* and Pope's association with Atterbury made early 1723 an inappropriate time for issuing proposals, as well as delaying Pope's part of the translation, but by the summer of 1723 Pope was hard at work with Broome and Fenton (*Correspondence*, ii. 182). By 1724 it was time to sign a contract, and as Fenton wrote to Broome on 9 January 1724, 'Tonson does not care to contract for the copy, and application has been made to Lintot, upon which he exerts the true spirit of a scoundrel, believing that he has Pope entirely at his mercy' (*Correspondence*, ii. 214). Certainly Lintot was bound to be wary after his difficulties with the *Iliad*, and the contract altered the terms which had annoyed him.

One fundamental change in arrangements was that whereas the *Iliad* was published in six annual volumes, the *Odyssey* was to be published in five volumes, the first three (as it worked out) in April 1725 and the last two in June 1726. (The text of the *Odyssey* and notes as published was shorter than the *Iliad*, so the volumes work out at about the same size—in folio, an average of 311 pages for the *Iliad* to 303 for the *Odyssey*.) I have suggested that the idea of publishing the *Iliad* in annual volumes was inspired by serial publications; by now the multi-volume format was accepted, but this time Pope knew he would have to deliver copies to subscribers and collect further instalments of the subscription without Lintot's help (unless he were to pay Lintot for the task, as he did with the *Iliad*), so it made sense to restrict the number of deliveries.[50] Moreover, with a team of three translators, the work could be completed more quickly; Lintot would have com-

[49] *Correspondence*, ii. 213; I suspect Sherburn's '[1724?]' is too late.

[50] Lintot did deliver some copies of the last two vols.; see Pope to Broome, 23 Aug. 1726: 'If you will send any correspondent you have for the two books of Homer, they shall be left directed for you at Mr. Jervas's. I did not send them before, knowing you corresponded with Lintot, who could at any time, on your order, have sent them, as he delivers many.' (*Correspondence*, ii. 390.)

plete sets for sale and the profits of a cheap edition not long after his capital expenditure. The collection of the second part of the subscription was made at the first delivery, so that the whole 5 guineas was collected between the publication of the proposals in January 1725 and the first delivery in April 1725; the deliveries were made from Jervas's house in Cleveland Court, St James's (*Early Career*, p. 270).

The contract of 18 February 1724 can be summarized fairly briefly.[51] The consideration was 50 guineas paid at the time of signing and further sums of 150 guineas paid 'at or before the Publication' of the first three and of the last two volumes. Pope made over the whole of the copyright 'for and during the Term of fourteen Years ... and for and during all and every Such farther ... Term or Termes as He ... by any Act or Acts of Parliament or otherwise Howsoever is Enabled to Grant and Assign the Same'. Lintot was to print 200 copies of each volume on the best writing royal paper and 550 on the best printing royal paper, specimens of which were annexed;[52] these were to be delivered a week or more before publication. (The *Iliad* contract had stipulated a month's delay, which Lintot resented deeply; it seems to have been changed to a week for the last two volumes.) Lintot undertook not to print any other volumes of the *Odyssey* in quarto, nor on the same sort of paper, nor with the same engravings as Pope's 750, within the next ten years. (With the *Iliad* Lintot had no right to reprint on royal paper or with engravings at any time.) Pope was to be responsible for the cost of the copper plates for the headpieces, tailpieces, and initial letters, though Lintot would pay for the cost of printing from them. (For the *Iliad*, Lintot had paid for the plates to be engraved under Pope's instructions.) Lintot was to keep the plates, but would permit Pope or his assigns to use them for other books, they being responsible for any 'touching up' of the plates that might be required.

In 1740 Pope, hearing that Gilliver had reprinted the first volume of the *Iliad* to complete sets, suspected there might be a plot to reprint the whole of Homer in quarto; and he then asked himself what use the plates were to Lintot unless he intended to print up sets of the *Odyssey* to go with sets of the *Iliad* he already had (*Correspondence*, iv. 224). I suppose Lintot's primary purpose was to be able to use the plates in other books he printed, though I have not yet identified any. Pope, however, exercised his rights to borrow the plates. Five of the tailpieces by Kent were lent to Samuel Buckley for his 1733 edition of Thuanus; they appear above the colophon in volumes I, and III–V (there was no space in volume II);

---

[51] Printed in *Early career*, pp. 313–16, from BL Egerton Charter 130. The counterpart is in the Houghton Library at Harvard, MS Eng. 232.2, and has the receipts for three payments: 50 guineas on 18 Feb. 1724, 150 on 21 May 1725, and 150 on 13 July 1726. The later receipts also include the two batches of 750 subscribers' copies.

[52] Two half-sheets are attached to the BL contract, one watermarked IV (as is the sheet attached to the Bodl. *Iliad* indenture), the other whiter and with no watermark, probably Genoa. Pope on Tonson's advice allowed Lintot to use a different paper 'so that the work need not to be retarded. Since you assure me that is as good, tho somewhat less Expensive' (*Correspondence*, ii. 217).

the fifth tailpiece is found at the end of the *History* in volume VI, p. 440.[53] Pope wrote to Buckley on 16 June 1732:

As One Instance (& I wish I could give you many) of my Desires to be serviceable to you, I have had the Articles examined betwixt Lintott & me, as to what I promisd of the use of the copper ornaments, Initials & Tailpieces, for your Work. I am very certain they are wholly in my power. Therfore I have written to him an order to deliver them to you; upon your going or order for them. But as he is a Grand Chicanneur, I would not have you tell him what Book: and as he is a great Scoundrell to me, I would willingly have him receive the small punishment of imagining I am printing with you Something of my own, for which he has (upon Rumours, for I never converse with Him) lately been importuning me, and receivd no other answer than a very true one, that I would never imploy him more. (*Correspondence*, iii. 294.)

The plates, especially the tailpieces, were subsequently used by John Wright for Pope's *Essay on man* (1734), his *Works* (1735), and other works in quarto; it seems possible that Buckley's request reminded Pope that they were at his disposal.

When Pope's proposals appeared in the press in January 1725, a year after the contract was signed, there was so astonishing a reply from Lintot that I cannot resist telling the story again. Tonson had advertised his edition of Shakespeare in the *Daily post* of 18 November (and 9 December) 1724:

Mr. Pope's Edition of Shakespear in Six Vols. in Quarto, is now very near finish'd, and there being but a small Number Printed, such Persons who are willing to subscribe for the same, are desir'd to send their Names and first Payments to Jacob Tonson in the Strand, before the 16th Day of December next, at which Time the List of Subscribers will be printed off. (*Early career*, p. 240.)

The wording is ambiguous as to whether the subscription was intended for the benefit of Tonson or of Pope, though later advertisements make it clear that it was for Tonson's benefit alone. Pope's proposals for the *Odyssey* are explicit about the Shakespeare:

I take this occasion to declare that the subscription for Shakespear belongs wholly to Mr. Tonson: And that the benefit of *this Proposal* is not solely for my own use, but for that of *two of my friends*, who have *assisted me in this work*.[54]

We know from a letter Pope sent to him at this time that Tonson had hoped that Pope would not make it clear that the Shakespeare was solely for his benefit, and also advised him that he should not refer to the assistance given by Fenton and Broome in the translation of the *Odyssey* (*Correspondence*, ii. 285–6). No doubt there was concern on the other side lest Pope's rich friends, seeing his name in Tonson's proposals, might subscribe for multiple copies of Shakespeare instead

[53] Vols. I–V were printed respectively by Henry Woodfall, Samuel Richardson, James Bettenham, James Roberts, and Thomas Wood. I have found no colophon in vols. VI and VII.

[54] No copy of the proposals is known to survive, but the relevant passage in its revised form is quoted in the *Dunciad variorum* (*Twickenham*, v. 31). There the proposals are said to have been 'printed by J. Watts, Jan. 10, 1724' [i.e. 1725].

of reserving that patronage for the *Odyssey*.[55] The same letter makes it clear that Lintot was incensed that Pope was engaged in a subscription publication for Tonson's benefit and had even listed Tonson among those who would take in subscriptions for the *Odyssey*; no doubt this was to be a factor in his subsequent actions.

The next step after the printing of Pope's proposals was that Tonson readvertised his Shakespeare on 18 January 1725, in hopes of getting more subscribers than those whose names were to have been printed off on 16 December. At the same time Pope advertised the *Odyssey*, and we have his letter to Samuel Buckley of 20 January 1725:

I shall take it as a favour of you to insert the inclosed advertisement both in the Gazette & Daily Courant, three times. What I particularly recommend to your care is to cause it to be distinguishd with proper dignity, & the title in Capitals, as here drawn. Also to stand at the head of the more vulgar advertisements at least rankd before Eloped wives, if not before Lost Spaniels & Strayd Geldings. Do not, I beseech you, grudge to bestow One Line at large in honour of my name, who wd bestow many to celebrate yours.   (*Correspondence*, ii. 285.)

Buckley gave the proposals the prominence Pope requested; in the *London gazette* of 23 January 1725 they appeared directly below a notice of a meeting of the directors of the South-Sea Company, and there and in the *Daily courant* the first line was 'PROPOSALS by Mr. POPE'. Figure 51 shows Pope's and Tonson's proposals as they appeared in the same number of the *Whitehall evening post* of 21–3 January. Three days later came Lintot's coup. He advertised in the *St James's evening post* of 23–6 January (Figure 51) that those who subscribed for his benefit could have copies of the *Odyssey* more cheaply than from Pope. Ironically his advertisement and Pope's appear on the same page of the *Post boy* on this day.

Apart from the shock effect of this advertisement, the practical effect was that the public could buy the large-paper folio (without the engravings of the subscribers' quarto) at a guinea less than the subscribers were paying, and that the second payment of 2 guineas would be made not on delivery of the first three volumes, but when volumes IV and V were published. Sherburn points out that 'Lintot's subscription was merely an unscrupulous advertisement of his trade editions',[56] but it had to be advertised in this way because by the agreement Lintot could not publish copies until a week after Pope's subscribers' copies were delivered. In this way he could take his money in advance, and the copies he sold himself, as opposed to those sold in the usual way through the trade, would not be subject to the trade discount. Perhaps this is why he was able to cut his prices to encourage their sale. The *Iliad* had cost a guinea a volume for the large paper;

[55] The lists show 411 subscribers for 417 copies of Shakespeare by contrast with 610 subscribers for 1,057 copies of the *Odyssey* (Griffith, i. 120–1).

[56] *Early career*, p. 257. He is, however, in error in saying that copies of the large or small folios could be trimmed by the binder to match the quartos of the *Iliad*. I think Sherburn exaggerates the effect that Lintot's advertisement had on the number of subscribers: if my calculations (n. 70 below) are right, Pope gained about 230 in Jan. and Feb.

PROPOSALS by JACOB TONSON for Mr. POPE's Edition of SHAKESPEAR.

I. This Work will confist of Six Volumes, printed upon Royal Paper, in Quarto, each Volume containing not lefs than 70 Sheets.

II. The Six Volumes are propofed until the 10th Day of February next, (at which Time the Subfcription will be clofed) at Five Guineas in Quires, Two Guineas to be paid in Hand, and the other Three on Delivery of the Book.

III. This Work will be delivered to the Subfcribers before the End of February next at fartheft, it being all printed off except the Preface, the Life of the Author, and the Index, which are now in the Prefs.

Note, If any Names are omitted in the Lift of the prefent Subfcribers now printed and made publick, they fhall be inferted upon the firft Notice to J. Tonfon in the Strand.

(a)

PROPOSALS by Mr. POPE, for a Tranflation of HOMER's ODYSSEY.

This Work confifts of the fame Number of Books as the ILIAD, and of as large a Body of Notes and Extracts. It is printed in the fame Manner, Size, Paper and Ornaments.

It is propofed to the Subfcribers at a Guinea lefs, namely, at Five Guineas.

The firft three Volumes, viz. 14 Books are already printed, in Confideration of which, three Guineas are to be now paid, and the remaining two upon Delivery of them.

The greater Number of the Impreffion being already fubfcribed for, thofe who would have the Book are defired to fend in their Names and Payments before the laft Day of February next, to Mr. Lintot at the Crofs-Keys between the Temple Gates in Fleetftreet, who will deliver Receipts for the fame; the Subfcription being then to be clofed.

(b)

PROPOSALS by Bernard Lintot, for his own Benefit, for printing a Tranflation of HOMER's ODYSSEY, by Mr. POPE.

This Work confifts of the fame number of Books as the Iliad, and of as large a body of Notes and Extracts. It is printed in five Volumes in large and fmall Paper Folio, in the fame Manner, Size and Paper. The large is propos'd to the Subfcribers at two Guineas per Sett lefs than the Iliad, namely, at four Guineas. The fmall Paper at ten Shillings lefs, namely, at fifty Shillings per Sett, half to be paid down, the Remainder upon the Delivery of the two laft Volumes. Two hundred and fifty Setts only are printed for thofe Gentlemen who have the Iliad in Folio, who are defir'd to fend in their firft Payments by the laft Day of February next, to the Proprietor Bernard Lintot, at the Crofs Keys between the Temple-Gates, in Fleet-ftreet, when this Subfcription will be clos'd. N.B. The Edition in 12mos. of Mr. Pope's Odyffey, with the Notes printed at the bottom of each Page, with Cuts, will be publifhed in one Month after the Subfcription Books are deliver'd.

(c)

FIG. 51. Proposals for the Shakespeare and *Odyssey*: (a) Tonson's proposals for Pope's Shakespeare in *Whitehall evening post*, 21–3 January 1725 (Bodl. N.N.; 49 × 77); (b) Pope's proposals for the *Odyssey* in the same issue (Bodl. N.N.; 46 × 77); (c) Lintot's proposals for the *Odyssey* for his own benefit in *St James's evening post*, 23–6 January 1725 (Bodl. N.N.; 76 × 75)

Lintot is offering the first three volumes of the *Odyssey* at the cost of two. The small-paper copies of the *Iliad* had cost 12s., so five volumes would have cost 60s., not 50s. as offered here. What the customers were not to know was that this special offer was not limited to those who subscribed before publication; on publication of the duodecimo Lintot advertised in the *Daily post* for 25 May 1725 that the prices per volume in quires were 17s. the large folio, 10s. the small folio, and 2s. the duodecimo.[57] Lintot had cut his prices for the same size of volume by nearly 20 per cent, and we are entitled to wonder why.

No doubt Lintot got some malicious satisfaction from the effect of his advertisement, but I hardly believe that he would jeopardize his profits for the sake of a gesture; if that had been his intention, he could have limited his special offer to those who had paid by the end of February and charged normal prices thereafter, thus making his effect but limiting his losses.[58] Another possibility is that the cost

---

[57] In 1726 the prices were 17s., 12s., and 3s. (Griffith, i. 130); these may include binding.

[58] In fact those who had bought their first 3 vols. for 2 guineas could (if they were unscrupulous enough) buy the second 2 at the standard price of 34s. rather than the 2 guineas of the proposals. Similarly with the small-paper folios.

of printing or paper had fallen, or that Watts could print more cheaply than Bowyer; as to this we have no direct evidence, but the cost of the quartos, discussed below, suggests this was not the case. A third possibility is that over the past year Lintot had come to know more about the collaboration in the translation, and feared its effect on his sales. Lintot had signed the contract with Pope on 18 February 1724, and printing started almost at once; by 22 December 1724 Pope could write to Harley, 'I have printed Eleven books, & have fourteen finishd' (*Correspondence*, ii. 279), while the proposals of 25 January 1725 are perhaps a little optimistic in saying that 'the first three Volumes (Viz. fourteen Books) are already printed'. It is clear that Lintot had plenty of opportunity to see how the collaboration worked, and now Pope had made it public in his proposals he may have feared repercussions and have lowered his prices in the hope of shifting his stock.[59] It is possible that the trade itself was sceptical, and unwilling to buy the large number of copies at special wholesale rates which had provided Lintot with some quick return of capital on the *Iliad*. My calculations below suggest that if the production costs of the *Odyssey* were the same per volume as the *Iliad*, then Lintot's possible profit on the edition (with a trade discount averaging 25 per cent) would be reduced by the price cuts from £1,350 to £800. If he could sell the large folios and 500 small folios direct to the public he would at once recover that £550; what is more, he would more than cover all his costs, and that may have been his main concern. Thereafter he would not need to sell off stocks wholesale, so his average rate of return would be somewhat higher.

The figures in Lintot's account-book for the *Odyssey* (Nichols, *Anecdotes*, viii. 300) have never been satisfactorily explained, but the solution is really quite simple.

Copy-money for the *Odyssey*, Volumes I. II. III.; and 750 of each Volume printed on Royal Paper, 4to	615	6	0
Copy-money for the *Odys[s]ey*, Volumes IV. V.; and 750 of each Volume, Royal	425	18	7½

If we take the copy-money in the first of these entries to be the initial payment of 50 guineas on signing the contract plus the 150 guineas paid at the time of publishing the first three volumes, it leaves £405. 6s. as the value of the 750 copies, or £135. 2s. per volume. Deducting 150 guineas from the payment in the second entry we get a cost for the last two volumes of £134. 4s. 3¾d. These are only average figures for volumes which differ in size, but the agreement is so close both with one another and my estimated average cost for the *Iliad* quartos of £133 that I think we can take this as the right solution.[60]

---

[59] Sherburn, *Early career*, p. 263 quotes the letter of 'Homerides' from the *London journal* of 17 July 1725, which compares the royal patents of the *Iliad* and the *Odyssey*. The first speaks of '*a Translation . . . by Mr. Alexander Pope*', the other of '*a Translation, UNDERTAKEN by . . . Alexander Pope, Esquire*'. Pope's contract with Lintot also uses the word 'undertaken', but its full implications may not have been clear.

[60] My figure of £133 for the *Iliad* does not include the cost of printing the engravings, and it is for 660 copies, not the 750 for the *Odyssey*. The cost of printing vol. II of the *Iliad* (which has 43 sheets against

I have tried to estimate Lintot's possible profits from the *Odyssey* and to compare them with the *Iliad* (Table 10). Again these figures rest on a number of assumptions and are merely intended to give an idea of the sort of figures involved.[61] I have taken the production costs per volume of the *Odyssey* as the same as the *Iliad*; they could well be a little lower. Lintot in his advertisements said that 250 large-paper folios were printed, but we can only guess that there were 1,000 small-paper folios by analogy with the *Iliad*. The question of trade prices and therefore of profits is more difficult; for the *Iliad* folios whose retail price was 12s., we know that the trade price was 10s. and the wholesale price to trade subscribers was 8s., and an average price of 9s. or three-quarters of retail price seemed reasonable. If my earlier hypothesis is right, that Lintot's 'subscription' offer was an attempt to sell copies direct to the public at the lower retail price of 10s. and to do away with a trade subscription, then it would be wrong to take an average trade price of three-quarters retail price. As it is, I have given the same discount as for the *Iliad* in the table, but have given below the overall profit figure that would have been achieved if Lintot had either kept prices and discounts at the *Iliad* level, or succeeded in selling a large number of copies direct to the public—the figures are roughly the same.[62]

Below in the table I have given Lintot's payment to Pope in copy-money and subscribers' copies using his own figures; we have seen that those for the *Iliad* seem to be rounded-up, but it is difficult to estimate how much the engravings for that work cost, so I have let the figures stand. (Pope paid for the engravings for the *Odyssey*, and again we don't know how much they cost; the difference in Lintot's figures for the two works gives a maximum figure of about £150.) Lintot's possible overall profit—that is, assuming that he sold all the copies he had printed at these prices—at £794 is little higher than the £669 for the *Iliad*, despite the fact that he had reduced his payment to Pope by more than half. Below I have given a reminder of what profit Lintot might have made if he had printed and sold his original order of 1,750 small folio copies of the *Iliad*, and what he might have made if he had not reduced the price of the *Odyssey*, or if he had succeeded in selling the copies at a lower discount. However, once again he failed to sell his stock of folios, for Thomas Osborne was able to offer them with the *Iliad* or separately at about half-price in May 1739;[63] in the following January he offered the *Iliad* alone, which suggests his stocks were more limited, either because Lin-

---

an average of 42¾, and so can be taken as typical) on *Iliad* papers but as 550 + 200 copies (the *Odyssey* number) would be £145. The average *Odyssey* vol. is 41½ sheets, and that would give a cost of £140, plus the cost of printing from the engravings. On the other side, we know that Lintot was able to economize on the paper (see n. 52 above). The important thing, of course, is the order of magnitude and not the precise figures.

[61] The figures for the *Iliad* sometimes differ slightly from those in Table 8 as a result of different methods of averaging and rounding up and down; the differences are insignificant.

[62] That is, a price of 9s. could represent either an average trade discount of 25 per cent on 12s. or a successful attempt to sell half the copies direct to the public at 10s. and half at a trade price of 8s.

[63] See above, p. 57.

TABLE 10. Lintot's Profits on *Odyssey* and *Iliad* Compared

	Iliad £	s.	d.	Odyssey £	s.	d.
Folio						
Demy: trade price		15	9		12	9
unit cost		4			4	
profit per vol.		11	9		8	9
profit per edn. of 250	145			109		
6 vols. (*Iliad*); 5 vols. (*Odyssey*)	870			545		
Pott: trade price		9	0		7	6
unit cost		2	4		2	4
profit per vol.		6	8		5	2
profit per edn. of 1000	333			258		
6 vols. (*Iliad*); 5 vols. (*Odyssey*)	2000			1290		
Profit on sale of folio edn.	2870			1835		
Payments to Pope (Lintot's figures)						
Copy-money	1275			367	10	
Subscribers' copies	926			673	15	
Total	2201			1041		
Profit on folio edn. less payment to Pope	669			794		
With 1750 pott *Iliad* and no loss from						
Odyssey subscription	2169			1350		
Duodecimos						
Profit per set		6			5	
Profit on edn. of 2500	750			625		
Profit on edn. of 5000	1500					

tot's 'subscription' had been successful or because he had printed fewer copies of the small folio.

This time, moreover, Lintot failed to make such a handsome return from the duodecimos. I have assumed that he printed the same number of copies as the first duodecimo of the *Iliad*, 2,500 copies, but since that edition sold out in six months he may have printed more. Whatever the cause, this edition was not reprinted for twenty years (Lintot's later editions are 1745, 1752, and 1758), so my figure of £625 is probably reasonable for his immediate returns. I have taken Lintot's profit on a set at the same figure per volume as the *Iliad*.[64] Bernard Lintot's only comfort, if he had been alive to see it, would have been that at the firm's trade sale in 1759 the copyrights of Homer, sold in shares of one-eighth, fetched nearly £1,000, a quite remarkable figure, comparable with the prices for Shakespeare, Milton, or the *Spectator*.[65]

---

[64] The *Odysssey* was advertised at 2*s*. a vol. in the *Daily post* of 25 May 1725; Lintot's proposals said they would be 'with cuts'. Vols. I–III were published within a month of the folios. In 1726 the duodecimos were priced at 3*s*., like those of the *Iliad* in the same advertisement (Griffith, i. 130); these copies may be calf gilt.

[65] See Ward trade sales catalogues, Bodl. John Johnson, and T. Belanger, 'Booksellers' trade sales, 1718–68', *Library*, 5th ser. xxx (1975), 295–6.

As for Pope, it is interesting to calculate Lintot's payments to him in terms of a royalty on the published price of the first edition in sheets. For the *Odyssey* it is about 29 per cent, but for the *Iliad* over 42 per cent; but if Lintot had sold 1,750 small folio copies of the *Iliad* it would have been 28 per cent. Pope in fact succeeded in driving nearly as hard a bargain for the *Odyssey* as for the *Iliad*, considering what Lintot expected to sell in each case. If one takes the duodecimo editions into account up to 1760, the payments to Pope work out as a royalty (paid in advance, of course) of 13 per cent for the *Iliad* and 12 per cent for the *Odyssey*.[66]

Pope's profits, of course, came mainly from the subscribers, who paid six times as much for their copies as they cost Lintot. I suggested when discussing the *Iliad* that Pope's subscription lists represent rather the efficiency of his influential friends in finding subscribers than a straight appeal to the public, and this is confirmed by Pope's willingness to share the public part of his subscription with his collaborators. I suggest below that this was less than a quarter of the whole. The extent of patronage is much more evident with the *Odyssey*, where there are many subscribers for five or ten copies who, if they only drew a single copy (as seems likely in many cases), were effectively presenting Pope with 25 or 50 guineas. Pope's correspondence with Edward Harley, Earl of Oxford, before the printing of the proposals is illuminating. Pope wrote on 12 December 1724 about his mother's recovery from a severe illness, and continues:

Tho if ever I attend my Subscription, I must do it now; the time of publication drawing so nigh, & I not having (thro this unfortunate accident) yet publish'd the Proposals to the Town. I am at last determined to do it, & to take no further care about it than to publish it: since I really cannot leave my poor Mother on any account whatever. I must desire to know in what manner to treat your Lordship & Lady Oxford in the printed List, which I am to annex to this Proposal? If I were to set you down for as many Subscriptions as You have procur'd me, half my List would lye at your door, & I might fairly make you a Benefactor of the greatest number. Yet as I am sensible you care not to be known for the Good you do, I am afraid to put you down in any distinguishing manner; & Yet again, I cannot bear but the world should know, that you do distinguish me. I have set down the Duchess & Duke of Buckingham for five Setts; will you allow me to do the same to your self & Lady Oxford? Mr Walpole & Lord Townshend are sett down for Ten, each: I would not deny my obligations; & tis all I owe Them. But to the Duchess & to your Lordship I would keep some measures; I am so much, & ever like to be so much, in hers & your debt, that I will never tell how much, without your absolute command or leave. Let but the whole world know you favor me, & let me enjoy to myself the satisfaction of knowing to what degree? I have kept back my Proposal from the press till I have the honour of your Commands on this subject.[67]

---

[66] Calculated on the basis of six edns. of 2,500 and one of 5,000 copies for the *Iliad*, three of 2,500 and one of 5,000 for the *Odyssey*.

[67] *Correspondence*, ii. 275–6. Sherburn makes the plausible suggestion that Walpole and Townshend were set down for 10 copies each because they were responsible for a royal grant of £200. He wrongly dates the treasury warrant 26 Apr. 1724 (at ii. 160 n. he dates it 29 Apr. 1725 and ascribes it to Carteret's goodwill). The warrant is reproduced in Maggs' catalogue no. 301, Plate xxiii; it is dated 26 Apr. 1725, refers to His Majesty's letters of Privy Seal dated 29 Sept. 1724 and a warrant of 19 Apr. to pay £200 to

Oxford replies from Wimpole, 17 December 1724:

as to the affair of your subscription I wish you had met with more success, I do not think it is at an end yet. pray why do you print the names of your subscribers with your proposals? I thought the names of your subscribers had been printed with the book that was first delivered, I think you are in the right to print proposals. as to my being set down I did forget to mention it to you when I was in town and saw you last, I would be for ten setts my Wife for five setts and peggy for one.   (*Correspondence*, ii. 277.)

The subscribers' list follows these instructions, but adds to the sixteen copies another for Robert Harley, who had died earlier in the year on 21 May. It seems worth listing those who subscribed for five copies or more: Duke of Argyll (5), Earl Arran (5), Lord Bathurst (10), Peter Bathurst (5), Lord Bingley (10), Hon. Martin Bladen (10), Duke of Buckingham (5), Duchess of Buckingham (5), Earl of Burlington (5), Lord Carleton (10), Lord Carteret (10), Duke of Chandos (10), Lord Viscount Cobham (5), Rt. Hon. Spencer Compton (10), Duke of Dorset (5), Lord Foley (10), Lord Gower (5), Duke of Grafton (5), Lord Viscount Harcourt (10), Earl of Kinnoul (5), Henrietta, Duchess of Marlborough (5), Duke of Newcastle (10), Mrs Newsham (10), Edward, Earl of Oxford (10), Countess of Oxford (5), Earl of Peterborough (5), Earl of Pontefract (5), Rt. Hon. William Pulteney (10), Duke of Queensborough and Dover (5), Earl of Scarborough (5), Lady Viscountess Scudamore (5), Earl of Sussex (5), Lord Viscount Townshend (10), Rt. Hon. Robert Walpole (10), Hon. Edward Wortley Mountague (5). These thirty-five subscribers are down for 250 copies; by contrast, there were 12 for 87 copies of the *Iliad*: Lord Ashburnham (5), Duke of Buckingham (10), Lord Viscount Bolingbroke (10), Earl of Burlington (6), Earl of Carnarvan (12), Earl of Clare (5), Duke of Devonshire (4), Earl of Halifax (10), Lord Harcourt (5), Hon. Simon Harcourt (5), Lord Lansdowne (10), Duke of Ormond (5).[68]

Apparently, these thirty-five multiple subscribers to the *Odyssey* presented Pope with 1,200 guineas. While there seems no doubt that such subscriptions are a form of patronage, the correspondence with Oxford suggests a possible converse: that Pope is using the subscription list in some cases as a way of acknowledging gifts received previously, or personal favours. Thus Pope appears to have given copies to Samuel Buckley and W. Aikman and included their names in the list of subscribers (*Correspondence*, ii. 286, 294); the two copies for Jervas are doubtless in the same class. An aspect of the subscription system which has been little explored is that friends of the author were given printed receipt forms to solicit subscriptions from their circle, as is implied by Pope's letter to Oxford quoted above. In some cases the friend might be expected to pay for a number of

Pope 'as his Matys Encouragement to the Work, he is now about, of Translating the Odysses of Homer into English Verse, and the subscriptions which are making for the same'. It is signed by Walpole, George Baillie, William Yonge, and George Dodington. The payment is dated 29 Apr. 1725; it bears Pope's receipt.

[68] One can contrast Gay's *Poems* of 1720, which has fewer generous subscribers, but Burlington is down for 50 copies and Pulteney for 25; see also the articles by Rogers and Hodgart (above, n. 25).

sets, and then try to find subscribers to take them up. This seems to be referred to in Pope's letter to Caryll of June 1725:

I also fancied that since 'twas possible you might not have got off the 4 subscriptions I sent you, you might prefer to have but 2 Sets, and so I would have taken the payment you made to Mrs Cope for the whole in full . . . I did purposely omit to set you down for four sets [in the list of subscribers], for the reason I just now mentioned to excuse you from being taxed too high, which I least of all Care my friends should be. (*Correspondence*, ii. 299.)

I have wondered (perhaps uncharitably) whether on these occasions the subscribers for five or ten sets paid five or ten times the first instalment of a subscription but did not necessarily pay the full amount of further instalments. In this case, where the second instalment was due within a few months, it seems likely that generous patrons paid the whole sum in advance.[69] For the same reason it seems unlikely that many would have died or lost interest as was possible over the six-year period of the *Iliad*.

The 1,057 payments from 610 subscribers of £5,549. 5s. and 350 guineas copy-money from Lintot make a total of £5,916. 15s. There were 140 more copies printed than there were subscribers, and though no doubt some multiple subscribers exercised their right to more than one copy, there were undoubtedly a substantial number of quartos left that could be sold, for Pope was able to give away copies as late as 1738. We can at least set these off against the cost of Kent's engravings. Pope's payments of £200 to Fenton for four books and of £400 to Broome for eight books, plus £100 for notes and another £100 for the subscribers he found amount to £800,[70] leaving Pope with some £5,000 profit.

It is perhaps worth trying to put these figures into modern terms. A conservative estimate, which we may soon be able to double, is that Pope made £100,000 each from his translations. The subscribers were paying 20 guineas a volume or 100 guineas a set for these works; the small folio *Odyssey* cost £50 a set. The cheap duodecimo was more reasonable at £10 unbound.[71]

---

[69] A partial draft of the proposals (not in Pope's hand) in BL Add. MS 4809, fo. 87ᵛ reads 'But if any Person subscribe for more than one set, one half only of such Subscription mony shall be paid down'. Normal subscribers paid down 3 out of 5 guineas.

[70] These are the accepted figures, from *Early career*, p. 259, and the payments to Broome are confirmed in *Correspondence*, iii. 507 and ii. 389, but O. Ruffhead, *The life of Alexander Pope* (1769), p. 205, gives the figure of £300 for Fenton, as does Spence, *Anecdotes*, no. 207a. The payments of £200 and £400 to Fenton and Broome presumably correspond to their share of the subscriptions received after Pope advertised to the public on 10 Jan. 1725. If we ignore the notes and assume that Pope's share for 12 books was £600, that suggests that about 230 subscriptions producing £1,200 were received in this period: a plausible figure.

[71] These are the figures for 1976; the prophecy that they could soon be doubled has been fulfilled, at least if we go by the General Index of Retail Prices, which indicates that they should be multiplied by 2.68 to update them to 1988.

3

THE PROBLEMS OF INDEPENDENCE

WRIGHT AND GILLIVER

With Pope's return to original composition with the *Dunciad* of 1728, we find a completely new relationship with the book trade, one in which the author takes charge, choosing his own printer and publisher and directing operations himself. The nature of the *Dunciad* itself called for a clandestine publishing operation; Pope may have hoped to conceal his authorship longer than proved possible, but apart from this any printer or bookseller associated with the work might have been subject to reprisals from those it attacked. The pattern of his subsequent career shows another, quite distinct, motive for his new independence: to extract the maximum profit from his publications and to keep the copyright of his works under his own control.

The author who feels he is not getting his proper share of the profits from his work is not an uncommon figure, but Pope seems to have been more than usually sanguine about the profits that could be made. Spence reports him as saying in 1739: 'An author who is at all the expenses of publishing ought to clear two thirds of the whole profit into his own pocket.' He meant that two-thirds of the retail price would be the author's profit, for Spence gives the following gloss:

For instance, as he explained it, in a piece of one thousand copies at 3*s*. each to the common buyer, the whole sale at that rate will bring in £150. The expense therefore to the author for printing, paper, publishing, selling, and advertising, should be about £50, and his clear gains should be £100.   (Spence, *Anecdotes*, no. 199.)

Pope may have got this idea from the quartos of the *Iliad* and *Odyssey*, which cost about 3*s*. 6*d*. a volume to print, or a sixth of their retail price of one guinea, leaving another 3*s*. 6*d*. to cover advertisement and distribution through the trade. From now on, when Pope prints a volume of his works in that format he always stipulates that the price is not to be less than 18*s*. to the trade or a guinea to gentlemen, leaving the retail bookseller 3*s*. profit and reserving 6*d*. to cover advertisement and distribution. While it is true that the trade price and the wholesale price of a sought-after work could be higher than normal—as the prices for the folio *Iliad* are, by comparison with other works in Henry Clements's Notebook—this calculation leaves no tolerable margin for the wholesaling which was the normal method of distribution within the trade. As we shall see, this led to a dispute with Gilliver when the *Works* of 1735 was ready for publication and Gilliver offered to take them all at a wholesaler's price of 13s.; Pope had to try to find other booksellers who would give 17s. and be content to pass them through the trade for a shilling's profit. Even this denied him sixpence of the profit he seems to have expected, and his later volumes were made thinner so that the production cost was smaller.[1]

Pope took much the same attitude as the Oxford University Press, which offered booksellers the *Catalogue of printed books in the Bodleian Library* (1738) at 5s. discount for a retail price of 3 guineas large paper and 2 guineas the small[2]— but Pope did at least have a willing buyer. Both were in need of Johnson's famous advice of 1776:

We will call our primary agent in London, Mr. Cadell, who receives our books from us, gives them room in his warehouse, and issues them on demand; by him they are sold to Mr. Dilly, a wholesale bookseller, who sends them into the country; and the last seller is the country bookseller. Here are three profits to be paid between the printer and the reader, or in the style of commerce, between the manufacturer and the consumer; and if any of these profits is too penuriously distributed, the process of commerce is interrupted. . . .

The deduction, I am afraid, will appear very great: but let it be considered before it is refused. We must allow, for profit, between thirty and thirty-five *per cent.* between six and seven shillings in the pound; that is, for every book which costs the last buyer twenty shillings, we must charge Mr. Cadell with something less than fourteen. We must set the copies at fourteen shillings each, and superadd what is called the quarterly-book, or for every hundred books so charged we must deliver an hundred and four.

The profits will then stand thus:

Mr. Cadell, who runs no hazard, and gives no credit, will be paid for warehouse room and attendance by a shilling profit on each book, and his chance of the quarterly-book.

Mr. Dilly, who buys the book for fifteen shillings, and who will expect the quarterly-book if he takes five and twenty, will sell it to his country customer at sixteen and sixpence, by which, at the hazard of loss, and the certainty of long credit, he gains the regular profit of ten *per cent.* which is expected in the wholesale trade.

The country bookseller, buying at sixteen and sixpence, and commonly trusting a considerable time, gains but three and sixpence, and, if he trusts a year, not much more than two and sixpence; otherwise than as he may, perhaps, take as long credit as he gives.[3]

Even if, like Boswell, we regard these figures as a little generous and advance them by a whole shilling in converting pounds to guineas, it still means that if

---

[1] Average size of *Iliad* vol., 43 sheets; of *Odyssey*, 41 ½ sheets; of *Works* II, 65 sheets; *Letters*, 1737, 46 sheets; *Works in prose* II, 52 sheets.

[2] Carter, *History of the Oxford University Press*, p. 274, and see pp. 370–3 for Blackstone's view that 10 per cent discount to their London agent was sufficient.

[3] Boswell, *Life of Johnson*, ed. G. Birkbeck Hill, rev. L. F. Powell (Oxford, 1934–50), ii. 425–6. From Ackers's accounts (below, n. 29) it seems that the arrangement for the quarterly book was that 96 books were charged for and 100 delivered.

Gilliver were to take Cadell's agent's role he could expect to pay 15s., the whole-sale bookseller 16s., and the retail or country bookseller 17s. 6d. The retail book-seller who paid 18s. for one of Pope's works was probably not suffering greatly, but the intermediaries were practically squeezed out. If Gilliver was being greedy, so was Pope.

Pope in this period was operating much as Bernard Shaw was to do;[4] he paid his own printer, and had his works distributed through booksellers—first Lawton Gilliver and then Robert Dodsley—both of whom (and here the parallel ends) he had helped to set up in business. Only in the last two years of his life did he turn back to the major figures in the trade, and even then they acted as his agents. There may have been another motive in common with Bernard Shaw, that of controlling the physical appearance and the styling of his works, though he seems to have had ample control over the appearance of his works before. As it turned out, his printer John Wright was not particularly distinguished; his main characteristic seems to have been a determination to follow Pope's copy at all costs. This was doubtless the only way he could keep his job, and it makes the work of the textual critic simpler; at the same time, as I shall show, it caused Pope problems when he forgot to give explicit directions.

Wright did not start printing for Pope until after the publication of the first *Dunciad*; he worked on the *Dunciad variorum* of 1729 which was published by Lawton Gilliver. The great period in which these two worked directly for Pope did not begin until the thirties, and it is perhaps best to say something of them before we turn to consider Pope's later works. Much of the information is based on work by my former pupil, Jim McLaverty, who identified Wright's printing by the ornaments he used and Gilliver's publications from his imprints and advertisements.[5]

John Wright had been foreman to John Barber, a printer who through his friendship with Bolingbroke became part of a Tory circle which included Harley, Atterbury, Swift, Pope, and Prior—not to mention Mrs Manley, who became his mistress. Swift was particularly close to Barber after the Tories came to power in 1710, and he became printer of the *London gazette* in 1711 and Queen's printer in 1713. He seems to have made his fortune in the South Sea speculation of 1720, and he became alderman of Castle-Baynard Ward in 1722. About this time he was advised by his doctors to go to Italy for his health; this he did, leaving Wright in charge of the printing house and also of a time bomb in the form of the *Works* of John Sheffield, Duke of Buckingham for which he had received a royal licence on 18 April 1722.

The Duke of Buckingham had been on friendly terms with Pope for some years before his death in February 1721, and the Duchess in consultation with

---

[4] Shaw's correspondence is rich in references to publishing arrangements; see, e.g., *Collected letters 1874–1897*, ed. D. H. Lawrence (1965), pp. 556–7, 598–9, 798, 811.

[5] J. McLaverty, *Pope's printer, John Wright* (Oxford, 1977), and 'Lawton Gilliver: Pope's bookseller', *SB* xxxii (1979), 101–24.

Atterbury (Bishop of Rochester, and at this time Pope's close friend) decided that Pope should edit and Barber publish his works. Pope seems to have completed his editorial work on the edition by the summer of 1722, though printing was not completed until the winter.[6] The fact that Sheffield's *Works* contained Jacobite passages might have been overlooked if those involved in the edition had not been implicated in Jacobite activity. The Duchess was an illegitimate daughter of James II, and according to a report from Sir Luke Schaub, ambassador in Paris, Barber on his visit to Italy was carrying a banker's draft for £50,000 to the Pretender. A biographer was to claim that Barber was sent on this errand by the Duchess.[7] Three months later, in August, Atterbury himself was sent to the Tower on a charge of high treason, and was still awaiting trial when Sheffield's *Works* were published on 24 January 1723 and seized by the King's Messenger three days later. Pope was clearly upset by this experience and by having to give evidence in Atterbury's trial in May that year.

There are uncertainties about the arrangements for publishing Sheffield's *Works*. Pope had written to Caryll that it 'will be a very beautiful book' (*Correspondence*, ii. 117); though the imprint is 'Printed by John Barber, alderman of London', the first volume (of poems) was the work of William Bowyer. The second volume may have been printed in Barber's own shop by Wright; the evidence of the ornaments is not clear. The copyright was entered to Barber in the Stationers' Register by Wright on 15 January 1723, but who was to get the profits? Lintot noted under 24 October 1722 'A Copy of an Agreement for purchasing 250 of the Duke of Buckingham's Works—*afterwards jockeyed by Alderman Barber and Tonson together*' (Nichols, *Anecdotes*, viii. 303). Bowyer's paper ledger confirms this division, recording that the hundred copies on fine paper were delivered to Tonson by 14 December 1722; 250 ordinary copies went to Wright on 15 January and 239 to Tonson on 19 January (these and one copy 'to Mr Pope in time of Printing' are added up with customary inaccuracy to 500).

The seizure must have been highly embarrassing to Tonson with his dependence on Whig patronage: Pope in his letter of 12 February 1723 to Buckley on the subject writes: 'Particularly remember me to Mr Stanyan, & desire him *not to forget his Servant Jacob*, who seems to me in tribulation, & not mindful of the text which admonishes Booksellers *patiently to Bear one anothers Burdens*' (*Correspondence*, ii. 158). A letter from Tonson to Pope deals somewhat obscurely with the accounting:

You have Inclosed the account of the D. of Bucks Works: for the Books sold I have allowed you all the mony I have rec'd, & the binding &c I have charged at the price it cost me. The Balance 197 lr 9s is ready when you will please to call & bring with you the agreement between Us which may be cancelled as I will do mine. & I will give you my Note to deliver the Books left when required. (*Correspondence*, ii. 211–12.)

---

[6] *Correspondence*, ii. 117, 123–5; Bowyer's paper stock ledger (fo. 20ᵛ) shows paper for the vol. of poems being delivered as late as 8 Nov. 1722.

[7] *An impartial history of . . . Mr. John Barber* (1741), p. 23 n.

It is probable that this agreement between Tonson and Pope was limited to the subscribers' books, but as there is no printed list of subscribers we have no way of knowing how many copies were involved. The fact that all the fine-paper copies were delivered to Tonson certainly suggests that he handled the subscribers' copies. But whatever the nature of the agreement between Pope and Tonson, this letter suggests they were glad to do away with the evidence of it.

Other letters between Pope and Tonson relate to later plans for republishing the copies held by Tonson, and those held by Wright. Pope shows his customary interest in the details of publication, and tries to advance the interests of Lintot, now settled as the bookseller for the *Odyssey*:

> Pray (as you know I am ignorant of the whole matter) tell me what Booksellers are concernd in the Shares of the Duke's Book, now Re-published, besides yourself, Mr Taylor & Innys? Why should not Lintot be admitted among 'em if there's any thing to be gott? You know I must be concerned in his interests, now he is again My Bookseller.[8]

Though Pope speaks of 'Shares', they can hardly be shares in the copyright—it seems more likely that Tonson arranged the republication by selling off the remaining copies to booksellers whom he could trust. He seems to have been acting on behalf of Wright (for Barber) as well as himself; but how many of the copies he held were his own and how many Pope's is obscure. What does emerge from all this is that Pope was in personal contact with Wright at this difficult time.

After Barber returned from Italy in 1724 he seems to have given up his printing house, possibly giving Wright the copyright of Freind's *History of physick* as a farewell present.[9] We know little more of Wright during the crucial period before he became Pope's printer, except that in 1728 he moved into premises in St Peter's Hill, and that when Pope returned to Bowyer as his printer fourteen years later, he wrote of Wright, 'It was Charity made me use him'.[10] He probably first started work there on the *Dunciad variorum*, largely printed in 1728, though his first book to be published was a reprint of the first volume of the Pope–Swift *Miscellanies* (Griffith 208). As far as we know, he printed almost nothing from then on but the works of Pope and his friends, and the probability is that Pope helped to set him up in business as he had formerly helped Henry Woodfall.[11]

---

[8] As Sherburn argues (*Correspondence*, ii. 218), this letter refers to the printing of the early sheets of the *Odyssey*, and must be later than the indenture of 18 Feb. 1724; it is presumably later than 29 Feb., when the republication of Buckingham's *Works* was advertised in the *Evening post* (Griffith 142a).

[9] Wright entered the two vols. in the Register on 28 Nov. 1724 and 12 Jan. 1726; the *Impartial history ... of Mr. John Barber*, p. 44, says Freind gave Barber the copy. Wright took over from Barber as printer to Christ's Hospital on 14 Feb. 1727 (Guildhall MS 12811/9).

[10] *Correspondence*, iv. 392. He did, however, continue to print edns. of Pope's *Works* in octavo until 1743, and vols. of the Pope–Swift *Miscellanies* as late as 1747.

[11] Nichols reports that Woodfall 'at the age of 40 . . . commenced master, at the suggestion, and under the auspices, of Mr. Pope, who had distinguished his abilities as a scholar whilst a journeyman in the employment of the then printer to this admired author' (*Anecdotes*, i. 300). As Woodfall took his first apprentice in 1719, the story may not be true; it might be a garbled version of the Wright story. For an account of Woodfall, see R. J. Goulden, *The ornament stock of Henry Woodfall 1719–1747* (1988).

My impression is that his printing office was on a very modest scale, mainly printing one work at a time; he did not take on any apprentices.

Lawton Gilliver, by contrast to Wright, was new to the trade in 1728. He came from Derbyshire, and was bound apprentice to the high Tory bookseller Jonah Bowyer on 6 March 1721; he was freed by Christiana Bowyer, widow and administratrix of Jonah Bowyer, on 7 May 1728.[12] Gilliver had already set up shop for himself in March 1728 at the expiry of his seven years apprenticeship, before he was formally made free; and I am convinced that Pope had already promised Gilliver his patronage, if not the future copyright of the *Dunciad*, for Gilliver as soon as he opened his shop chose the sign of Homer's Head. There was nothing in Gilliver's training or in the type of books he dealt in to make such a sign relevant, but as bookseller to Pope, whose last publication had been his translation of the *Odyssey*, it was the perfect choice.

From publicaton of the *Dunciad variorum* in 1729 until 1733 Pope had little work of his own for Gilliver and Wright, but apart from the everyday work of the trade, there is a substantial remainder of their work which has connections with Pope. To start with, the *Grub Street journal* started publication on 8 January 1730. However ambiguous Pope's relationship with the *Journal* may be, we do know from the minute book of the partners in the *Journal* that Gilliver was treasurer for the first six months, had the duty of correcting the press, and at first held six of the twelve shares in the paper.[13] His connection with it continued until its demise at the end of 1737—the time when his connection with Pope also ended.

The other work that occupied Gilliver and Wright was a number of poems by Pope's circle of friends and protégés, starting with James Bramston's *Art of politicks* in December 1729 and followed by Edward Young's *Two epistles to Mr. Pope*, George, Lord Lyttelton's *Epistle to Mr. Pope*, and Richard Savage's *The gentlemen of the Dunciad* in 1730, Walter Harte's *Essay on satire* and James Miller's *Harlequin Horace* in 1731, Lyttelton's *The progress of love* and Gilbert West's *Stowe* in 1732. Pope's letter to Caryll of 6 February 1731 hints at his connection with this activity:

The Art of Politicks is pretty. I saw it before 'twas printed. There is just now come out another imitation of the same original, *Harlequin Horace*: which has a good deal of humour. There is also a poem upon satire writ by Mr Harte of Oxford, a very valuable young man, but it compliments me too much: both printed for L. Gilliver in Fleet Street. (*Correspondence*, iii. 173.)

It must also be due to Pope's influence that the copyright of *The genuine works* of George Granville, Lord Lansdowne, was shared between Tonson and Gilliver in

---

[12] McKenzie (ed.), *Stationers' Company apprentices 1701–1800*, no. 987. M. Treadwell tells me that administration was granted to the widow on 2 May 1727 (PRO PROB 6/103/108).

[13] The minute book of the partners is in the library of The Queen's College, Oxford (MS 450); see M. L. Turner, 'The minute book of the partners in the *Grub Street journal* [with transcription]', *PubH* iv (1978), 49–94. The most recent discussion of Pope's involvement is by B. A. Goldgar, 'Pope and the *Grub-street journal*', *MP* lxxiv (1977), 366–80.

1732. Other authors published by Gilliver such as Henry Brooke, Aaron Hill, George Jeffreys, and David Mallet had their work seen by Pope before publication, while others again such as William Brownsword, William Burscough, Robert Dodsley, John Lockman, and Thomas Newcomb had relations of various kinds with Pope. While we have a number of cases where we know that Pope was responsible for Gilliver's being employed, in others the mere fact that Gilliver was Pope's bookseller may have led to authors approaching him; this sort of prestige must have been one of the attractions of the relationship for Gilliver.

If I seem to be building a special relationship between Gilliver and Pope in this early period on circumstantial evidence,[14] there is good contractual evidence for it which I shall deal with after considering the *Dunciad*.

THE DUNCIAD

We can now turn back to a chronological survey of Pope's publications, and are faced by the complex problems of the *Dunciad*. A series of law suits should clarify the transactions involved, but the testimony is often contradictory; nor can we trust Pope's memory. I shall pass over the details, legal and bibliographical, as rapidly as I can.

The original *Dunciad* was advertised on 18 May 1728,[15] and I have already shown that its typography is modelled on the duodecimo *Essay on criticism* of 1713 (Figure 52). The imprint can, I hope, be seen more clearly in the light of what has been said about mercuries, or pamphlet-sellers; it cannot have been printed *for* Anne Dodd, since she lacked the right (whether by rule or custom of the Stationers' Company) of holding copyrights; her name was put on these editions, either as an equivalent (made necessary by the Stamp Act) to saying 'sold by the pamphlet-sellers of London & Westminster', or because she was in fact the distributor to the pamphlet shops. Anne Dodd was to testify in the following year that she had never sold any copies of the *Dunciad variorum* on which her name also appeared,[16] but we cannot be sure that she did not handle the original *Dunciad* as she did so many other pamphlets. Savage writes 'On the Day the Book was first vended, a Crowd of Authors besieg'd the Shop . . . On the other Side, the Booksellers and Hawkers made as great Efforts to procure it' (*Twickenham*, v, p. xxii). This suggests there was a single shop involved, and whose but Dodd's? The statement of the imprint that the work was first printed in Dublin is now accepted as false, and as probably an attempt to throw responsibility for the work

---

[14] In the good company of R. H. Griffith; see his review of the *Twickenham* edn. of the *Dunciad* in *PQ* xxiv (1945), 154–5.

[15] For a detailed account of the printing of the duodecimo and octavo, which were printed together, see D. L. Vander Meulen, 'The printing of Pope's *Dunciad* 1728', *SB* xxxv (1982), 271–85.

[16] On 3 June 1729; PRO C 41/43 No. 570, referred to but not quoted by J. Sutherland, 'The Dunciad of 1729', *MLR* xxxi (1936), 351 n.

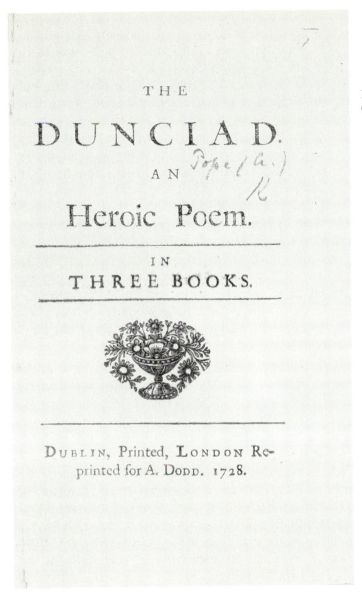

FIG. 52. Title-page of duodecimo *Dunciad* (1728) (BL C. 59. ff. 13(4); 175 × 107)

on Swift. It is the first example of this imprint I have found in a verse work (there may be examples in prose). There are occasions when it seems to be used on libellous or seditious works as an attempt to misdirect any pursuer, and that may be the case here.

Pope also had to find a discreet printer, and chose James Bettenham, Bowyer's son-in-law, who had helped with the printing of the second duodecimo printing of the *Iliad* in 1720. Daniel Prince, the Oxford bookseller and overseer of the learned side of the Oxford University Press from 16 February 1758, recorded that he was 'the apprentice trusted to go to the Author with the proofs in great secrecy. I had the wit to keep the sheets with some of his marks to correct' (Nichols, *Anecdotes*, iii. 705). It was Bettenham who entered the book in the Stationers' Register on 30 May, and deposited copies of the octavo printing there. Printers did sometimes make entries in the Register, but this is the only entry Bettenham ever made, and there can be little doubt that he made it on Pope's behalf; this is a pattern which will recur, and we can only speculate on whether he had an assignment of copyright from Pope and then reassigned it to the author.[17] What does seem very likely is that Pope (like many authors of controversial pamphlets) paid the printing costs and recovered them, together with the profits, from the pamphlet-sellers; Sutherland quotes the *Daily journal* of 9 September 1728: 'The Bookseller is obliged to the incensed Writers of the Town . . . or rather the *Author*, if it be true, as is said, that he makes a little Profit by the *retailing* of it' (*Twickenham*, v, p. xx n.).

Within a month of the *Dunciad*'s publication, and while it was still being reprinted and pirated, Pope had started work on the annotated edition; the full details of the edition 'to be printed in all pomp' are given in his letter to Swift of 28 June 1728 (*Correspondence*, ii. 503). This is where John Wright was brought in as printer, and John Dennis gives us what is no doubt an exaggerated but also a significant account when he writes:

Does not half the Town know, that honest *J.W.* was the only Dunce that was persecuted and plagu'd by this Impression? that Twenty times the Rhapsodist alter'd every thing that he gave the Printer? and that Twenty times, *W.* in his Rage and in Fury, threaten'd to turn the Rhapsody back upon the Rhapsodist's Hands? (*Twickenham*, v. p. xxvii.)

There is no doubt that Wright was to become only too familiar with Pope's rewriting in proof.

It seems likely that Pope had hoped to publish the *Dunciad variorum* by the end of 1728 (see Swift's letter of 16 July, *Correspondence*, ii. 504), and this may be why he was to claim in his 1743 suit against Lintot (who had purchased the copyright in the mean time) that Gilliver printed or published the *Dunciad* in December 1728.[18] But when 'after many unforseen delays' (which may well include the illness of Pope and his mother) the book was nearing completion at the end of the year, Pope, Wright, and Gilliver began to fear 'lest if the Printer & Publisher be found, any [legal] Action could be brought'.[19] Pope asked Burling-

---

[17] Sutherland is wrong in suggesting in the *Twickenham* edn. (v, p. xx) that the entry would be invalid because the Copyright Act required the author's name to be given; it did not. The problem in Chancery was that Gilliver had not claimed that he had acquired the copy from the author; see C. Viner (ed.), *A general abridgment of law and equity* (1751–8), iv. 278–9 (cited by Sutherland, 'The Dunciad of 1729', pp. 350–1).

[18] Sutherland, 'The Dunciad of 1729', p. 351, quoting PRO C 11/549/39. Pope may at this time have given Gilliver an assignment which was later superseded; it would in that case rightly have been destroyed.

[19] *Correspondence*, iii. 4, Pope to Burlington. Pope explicitly says that he is safe, but he *may* have had apprehensions on his own account.

ton for the advice of his lawyer, Nicholas Fazakerley. His advice cannot have been encouraging, for after the poem had been presented to the King and Queen by Sir Robert Walpole on 12 March, the book was circulated privately by three of Pope's noble friends, Burlington, Oxford, and Bathurst, on the assumption that no Dunce would dare to bring their lordships to a court of law for publishing it. A fortnight later Pope asked Oxford 'to send about 20 books to Cambridge, but by no means to be given to any Bookseller, but disposed of as by your own Order at 6s. by any honest Gentleman or Head of a House' (*Correspondence*, iii. 26). On 8 April Pope wrote to Caryll to say 'I understand that now the booksellers have got 'em by the consent of Lord Bathurst' (*Correspondence*, iii. 31). Two days later the edition was advertised in the press as 'printed for Lawton Gilliver ... and A. Dodd', and two days after that, on 12 April, Gilliver entered the copyright in the Stationers' Register and deposited nine copies for the deposit libraries.

We know from a Chancery action Gilliver brought the next month against four pirates (James Watson, Thomas Astley, John Clarke, and John Stagg) that at this moment Gilliver did not own the copyright, but he said that he would soon have the assignment.[20] He was, however, anxious to establish his claim by an entry in the Register, because in five days' time he was to publish the first octavo edition with his own imprint. (The story is here complicated by the fact that this edition, as the defendants pointed out, had originally carried the imprint of A. Dod, but I am satisfied that it was never issued in that form; the only surviving copy has the title slit as a sign that it should be cancelled.[21]) It can scarcely be a coincidence that the day after Gilliver's publication of the octavo, on 18 April 1729, Pope wrote to Oxford:

I did not think so very soon to trouble you with a Letter: But so it is, that the Gentlemen of the Dunciad intend to be vexatious to the Bookseller & threaten to bring an action of I can't tell how many thousands against him. It is judged by the Learned in Law, that if three or four of those Noblemen who honour me with their friendship would avow it so openly as to suffer their Names to be set to a Certificate of the nature of the inclosed, it would screen the poor man from their Insults. If your Lordship will let it be transcribd fair, & allow yours to be subscribd with those of Lord Burlington, Lord Bathurst & one or 2 more, I need not say it will both oblige & honour me vastly. I beg a Line in answer, I cannot say how much I am Ever

My Lord your obligd affect: Servant A. Pope

Whereas a Clamor hath been raisd by certain Persons, and Threats uttered, against the Publisher or Publishers of the Poem calld the *Dunciad* with notes Variorum &c. We whose names are underwritten do declare ourselves to have been the Publishers and Dispersers thereof, and that the same was deliverd out and vended by our Immediate direction. (*Correspondence*, iii. 31–2.)

No such action seems to have been taken, but at some point Pope assigned the copyright to the three lords, who in their turn assigned it to Gilliver for £100 shortly before he published the second edition; his entry for that in the

---

[20] Sutherland, 'The Dunciad of 1729', p. 352.
[21] Ibid. The surviving copy is in the BL, 12274. i. 10.

Stationers' Register on 21 November specifically refers to this assignment, which was drawn on 16 October 1729, and subsequently signed by the three noblemen. This later assignment survives in the British Library;[22] the former does not. Pope wrote to Lintot on 31 January 1741:

When you purchasd the Shares in the Dunciad, I hope Mr Gilliver deliverd you his Title under the Hands of the Lords as well as mine to them . . . for he told me he coud not find it, and without it yours would be (I apprehend) insufficient.[23]

Two problems remain: when was Pope's assignment to the lords made, and under what terms did Gilliver sell the quarto? We can, I think, take it that the profits of the first octavo were his, despite the copyright situation.

To the first question, Sutherland seems to have given different answers at different times, which is not surprising; I have been in the same state. At present I feel that the crucial evidence is the sudden panic after the octavo was published with Gilliver's imprint on 17 April. If we look at the alternative dates given for the assignment of the copyright, they all have their place. It was originally intended that Gilliver should publish the quarto, and hence comes Pope's evidence of an assignment in December 1728; everyone (or else Gilliver and Wright alone) got cold feet, and it was decided that the quarto should be circulated privately at first. The octavo must by this time have been nearly ready, and it is plausible, as Gilliver claimed in his suit against the pirates, that he acquired the copyright on or about 31 March and subsequently proceeded to advertise (10 April) and enter (12 April) the quarto and then to issue the octavo. It was only when the Dunces threatened him with an action, once the octavo had appeared on 17 April with his imprint, that he withdrew. It seems to me likely that, instead of the lords publishing the declaration that Pope suggested to them on 18 April, they instead bought the copyright—thus achieving the same end without exposing themselves to publicity. I would therefore date the assignment to the lords soon after 18 April; the object was to protect Gilliver and not, as Sutherland suggests, the copyright.

As to Gilliver's position in regard to the profits from the quarto, the defendants in the piracy case gave interesting evidence. John Stagg said that the quarto *Dunciad* 'was published on or before April 5, and on either April 7 or 8 he bought a number of copies at five shillings apiece'. At this time Gilliver 'came to Stagg and said that he too had bought some at the same price, and begged Stagg not to

---

[22] MS Egerton 1951, fo. 6; printed by R. W. Rogers, *The major satires of Alexander Pope* (Urbana, 1955), p. 116.

[23] *Correspondence*, iv. 333. Sherburn's n. 2, referring to Lintot's letter of 29 Jan. 1740, seems mistaken. There is no reason to think an earlier assignment has been lost; Pope's memorandum on Lintot's letter of 29 Jan. 1740, 'see Gillivers assignment Lords', is related to his questioning when the copyright expired. It could well be that he borrowed the assignment from Gilliver to find out, and this is how it came to be with his other contracts, and not in Gilliver's possession. Perhaps it is indelicate to suggest that when Pope wrote that he hoped Gilliver had given Lintot the assignment, he knew he had it himself and was trying to make Lintot believe his claim to copyright would not stand up in court.

THE
DUNCIAD.

BOOK the FIRST.

BOOKS and the Man I ſing, the firſt who brings
The Smithfield Muſes to the Ear of Kings.

REMARKS on BOOK the FIRST

* THE *Dunciad*, *Sic* M. S. It may be well diſputed whether this be a right Reading? Ought it not rather to be ſpelled *Dunceiad*, as the Etymology evidently demands? *Dunce* with an *e*, therefore *Dunceiad* with an *e*. That accurate and punctual Man of Letters, the Reſtorer of *Shakeſpeare*, conſtantly obſerves the preſervation of this very Letter *e*, in ſpelling the Name of his beloved Author, and not like his common careleſs Editors, with the omiſſion of one, nay ſometimes of two *ee*'s [as *Shak'ſpear*] which is utterly unpardonable. Nor is the neglect of a *Single Letter* ſo trivial as to ſome it may appear; the alteration whereof in a learned language is an *Atchievment that brings honour* to the Critick who advances it; and Dr. *B.* will be remembered to poſterity for his performances of *this ſort*, as long as the world ſhall have any Eſteem for the Remains of *Menander* and *Philemon*.

THEOBALD.

I have a juſt value for the Letter E, and the ſame affection for the Name of this Poem, as the forecited Critic for that of his Author; yet cannot it induce me to agree with thoſe who would add yet another *e* to it, and call it the *Dunceiade*; which being a French and foreign Termination, is no way proper to a word entirely Engliſh, and Vernacular. One *E* therefore in this caſe is right, and two *E*'s wrong; yet upon the whole I ſhall follow the Manuſcript, and print it without any *E* at all; mov'd thereto by Authority, at all times with Criticks equal if not ſuperior to Reaſon. In which method of proceeding, I can never enough praiſe my very good Friend, the exact Mr. *Tho. Hearne*; who, if any word occur which to him and all mankind is evidently wrong, yet keeps he it in the Text with due reverence, and only remarks in the Margin, *ſic* M. S. In like manner we ſhall not amend this error in the Title itſelf, but only note it *obiter*, to evince to the learned that it was not our fault, nor any effect of our own Ignorance or Inattention.

SCRIBLERUS.

VERSE I. *Books and the Man I ſing, the firſt who brings*
The Smithfield *Muſes to the Ear of Kings.*

Wonderful is the ſtupidity of all the former Criticks and Commentators on this Poem! It breaks forth at the very firſt line. The Author of the Critique prefix'd to *Sawney*, a Poem, *p.* 5. hath been ſo dull as to explain *The Man who brings,* &c. not of the Hero of the Piece, but of our Poet himſelf, as if he vaunted that *Kings* were to be his Readers (an Honour which tho' this Poem hath had, yet knoweth he how to receive it with more Modeſty.)

FIG. 53. Abandonment of italic for proper names in text but not notes of *Dunciad variorum* (1729), p. 1 (Bodl. CC 76(1) Art; 249 × 188)

sell any copies to the booksellers at less than six shillings each'.[24] This attempt to maintain small profit margins (the retail price was 6*s.* 6*d.*) runs so closely parallel to Pope's negotiations with his *Works* in 1735 and later that one is tempted to feel that Gilliver was acting on Pope's instructions and as his agent. That is not to say that, after testing the temperature of the water by releasing copies of the *Dunciad* to several booksellers, Pope would not allow Gilliver to sell and take his profits on all the copies that were left. But I suspect that whatever may have been Pope's original intention (and whether or not set out in a superseded contract), in the end he paid Wright's bill for the printing and handled the distribution himself, letting Gilliver act as his agent in the later stages.

The pirates' other evidence impugned Gilliver's rights under the Copyright Act by pointing out that he entered the book in the Stationers' Register two days after he had advertised it in the *Daily post*, whereas the Act laid down that books must be entered before publication. Moreover, the nine deposit copies which Stagg had examined were 'on a worse paper'; Pope had had fine-paper copies printed, and the Act said that deposit copies should be on fine paper when there were such printed. Pope was to pay close attention to both these matters in the future; there are cases where the only fine-paper copies known are those which were deposited at Stationers' Hall.

As to the printing of the quarto *Dunciad*, it is noteworthy for two typographic features. The first, which we can see on the first page of text with its engraved headpiece, probably the work of Kent, is the abandonment of italic for proper names—here 'Smithfield'—in the text, but not in the notes (Figure 53). The other is the abandonment of catchwords; on this page and on page 3 they are found where a footnote is continued on the following page, but thereafter they vanish completely so far as the text is concerned (Figure 54). Here, however, we begin to note the literal-mindedness of John Wright as a printer. The Prolegomena to the poem have catchwords, perhaps because Pope had not yet made clear his wishes. But after the poem, when we come to the appendices, the catchwords reappear (except on pp. 99–102) together with italics (Figure 55). One might have expected Wright to be consistent, but he seems to have been satisfied to do as he was told. By the time we reach the second octavo edition at the end of the year, Wright has removed the catchwords from the appendices, but not yet from the Prolegomena. Hence my earlier suggestion that he only did what he was specifically ordered to do; and if anyone doubts that Pope was responsible for the decision to make these typographical changes, they certainly cannot attribute them to Wright's 'house style'. When we find small differences in capitalization between the successive editions of the *Dunciad*, therefore, we can assume with some confidence that we do not owe them to Wright's own initiative; some, indeed, may be due to careless corrections by Pope.

---

[24] Sutherland, 'The Dunciad of 1729', pp. 350–1, paraphrasing PRO C 41/43, no. 567.

135 A shaggy Tap'stry, worthy to be spread
   On Codrus' old, or Dunton's modern bed;
   Instructive work! whose wry-mouth'd portraiture
   Display'd the fates her confessors endure.

REMARKS.

" Reflions upon his *Genius*. An honeſt mind written theſe notes (as was my intent) in
" will love and eſteem a *man of merit*, tho' he the learned language, I might have given him
" be deform'd or poor. Yet the author of the the appellations of *Balatro*, *Calceatus capra*, or
" *Dunciad* hath likell'd a perſon for his *rueful Sciurus in trevis*, being pirates in good eſteem,
" *length of face*." MR. S. JOHN. *June* 8. and frequent uſage among the beſt learned. But
This *Genius* and *man of worth* whom an honeſt in our mother-tongue were I to tax any Gentle-
mind ſhould love, is Mr. *Curl*. True it is, he man of the *Dunciad*, ſurely it ſhould be in words
ſtood in the Pillory, an accident which will not to the vulgar intelligible, whereby civilian
lengthen the face of any man tho' it were ever charity, decency, and good accord among au-
ſo comely, therefore is no reflection on the na- thors, might be preſerved. SCRIBLERUS.
tural beauty of Mr. *Curl*. But as to reflections VERSE 135. *A ſhaggy Tap'ſtry.*] A ſorry
on any man's Face, or Figure, Mr. *Dennis* ſaith kind of Tapeſtry frequent in old Inns, made of
excellently; " Natural deformity comes not by worſted or ſome coarſer ſtuff; like that which is
" our fault, 'tis often occaſioned by calamities ſpoken of by Doctor *Donne* — *Faces as frightful*
" and diſeaſe, which a man can no more help, *as theirs who whip Chriſt in old hangings.* The
" than a monſter can his deformity. There is imagery woven in it alludes to the mantle of
" no one misfortune, and no one diſeaſe, but *Columbus in Æn.* 5.
" what all the reſt of men are ſubject to. VERSE 136. *On Codrus' old*, or *Dunton's*
" But the deformity of this Author is viſible, *modern bed.*] Of *Codrus* the Poet's bed ſee *Ju-*
" preſent, laſting, unalterable, and peculiar to *venal*, deſcribing his poverty very copiouſly. *Sat.*
" himſelf; it is the mark of God and Nature 3. v. 203, &c.
" upon him, to give us warning that we ſhould
" hold no ſociety with him, as a creature not   *Lectus erat Codro, &c.*
" of our original, nor of our ſpecies: And they
" who have refuſed to take this warning which Codrus had but one bed, ſo ſhort to boot,
" God and Nature have given them, and have That his ſhort Wife's ſhort legs hung dangling out:
" in ſpite of it by a ſenſeleſs preſumption, ven- His cupboard's head ſix earthen pitchers grac'd,
" tur'd to be familiar with him, have ſeverely Beneath them was his truſty tankard plac'd;
" ſuffer'd, &c. 'Tis certain his original is not And to ſupport this noble Plate, three Lay
" from *Adam*, but from the Devil, &c. *Dennis* A bending Chiron, caſt from honeſt clay.
" and *Gildon*: *Charact. of Mr. P.* 8°. 1716. His few Greek books a rotten cheſt contain'd,
It is admirably obſerv'd by Mr. *Dennis* againſt Whoſe covers much of mouldineſs were fed,
Mr. *Law*, p. 33. " That the language of *Bil-* Where mice and rats devour'd a poetic bread,
" lingſgate can never be the language of Charity, 'Tis true, poor *Codrus* nothing had to boaſt,
" nor conſequently of Chriſtianity." I ſhould And on Heroic Verſe luxuriouſly were fed.
elſe be tempted to uſe the language of a Critick: Dryd.
For what is more provoking to a Commentator, But Mr. *C.* in his dedication of the Letters,
than to behold his author thus pourtray'd? Yet Advertiſements, &c. to the Author of the *Dun-*
I confeſs it really hurts not *Him*; whereas ma- *ciad*, aſſures us, that " *Juvenal* never ſatyrized
liciouſly to call ſome others dull, might do them the poverty of *Codrus*.
prejudice with a world too apt to believe it. *John Dunton* was a broken Bookſeller and abu-
Therefore tho' Mr. *D.* may call another a *little* ſive ſcribler: he writ *Neck or Nothing*, a vio-
aſs or a *young aſs*, for be it from us to call him lent ſatyr on ſome Miniſters of State; the dan-
a *turdball lion*, or an *old ſerpent*. Indeed, had I

   Earless on high, stood un-abash'd Defoe,
140 And Tutchin flagrant from the scourge, below:
   There Ridpath, Roper, cudgell'd might ye view;
   The very worsted still look'd black and blue:
   Himself among the storied Chiefs he spies,
   As from the blanket high in air he flies,
145 And oh! (he cry'd) what street, what lane, but knows
   Our purgings, pumpings, blanketings and blows?
   In ev'ry loom our labours shall be seen,
   And the fresh vomit run for ever green!
   See in the circle next, Eliza plac'd;
150 Two babes of love close clinging to her waste,

REMARKS.

ger of a death-bed repentance, a libel on the late
Duke of *Devonſhire*; and on the Rt. Rev. Biſhop of *Peterborough*, &c.
  VERSE 140. *Himſelf among the ſtoried thiefs*
he flies, &c.] The hiſtory of *Curl's* being toſs'd
in a blanket, and whipp'd by the ſcholars of
Weſtminſter, is ingeniouſly and pathetically related in a poem entituled *Neck or Nothing*. Of
his purging and vomiting, ſee *A full and true*
account of a horrid revenge on the body of Edm.
Curl, &c.
  VERSE 149. *See in the circle next*, *Eliza*
plac'd.] In this game is expos'd in the moſt
contemptuous manner, the profligate licencious-
neſs of thoſe ſhameleſs ſcriblers (for the moſt
part of That ſex, which ought leaſt to be capable of ſuch malice or impudence) who in li-

equally and alternately were cudgell'd, and de-
ſerv'd it.
  VERSE 143. *Himſelf among the ſtoried thiefs*
he ſpies, &c.] The hiſtory of *Curl's* being toſs'd
in a blanket, &c.
  VERSE 140. *And Tutchin flagrant from the*
ſcourge.] *John Tutchin*, author of ſome vile verſes, and of a weekly paper call'd the *Obſervator*:
He was ſentenc'd to be whipp'd thro' ſeveral
towns in the weſt of *England*, upon which he
petition'd King *James* II. to be hanged. When
that Prince died in exile, he wrote an invective
againſt his memory, occaſioned by ſome humane
Elegies on his death. He liv'd to the time of
Queen *Anne*.
  VERSE 141. *There* Ridpath, Roper.] Authors of the *Flying-Poſt* and *Poſt-Boy*, two ſcan-
dalous papers on different ſides, for which they

IMITATIONS.

VERSE 143. *Himſelf among the ſtoried thiefs*
he ſpies, &c.] A parody on theſe of a late noble
author,
  His bleeding arm had furniſh'd all their room,
  And run the purple in the loom.
  VERSE 150. *Two babes of love cloſe clinging to*
her waſte,] Virg. Æn. 5.
  Creſſa genus, Pholöe, geminiſque ſub ubere nati.

VERSE 143. *Himſelf among the ſtoried thiefs*
he ſpies, &c.] Virg. Æn. 1.
  Sæpius prædipuos permaturos agentes Achivos —
  Conſpicit & lacrymans. Quis jam locus, inquit,
                                    Achati,
  Quæ regio in terris noſtri non plena laboris?
  VERSE 148. *And the freſh vomit run for*

FIG. 54. Abandonment of catchwords in *Dunciad variorum* (1729), pp. 34–5 (Bodl. CC 76(1) Art; 249 × 358)

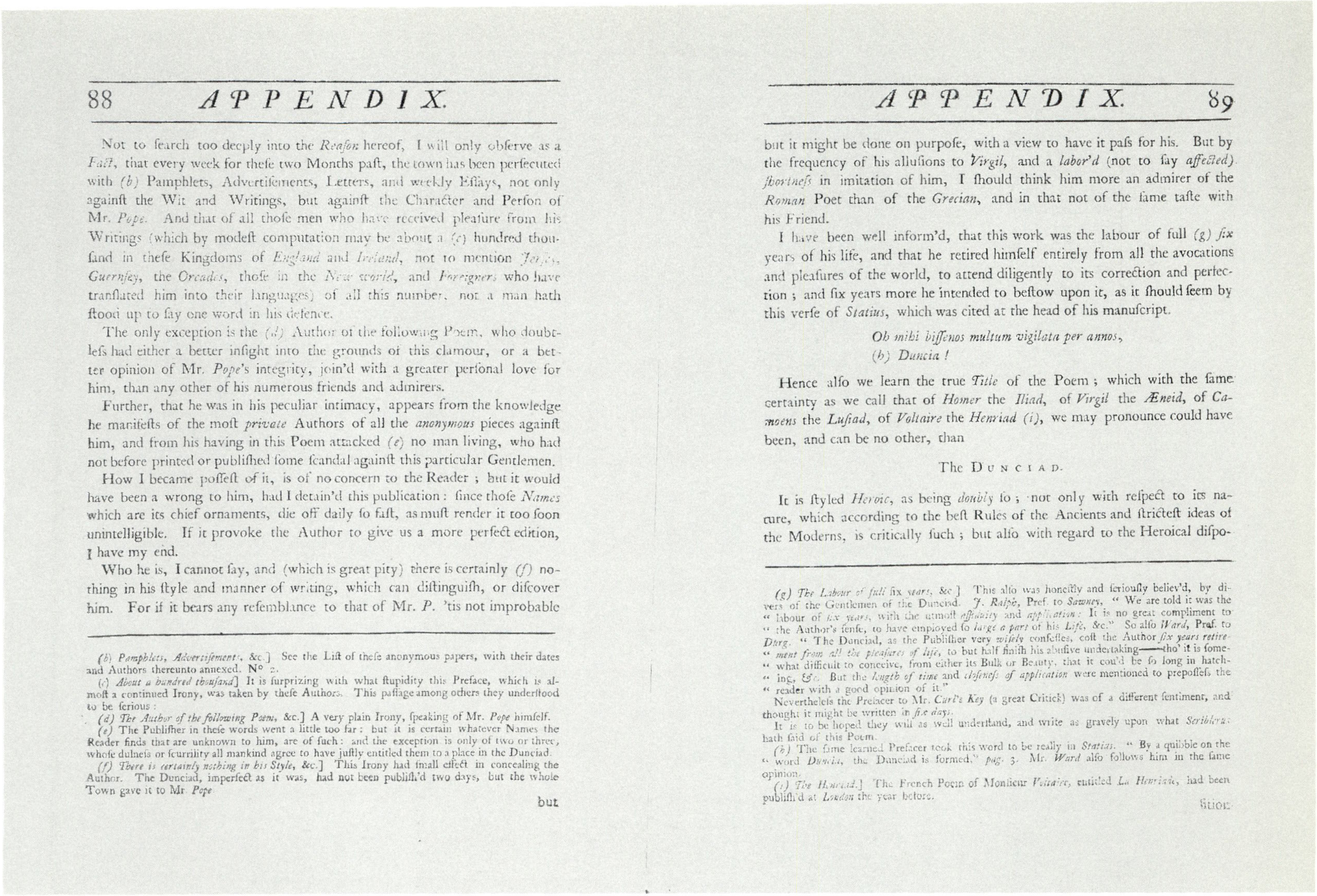

Not to fearch too deeply into the *Reafon* hereof, I will only obferve as a *Faft*, that every week for thefe two Months paft, the town has been perfecuted with (*b*) Pamphlets, Advertifements, Letters, and weekly Effays, not only againft the Wit and Writings, but againft the Character and Perfon of Mr. *Pope*. And that of all thofe men who have received pleafure from his Writings (which by modeft computation may be about a (*c*) hundred thoufand in thefe Kingdoms of *England* and *Ireland*, not to mention *Jersey*, *Guernfey*, the *Orcades*, thofe in the *New world*, and *Foreigners* who have tranflated him into their languages) of all this number, not a man hath ftood up to fay one word in his defence.

The only exception is the (*d*) Author of the following Poem, who doubt-lefs had either a better infight into the grounds of this clamour, or a bet-ter opinion of Mr. *Pope's* integrity, join'd with a greater perfonal love for him, than any other of his numerous friends and admirers.

Further, that he was in his peculiar intimacy, appears from the knowledge he manifefts of the moft *private* Authors of all the *anonymous* pieces againft him, and from his having in this Poem attacked (*e*) no man living, who had not before printed or publifhed fome fcandal againft this particular Gentlemen.

How I became poffeft of it, is of no concern to the Reader ; but it would have been a wrong to him, had I detain'd this publication: fince thofe *Names* which are its chief ornaments, die off daily fo faft, as muft render it too foon unintelligible. If it provoke the Author to give us a more perfect edition, I have my end.

Who he is, I cannot fay, and (which is great pity) there is certainly (*f*) no-thing in his ftyle and manner of writing, which can diftinguifh, or difcover him. For if it bears any refemblance to that of Mr. *P.* 'tis not improbable

but

(*b*) *Pamphlets, Advertifements, &c.*] See the Lift of thefe anonymous papers, with their dates and Authors thereunto annexed. № 2.

(*c*) *About a hundred thoufand.*] It is furprizing with what ftupidity this Preface, which is almoft a continued Irony, was taken by thefe Authors. This paffage among others they underftood to be ferious.

(*d*) *The Author of the following Poem, &c.*] A very plain Irony, fpeaking of Mr. *Pope* himfelf. The Publifher in thefe words went a little too far : but it is certain whatever Names the Reader finds that are unknown to him, are of fuch : and the exception is only of two or three, which dulnefs or fcurrility all mankind agree to have juftly entitled them to a place in the Dunciad.

(*f*) *There is certainly nothing in his Style, &c.*] This Irony had fmall effect in concealing the Author. The Dunciad, imperfect as it was, had not been publifh'd two days, but the whole Town gave it to Mr *Pope*

---

but it might be done on purpofe, with a view to have it pafs for his. But by the frequency of his allufions to *Virgil*, and a *labour'd* (not to fay *affected*) *heartinefs* in imitation of him, I fhould think him more an admirer of the *Roman* Poet than of the *Grecian*, and in that not of the fame tafte with his Friend.

I have been well inform'd, that this work was the labour of full (*g*) *fix* years of his life, and that he retired himfelf entirely from all the avocations and pleafures of the world, to attend diligently to its correction and perfec-tion ; and fix years more he intended to beftow upon it, as it fhould feem by this verfe of *Statius*, which was cited at the head of his manufcript.

Ob mihi biffenos multum vigilata per annos,
(*b*) *Duncia !*

Hence alfo we learn the true *Title* of the Poem ; which with the fame certainty as we call that of *Homer* the *Iliad*, of *Virgil* the *Æneid*, of *Ca-mœns* the *Lufiad*, of *Voltaire* the *Henriad* (*i*), we may pronounce could have been, and can be no other, than

The D u n c i a d.

It is ftyled *Heroic*, as being *done by fo* ; nor only with refpect to its na-ture, which according to the beft Rules of the Ancients and ftricteft ideas of the Moderns, is critically fuch ; but alfo with regard to the Heroical difpo-

(*g*) *The Labour of full fix years, &c.*] This alfo was honeftly and ferioufly believ'd, by di-vers of the Gentlemen of the Dunciad. *J. Ralph*, Pref. to *Sawney.* " We are told it was the " labour of *fix years*, with the utmoft *affiduity* and *application:* It is no great compliment to " the Author's fenfe, to have employed *fo large a part* of his Life, &c." So alfo *Ward,* Pref. to *Durg.* " The Dunciad, as the Publifher very wifely confeffes, coft the Author *fix years retire-* " *ment from all the pleafures of life,* to but half finifh his abufive undertaking——tho' it is fome-" what difficult to conceive, from either its Bulk or Beauty, that it cou'd be fo long in hatch-" ing, &c. But the *length of time* and *difmay of application* were mentioned to prepoffefs the " reader with a good opinion of it."

Neverthelefs the Preface to Mr. *Curl's* Key (a great Critick) was of a different fentiment, and thought it might be written in *fix days.*

It is to be hoped they will as well underftand, and write as gravely upon what *Scriblerus* hath faid of the Poem.

(*b*) The fame learned Prefacer took this word to be really in *Statius.* " By a quibble on the " word *Duncia,* the Dunciad is formed." pag 3. Mr. *Ward* alfo follows him in the fame opinion.

(*i*) *The Henriad.*] The French Poem of Monfieur *Voltaire,* entitled *La Henriade,* had been publifh'd at *London* the year before.

Siuce

FIG. 55. Italics and catchwords in appendix to *Dunciad variorum* (1729), pp. 88–9 (Bodl. CC 76(1) Art; 249 × 358)

POEMS LEADING TO *WORKS* II, 1735

I cannot help regarding the period from 1730 to the publication of the second volume of *Works* in 1735 as something of a tragi-comedy; indeed, the bibliographical results of the tragi-comedy lasted until 1741. In 1730 Pope spoke to Spence of his plans for what he was later to call (writing to Swift on 6 January 1734) his *Opus magnum*, dealing with 'The nature of man' and 'The use of things'.[25] The *Epistle to Burlington*, published at the end of 1731, was the first instalment of the second category, which (in a much reduced form) was to become the second book of 'Ethic epistles'. By the end of 1732, Pope was ready for a period of intense publishing activity, which he no doubt hoped would see the completion of the *Opus magnum*, and on 1 December 1732 he signed a contract with Gilliver as an important step towards putting these plans into effect. The contract says that Pope intends to publish 'certain Poems or Epistles in Verse' and that it is also his intention that 'Lawton Gilliver shall have the sole liberty of printing & selling' as many of them as Pope 'shall think fit for the space of one year only from the time of entring the same' in the Stationers' Register. Gilliver is to pay £50 for each poem or epistle. He binds himself to accept every poem that Pope permits him to print and to pay the £50 at or before the delivery of each poem. In return Gilliver gains the sole rights to print and sell in what volume or size he thinks fit. He also agrees to enter each poem in the Stationers' Register in his own name, and at the expiration of one year transfer his rights to Pope, or anyone else Pope may appoint.[26]

At about this time Pope and Gilliver must have reached an understanding about the publication of the second volume of Pope's *Works* in formats to match the first volume of 1717; it would consist of the *Dunciad*, of which Gilliver owned the copyright, and the new poems by Pope, presumably after Gilliver's copyright in them had expired and been transferred back to Pope. We have some idea of the date of this agreement because the new edition of the *Dunciad* which forms part of the *Works* of 1735 has a certificate between the text of the poem and the commentary (Figure 56) in which (in legal black letter) Pope declares that '*We*, having carefully revised this our Dunciad ... do declare every Word, Figure, Point, and Comma of this Impression to be Authentic', and it is dated 3 January 1733 (in fact 1732 in the legal style) in the presence of the Lord Mayor, John Wright's retired master, John Barber. While this is clearly a light-hearted bit of nonsense, there is no reason to doubt that the text of the poem had been printed by this date, five weeks after Pope's contract with Gilliver.

The position is further clarified by an undated declaration by Pope, appended to the Gilliver contract. In it Pope asks that if he die before all his epistles are printed, his executors shall offer Gilliver the refusal of all he leaves fit for the press

---

[25] Spence, *Anecdotes* no. 299, and *Correspondence*, iii. 401. For a full account of Pope's plans, see M. Leranbaum, *Alexander Pope's 'Opus magnum' 1729–1744* (Oxford, 1977).

[26] BL MS Egerton 1951, fos. 8–9; printed by Rogers, *Major satires of Pope*, pp. 116–19.

55

By the Author

A

DECLARATION.

WHEREAS certain Haberdashers of Points and Particles being instigated by the Spirit of Pride, and assuming to themselves the Name of Criticks and Restorers, have taken upon them to adulterate the common and current Sense of our Glorious Ancestors, Poets of this Realm; by clipping, coining, defacing the Images, or mixing their own base Allay, or otherwise falsifying the same, which they publish, utter, and vend as genuine: The said Haberdashers having no Right thereto, as neither Heirs, Executors, Administrators, Assigns, or in any sort related to such Poets, to all, or any of Them: Now We, having carefully revised this our Dunciad, beginning with the Word Books, and ending with the Word flies, containing the entire Sum of one thousand and twelve Lines; do declare every Word, Figure, Point, and Comma of this Impression to be Authentic: And do therefore strictly enjoin and forbid any Person or Persons whatsoever, to rase, reverse, put between hooks, or by any other means directly or indirectly change or mangle any of them. And we do herein earnestly exhort all our Brethren to follow this our Example, which we heartily wish our Great Predecessors had heretofore set,

as a Remedy and Prevention of all such Abuses. Provided always, that nothing in this Declaration shall be construed to limit the lawful and undoubted Right of every Subject of this Realm, to judge, censure, or condemn, in the whole or in part, any Poem or Poet whatsoever.

Given under our hand at London, this third day of January, in the Year of our Lord one thousand seven hundred thirty and two.

Declarat' cor' me.   JOHN BARBER, Mayor.

FIG. 56. Author's declaration: Pope, *Works II* (1735), pp. 55–6 (Bodl. Vet. A4 d. 142; 289 × 225)

in order to publish them all together with what were before printed, and with the Dunciad (of which he already has the property and of which he hath lying by an Edition in Quarto and Folio) And if the said Mr Gilliver shall not agree with my Executors for the perpetuity of the Epistles, That whosoever shall purchase the perpetuity shall make it a Condition to cause the said Epistles to be printed in the same manner as the abovesaid Edition of the Dunciad, to be bound up, and to go therewith in one Volume: which shall be intituled, The Works of Mr Alexander Pope. Volume the Second.[27]

When the *Works* were entered in the Stationers' Register on 11 April 1735, the copyright was given as one-half Pope's and one-half Gilliver's. That this was no formality can be seen if we examine what share of the individual copyrights each owned. If we regard the preliminaries as common property, we find we can attribute 258 pages of the quarto to Gilliver and 256 to Pope. To my mind it is clear that Pope was committed to producing as much material as Gilliver already owned, and as his plans (or his capabilities) for executing the *Opus magnum* declined, he was faced with the problem of making up his quota. It will be worth examining Pope's new poems between 1732 and 1735 with this in mind, before considering the actual make-up of the *Works*.

Another consideration concerns the printer John Wright. If I am right in my assumption that he had only a small shop, it would be sensible to print off sections of the *Works* as they were completed, as had already been done with the *Dunciad*. We find evidence of this being done as well, but not without false starts and confusion.

Finally, we should consider how profitable Gilliver's contract for the individual epistles was likely to be (Table 11). He was to pay £50 for a year's rights to each poem, but the decision on what poems he was to print was entirely in Pope's hands. While we have no figures at all for Wright's printing (let alone any accounting between Pope and Gilliver, which must have become complex), we have comparable figures for Woodfall's printing *The first epistle of the second book of Horace* for Dodsley in 1737.[28] From a comparison of similar figures, I think the cost without the fine-paper copies would be 25 instead of 27s. a sheet; although Pope often had fine-paper copies printed, he may have paid his bookseller's expenses for them. The paper size is probably crown, not notably good; from a range of prices from 8 to 12s. a ream, I have taken a figure of 9s. We know the work was sold retail at 1s. and I have assumed a trade price of 9d., which is in line with the discount given to the pamphlet-sellers on the pirate *Dunciad* of 1742, and to Wilford by Ackers for pamphlets, but it does not include allowances for the quarterly books or for publishing charges.[29]

If Gilliver were working on this basis, the first edition of 2,000 copies would have left him a profit of only £3. 13s. 0d. after he had paid Pope £50; the second

[27] BL MS Egerton 1951, fo. 12; printed by Rogers, ibid.

[28] P.T.P., 'Pope and Woodfall', p. 377.

[29] For the *Dunciad*, see H. P. Vincent, 'Some *Dunciad* litigation', *PQ* xviii (1939), 289, and for Ackers's account with Wilford, see *A ledger of Charles Ackers*, p. 126, and discussion, p. 14.

TABLE 11. Gilliver's Profits on Single Epistles

Calculation on Woodfall's account to Dodsley:
   'May 12, 1737
   Printing the first Epistle of the Second Book of
   Horace imitated, folio, double [pica], Poetry
   No. 2000 and 150 fine [7] shts. at 27s. pr. sht.   9. 09. 0.'

   [less fine, at 25s. per sheet]   8. 15. 0.
   Paper: 7 sheets × 4 reams at 9s.   12. 12. 0.
   (Price per copy 2.56d.)   21. 7. 0.

   'July 21, 1737
   Printing 500 Mr. Pope's first Epistle of Second
   Book of Horace, 7 shts. at 11s. pr . sht.   3. 17. 0.'
   Paper: 7 sheets × 1 ream at 9s.   3. 3. 0.
   (Price per copy 3.36d.)   7. 0. 0.

Sold retail at 1s.   Take trade price as 9d. (no
allowance for quarterly copies)
   Profit on 2000   53. 13. 0.
   Profit on next 500   11. 15. 0.

Calculation on smaller pamphlet:
   Pamphlet of 5 sheets at 25s. per sheet,
   2000 copies   6. 5. 0.
   Paper: 5 sheets × 4 reams at 9s.   9. 0. 0.
   (Price per copy 1.83d.)   15. 5. 0.

   profit on 2000   59. 15. 0.

edition would have been pure profit, and in the case of some poems which went into several reimpressions rather than one new edition, this figure could have been greater. In fact, this pamphlet of seven sheets is larger than usual; the average size of Pope's poems of the Gilliver period is about five sheets, and in the table I have recalculated costs for a five-sheet pamphlet. Although the price per copy is reduced, Gilliver's profit is still only £9. 15s. 0d. on the first two thousand copies. By the time Dodsley ordered two thousand copies of the Horace epistle in 1737 the demand for Pope's epistles should have been fairly well established; and though Gilliver may sometimes have sold more, he cannot in general have gained much more than prestige by buying copyrights on these terms.

   If we now look at Table 12, listing Pope's poems, we can see how matters developed. We see that the *Epistle to Burlington*, which was issued before the contract, had fine-paper copies and went into three reimpressions and two new editions.[30] The Timon's villa passage in the poem provoked one of the major controversies of Pope's career—seven pamphlet attacks within a few months of publication[31]—and this would have stimulated sales sufficiently to encourage Gilliver to sign his agreement with Pope. (Perhaps it gave him false expectations, as the *Rape of the lock* may have given Lintot false hopes for sales of the *Iliad*.) The

---

[30] For an account of four poems of this period, including *Burlington*, see W. B. Todd, 'Concealed Pope editions', *BC* v (1956), 48–52.

[31] See J. V. Guerinot, *Pamphlet attacks on Alexander Pope 1711–1744* (1969), p. xxiv.

TABLE 12. Pope's Poems, 1731–1735

Publication	title-page date	pub date	imprint	copyright	printer
Epistle to Burlington (+FP, 3 reimpr., 2 edns.)	1731	13 Dec	for L. Gilliver	Gilliver SR 7 Dec	Wright
Epistle to Bathurst (+FP, 2 edns.)	1732	15 Jan 1733	by Wright for Gilliver	Gilliver SR 13 Jan 1733	Wright
1st Satire, 2nd book Horace (+2 reimpr. 1 edn.)	1733	15 Feb	by L.G.; sold by A. Dodd, E. Nutt	Gilliver SR 14 Feb	
Essay on Man I ($2°$+reimpr.+LP+2 edns. $4°$+FP)	[1733]	20 Feb	for J. Wilford	'Wilford' SR 10 Mar	Huggonson
Essay on Man II (+2 edns.)	[1733]	29 Mar	for J. Wilford	'Wilford' SR 29 Mar	E. Say
Essay on man III (+1 edn.)	[1733]	8 May	for J. Wilford	'Wilford' SR 9 May	S. Aris
The impertinent $4°$	1733	5 Nov	for John Wileord		
Epistle to Cobham (+2 edns. 1737)	1733	16 Jan 1734	for Lawton Gilliver	Gilliver SR 5 Feb 1734	Wright
Essay on man IV	[1734]	24 Jan	for J. Wilford	'Wilford' SR 25 Jan	Wright
Essay on man [complete] ($2°$+LP, $4°$+FP)	1734	20 Apr	by Wright for Gilliver		Wright
1st & 2nd Satires Horace ($2°$+LP, $4°$+FP)	1734	4 Jul	for L. G.	'Gilliver' SR 3 Jul	Wright
2nd Satire 2nd book Horace	1735	3 Oct 1734	by Wright for Gilliver		Wright
Sober advice from Horace (+1 edn., reissue 1738)	[1734]	21 Dec	For T. Boreman sold bkslrs	Brindley, Payne, Boreman, Corbett	
Epistle to Arbuthnot	1734	2 Jan 1735	by Wright for Gilliver	Gilliver SR 2 Jan 1735	Wright
Epistle to a lady	1735	7 Feb	by Wright for Gilliver	Gilliver SR 7 Feb	Wright
Works II ($2°$+LP, $4°$+FP)	1735	24 Apr	by Wright for Gilliver	Gilliver $\frac{1}{2}$ Pope $\frac{1}{2}$ SR 11 Apr	Wright

FP: fine paper; LP: large paper; SR: Stationers' Register. Poems are in folio unless otherwise indicated.

clamour may also have caused Pope to delay publication of other epistles for a
while in the hope of a better reception for them later. Then there is a rush of five
publications in the first five months of 1733. Pope was anxious, as we know, that
the anonymous *Essay on man* should not be recognized as his, and his stratagem
was highly successful;[32] to help this plan it bore the imprint of John Wilford,
rather than that of Gilliver. Wilford also entered each part in the Stationers'
Register, though Pope's copy of the contract with Gilliver shows that Gilliver
paid £150 copy-money for the first three parts on 23 March. Wilford had become
a shareholder in the *Grub Street journal* before 7 September 1732, and so would
have been in close contact with Gilliver.[33] In what seems an excessive desire for
concealment (perhaps Wright could not be trusted to hold his tongue, though he
certainly did so in the matter of Pope's letters and Bolingbroke's *The idea of a
patriot king*), these first three parts were printed by John Huggonson, Edward
Say, and Samuel Aris.[34] Of these, Huggonson was also associated with Gilliver as
a shareholder in the *Grub Street journal* from its inception, and became its printer
in October 1733 in place of Sam Aris. Pope's authorship became known in the
summer of 1733, so Wright was allowed to print the fourth part as well as to reprint the second part (Foxon P834). *The first satire of the second book of Horace*
was only obliquely ascribed to Pope, in such a way that he could disclaim responsibility, and the presence of the mercuries Mrs Dodd and Mrs Nutt in the
imprint may reflect that; Wright's absence as printer seems unnecessary, but he
may have been too busy completing the *Dunciad* for the *Works* to undertake it.

The fact that the second volume of Pope's *Works* was already planned explains,
I think, why the first part of the *Essay on man* was reimpressed (with corrections)
in large and small folio and also in ordinary and fine-paper quarto (Foxon P823–
6). It seems almost beyond belief that Pope should have hoped to get a poem fully
to his liking on first publication, but I can see no other explanation. Moreover,
this reprint has the running-title 'Epistles' which is found both in Pope's final
manuscript and in the collected editions, but not in the separate folios. In fact
Pope abandoned this idea, published a heavily revised version of the first part in
April 1733 (Foxon P827), and revised the text again for the collected edition of
April 1734 (Foxon P850). This was indeed used for the folio edition of the *Works*
according to plan, but was reprinted with further corrections for the quartos,
probably because the number of copies printed was exhausted.[35]

On 7 June 1733 Pope's mother died, and this event may be responsible not
only for the delay in completing the *Essay on man* but also for thoughts of excluding the rest of the *Opus magnum* from his forthcoming volume of works. Pope
wrote to Bethel on 9 August 1733, 'I have now but too much melancholy leisure,

[32] See Maynard Mack's account, *Twickenham*, iii/1, pp. xv–xvi.

[33] See Turner, 'The minute book of the partners in the *Grub Street journal*', p. 59.

[34] Huggonson can be identified by an ornament he used in printing Bridges' *Divine wisdom and provid-
ence* in 1736, Say from an ornament in Henry Carey's *Poems*, 1729, while Aris uses a block with his name
on it.

[35] The fine-paper quartos at Bodl. and Harvard use the 1734 edition.

and no other care but to finish my Essay on Man' and that he will print it 'in a fortnight in all probability' (*Correspondence*, iii. 381). What we find printed after his summer travels, however, is the ambiguous *The impertinent*, an adaptation of Donne's fourth satire, published as an undistinguished quarto without an author's name, without printer's ornaments, and with Wilford's name misspelt 'Wileord'—normally the mark of a piracy, just as the *Dunciad* piracy of 1729 had the imprint of 'A. Dob' in place of 'Dod'.[36] Despite its surreptitious appearance, the text is clearly a recent revision by Pope of an earlier version that may well have first been written—as Pope was later to claim—before 1714.[37] It was reprinted with Donne's second satire in the 1735 *Works*. To my mind, this is the first clear sign that Pope was beginning to doubt his capacity to fill his half of that volume with the *Opus magnum*, and was looking for alternative materials. Imitations, whether of Horace or Donne, were well adapted to filling space, for by printing the original text on facing pages one doubled their bulk. Thus the two satires of Donne occupied 48 pages in the quarto *Works*, and the two satires of Horace another 42—taking up, between them, more than the space finally occupied by the seven epistles. I see this publication, therefore, as a trial by Pope to see how the work looked in print, and perhaps to get some critical reaction to it.

After this interlude, things appear to move forward according to plan with the *Epistle to Cobham* and the last part of the *Essay on man*, followed by the collected *Essay on man*, printed on the four sorts of paper needed for the *Works* and with engravings in the quarto and large folio copies. I have left the copyright column blank for this in Table 12, since the delay in publishing the fourth part has complicated what would already have been a complex situation. By this time Gilliver's year's right to parts I and II has expired and part III has only three weeks to run; part IV, though, has nine months left. One wonders how the accounting was done, especially since this printing was intended to be part of Pope's half of the *Works*. Note, however, that another full year has expired by the time they are finally issued in the *Works*, and that Gilliver's contract with Pope prevented him from selling copies once a year had expired. If Pope enforced this provision, it would have been possible (if complicated) to divide the profits between them as the rights returned to Pope.[38]

Care was taken to avoid this problem with the parallel edition in various formats of *The first satire of the second book of Horace. To which is added the second satire of the same book.* Gilliver's rights to the first satire had already expired, but though he entered the composite work in the Stationers' Register to protect the second half, there is no record of his paying his £50. The copyright was wholly

---

[36] It was advertised with Wilford's name correctly spelt in the *Grub Street journal*, 8 Nov. 1733.

[37] See J. Butt's comments on the 'advertisement' in *Works* II of 1735 (*Twickenham*, iv, pp. xli–xlii).

[38] There is a further question about the folio poems repr. after Gilliver's rights had expired. The only unambiguous example is the 'second edition' of *Essay on man* I, dated 1735 (Griffith 409), but parts II and III (Griffith 313 and 315) may also be late reprints. It is possible that the latter were printed surreptitiously by Gilliver, but equally possible that Gilliver published them as Pope's agent. (Cf. also Foxon P913–14 for possible late reprints of the *Epistle to Burlington*.)

Pope's, and Gilliver presumably sold copies—mainly of the quarto—as publisher for Pope.

This summer of 1734 is, I think, the time when Pope began to plan the final contents of the *Works*. In presentation copies of the fine-paper quarto of the *Essay on man*, he had outlined the contents of the second book of the *Opus magnum*; he recalled these, but one copy survives in the collection of Harry Forster, now in the Cambridge University Library (Figure 57). I think there is no doubt that when this was printed in the spring Pope must have known that his epistles could not be brought into this pattern without extensive additions and reshaping, and that he therefore planned to hold them back for further work. Soon afterwards, in May or June, the place of the epistle 'Of the limits of human reason' seems to have been taken by Walter Harte's *Essay on reason* which Pope praised so highly; he writes to Mallet, 'Pray tell Mr Harte I have given Gilliver his Poem to print, but whether he would chuse to publish it now, or next winter, let himself judge' (*Correspondence*, iii. 408). It was in fact published about 7 February the following year. Pope himself clearly needed time to reconsider what he was to do with his epistles and to look round for other materials to make up his half of the *Works*. Vinton Dearing and Maynard Mack have shown that a group of Pope's prose works, which I shall call the Scriblerus pieces—the *Art of sinking in poetry*, *Virgilius restauratus*, and the *Key to the lock*—were printed to follow Gilliver's new edition of the *Dunciad*; they are so bound in fine-paper quarto copies in the Bodleian and Harvard.[39] The *Dunciad* collates B–2F^4 (pp. 1–204); these pieces continue as 2G–2P^4 2Q^2 (pp. 205–80). In the ordinary quartos (Griffith 531) they have a watermark of the Strasburg bend with the initials RVH (R∧ H) as a countermark. This watermark is also found in the satires of Donne which were printed to follow the satires of Horace; it seems likely that these would have been Wright's next printing job after Horace was finished in July 1734, or perhaps more probably after Pope returned from his summer visits in October. I would hypothesize that Wright first printed the Donne satires, then these prose works, and then went on to reprint the *Essay on man* for ordinary-paper quartos, where the watermark occurs only in the first five sheets, B–F^4.[40] If we look at the totals of pages which would be available at this point (Table 13), we can see that all Pope's problems would be solved by including the prose. What is more, Pope seems to have completed the *Epistle to Arbuthnot* by 3 September[41] and if this were also to be separately published, giving Gilliver a year's copyright, its 22 pages would still only make Gilliver's total 246 pages against Pope's 242. The Scriblerus pieces would follow the long notes at the end of the *Dunciad* with some

[39] V. A. Dearing, 'The Prince of Wales's set of Pope's Works', *HLB* iv (1950), 320–38; M. Mack, 'Two variant copies of Pope's *Works . . . Volume II*: Further light on some problems of authorship, bibliography, and text', *Library*, 5th ser. xii (1957), 48–53 (*Collected in himself*, pp. 139–44).

[40] The watermark appears in the prelims (A2.3 and A4, not the same sheet); *Essay on man* ($^\pi$B^4, B–F^4); *Bathurst* (the cancel B3); *Addison* and *Oxford* (both half-sheets in G); *Arbuthnot* (H–I^4); *Satires of Donne* (F–L^4).

[41] *Correspondence*, iii. 431. Gilliver paid his £50 on 1 Dec. 1734.

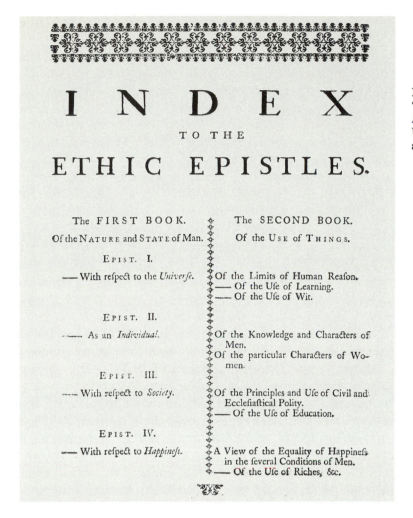

FIG. 57. Recalled plan of Pope's *'Opus magnum'* in unique copy of *Essay on man* (1734) (Cambridge University Library, Forster b. 129; approx. 257 × 204)

plausibility: indeed, Pope was to write a couple of years later, referring to these notes: 'There having been formerly published together with the *Dunciad*, Part of the *Works* in *Prose* of Mr. Pope . . . the Proprietors . . . thought it would be agreeable to the Purchasers . . . to collect the same in the like Form.'[42] This was when the original plan had fallen through and Pope was trying to dispose of these printed sheets.

The reason why the plan fell through was very simple: these three prose works had been printed in the Pope–Swift *Miscellanies*; part of the copyright belonged to Benjamin Motte, and Pope had been on bad terms with him since Motte's failure to make the payments due in 1729 (*Correspondence*, iii. 322). The copy-

---

[42] Quoted from the 1737 'The booksellers to the reader' by Mack, 'Two variant copies', p. 51 (*Collected in himself*, pp. 141–2).

TABLE 13. Quarto *Works*, 1735: Possible Shares of Copyright (in pages)

	Scriblerus Pieces Plan		Revised Plan	
Pope:	*Essay on man* πB⁴ B–I⁴ K²	76		76
	Satires (Horace and Donne) A–L⁴	90		90
	Scriblerus pieces Gg–Pp⁴ Qq²	76		—
			three epistles	48
			Addison and *Oxford*	8
			Epitaphs	16
			prelims to Epistles and Postscript	20
	TOTAL	242	TOTAL	258
Gilliver:	*Dunciad* B–Ff⁴	224		224
	? *Arbuthnot*ᵃ	22		22
			To a lady	12
	TOTAL	246	TOTAL	258

ᵃ Pagination of *Arbuthnot* in the quarto is 57–78. If this was first printed (as watermarks suggest) at the end of Epistles, these 78 would match the Scriblerus total of 76.

right situation was in fact quite complicated. *The art of sinking* published in the 'last' volume was indeed Motte's property; *Virgilius restauratus* was originally published in the *Dunciad variorum* of 1729 and reprinted in the third volume of *Miscellanies*, 1732, under the joint imprint of Motte and Gilliver—Pope had arranged for Gilliver to have the copyright after dismissing Motte in 1729, but Swift prevailed on him to give Motte a half-share; while the *Key to the lock* had been purchased by Lintot in April 1715, and he was to complain of the infringement of his copyright by its inclusion in *Miscellanies* II.[43] Nevertheless, by putting a veto on *The art of sinking*, Motte could prevent these sheets being used. Pope may have assumed that Motte would have allowed these copyrights to be used in exchange for an appropriate number of copies, as Tonson had done with the first volume of *Works* in 1717; but even if he had, this would have upset the fifty–fifty balance of copyright between Pope and Gilliver. Pope had then no option but to use the epistles he had completed to fill up the volume.[44]

The make-up of the section of the *Works* containing the epistles is full of anomalies in signatures and pagination, and includes a series of cancels, two of whole signatures in the quarto; the problems are too intricate to be dealt with

[43] See *Correspondence*, iii. 306 for *Miscellanies* III, and *Correspondence*, iv. 324 for Lintot's claim to the *Key to the lock*.

[44] Negotiations with Gilliver must have been complex. Pope would have been going back on his declaration to him (BL MS Egerton 1951, fo. 12, discussed above) if he had not included the £50 epistles in *Works* II, and the vol. would have been less attractive without them. The 'Directions to the Binder' found in some copies of *Works* II lists '*The Dunciad, with Notes, and Pieces in prose*', and the advertisement leaf to *Epistle to a lady* says section III of *Works* II, The Dunciad, will contain 'some additional pieces'. These look like references to the Scriblerus pieces; but it is not clear whether this was the original or a subsequent plan.

here, but something must be said of the *Epistle to a lady*. It looks as though Pope started to print the three epistles previously published, *Bathurst*, *Burlington*, and *Cobham*, in that order, and that the inclusion of the new *Epistle to a lady* (Gilliver paid his £50 on 4 January 1735) was an afterthought.[45] Even when he decided to include the poem, he remained uncertain which version of the text to print. At a late stage he removed 86 lines from the text of the folio (causing the line number-ing to leap from 105 to 195 within 4 lines) by cancelling F–G² and replacing them with a new F².[46] Bateson has calculated that the missing 86 lines can be accounted for by the controversial material—the character of Philomedé, the character of Atossa, and the lines on Queen Caroline—that Pope subsequently printed in the death-bed edition of 1744 (*Twickenham*, iii/2, p. 43), and the strongest support-ing evidence for this hypothesis comes from a thick-paper quarto at Harvard reported by Vinton Dearing.[47] The unusual feature of this quarto is a complete sheet I⁴ which replaces the two leaves I1 and I2 in the ordinary quartos and con-tains the additional lines on Atossa, Philomedé, and the Queen. Because the Har-vard volume was put together for presentation to the Prince of Wales in 1738 Dearing assumes this sheet represents a later state of the text, a revision of the 1735 poem; but the evidence suggests the contrary—that this is the earlier ver-sion. When we compare the ordinary 1735 quarto and folio, it is clear that the quarto has the later text of this part of the poem: it contains an additional couplet which disturbs the line numbering (lines 107–8 of *Twickenham*); it normalizes 'rome' to 'roam' (line 223); and it incorporates a reading from the folio's (and its own) errata (line 272). But Dearing's Harvard text sometimes agrees with the folio against the quarto, and so it is probably earlier than the usual quarto text. It is even possible that the Harvard quarto presents us with a rare example of the first setting of the sheet (the equivalent of F–G² removed from the folio) before Pope decided to omit the controversial lines.

The addition of the *Epistle to a lady* must have caused new problems in balanc-ing Pope's section of the *Works* with Gilliver's. Without it, but adding the early epistles to Oxford and Addison and the section of epitaphs, Pope's total of pages creeps up to about 238 against Gilliver's 246. But if the *Epistle to a lady* were to be included after Gilliver had paid for his year's copyright—as it was—Gilliver's side of the balance would go up to 258. It was roughly compensated for by 16 pages of postscript, containing notes on the epistles to Bathurst and Arbuthnot and the collection of variant readings from earlier editions made by Jonathan Richardson the younger. The introduction, which begins 'It was intended in this

[45] The original order in the quarto may have been A–G⁴ H² (pp. 1–60); *Bathurst* now begins on K2ʳ but its second leaf is B1, while *Cobham* ends on p. 60. But the leap in pagination between *Bathurst* and *Burlington*, representing a missing sheet E, suggests early disruption of this order. The new order, now canonical, was *Cobham*, *Lady*, *Bathurst*, *Burlington*.

[46] The inner pages of the new F² have an additional 2 lines each and F2ᵛ is unusual in having a catch-word (to make sure the binder goes straight on to H1ʳ).

[47] Dearing, 'The Prince of Wales's set of Pope's works'. F. Brady has subsequently corrected Bateson's arithmetic to give 82 lines· 'The history and structure of Pope's *To a lady*', *SELit* ix (1969), 439–62.

Edition, to have added *Notes* to the *Ethic Epistles* as well as to the *Dunciad*, but the book swelling to too great a bulk, we are oblig'd to defer them', is disingenuous, though it is doubtless true that Pope had intended to annotate the epistles if they had been published at leisure.

The scenario I have written lends itself to a number of variations, but in essentials it explains the curious construction of the *Works* and also the premature codification of the epistles and with it abandonment of plans for a larger *Opus magnum*. Pope may well not have been sorry to have been brought to a halt; he wrote to Swift on 19 December 1734, perhaps while licking the *Epistle to a lady* into shape, 'I am almost at the end of my Morals, as I've been, long ago, of my Wit; my system is a short one, and my circle narrow'.[48]

Even when the volume was completed Pope's troubles were not at an end. The logical consequence of equal shares in the copyright for Pope and Gilliver would be equal payments of the bills for printing and paper, and equal shares in the copies produced.[49] We do not know how many copies were printed; the first advertisement in the *Grub Street journal* of 24 April 1735 is merely negative: 'no greater Number of this Volume is printed in large Folio and Quarto with *expensive Ornaments*, than to answer the like Impressions of *the first Volume* of his *Works*, and of the *Iliad* and *Odysseys*.' The first volume of the *Works* consisted of 500 copies in quarto plus another 250 on fine paper, and 250 copies in large folio;[50] but Pope writing to Allen about plans for his *Letters* on 30 July 1736 says of the quartos, 'there were 500 of the former volumes, of which above 100 remain on my hands' (*Correspondence*, iv. 24). This (if not a slip of the pen in writing 'volumes' in the plural) is incorrect for the numbers of the first volume but it seems plausible to assume that there were 500 copies of the second volume in quarto, with only comparatively few fine-paper copies, for they seem much scarcer than those of the first volume—and this time Pope would have had to bear all the cost himself; there may have been 250 large-paper folios in addition.[51] Since Lintot had difficulty in disposing of the thousand small-paper folios of the Homer, there may well have been less than a thousand of the *Works* II. Of these uncertain numbers, Pope certainly took over part of the quartos and large-paper folios, since two years later he wrote to ask 'what number of Second Vols. of my

[48] *Correspondence*, iii. 445. Cf. his letter to Swift of 25 Mar. 1736 where he speaks of a possible new group of four epistles, 'But alas! the task is great, and *non sum qualis eram*!' (*Correspondence*, iv. 5).

[49] There would be complications resulting from the separate sale, before publication of the *Works*, of the collected *Essay on man* and the *Satires of Horace* in folio and quarto, and the consequent offer of 'parts singly to compleat former setts' once the volume was published; see Griffith, i. 258, 260; ii. 282, 287–8.

[50] See Fleeman, '18th-century printing ledgers'.

[51] Pope's letter to Allen of 5 June 1736 (*Correspondence*, iv. 20) also suggests that 500 in quarto are in Pope's mind. Pope received 120 copies of *Works* I in fine-paper quarto under the terms of his agreement with Lintot and many of these may have remained on his hands. The bookseller's advertisement in the *Daily courant* of 13 May 1735 says 'whoever is willing to part with their first Volumes in large Folio or Quarto, may receive Fifteen Shillings for the same', which suggests a problem in finding customers for a second vol. when they did not own the first. Gilliver's trade sales of 1742 contained a few copies in quires of vol. I in quarto and folio (see n. 55 below).

Works, Quarto or folio' were in Knapton's hands so that he could print an appro-priate number of additions (*Correspondence*, iv. 66).

I think it is clear that Pope expected to sell his share of the copies through Gilliver, without taking any further responsibility, but what went wrong is seen from a letter he wrote to Buckley on 9 April, a fortnight before publication:

It was meerly an Unwillingness to give you Trouble, that hinderd my doing myself the Service of desiring your Assistance in printing this book. As it is, it has cost me dear, & may dearer, if I am to depend on my Bookseller for the Re-imbursement. If it lye in your way to help me off with 150 of them, (which are not to be sold to the Trade at less than 18s or to Gentlemen than a Guinea) it would be a Service to me, a Bookseller having had the Conscience to offer me 13s a piece, & being modestly content to get 8s in the pound himself, after I have done him many services. Another, quite a Stranger, has taken 100 at 17s but I want to part with the rest.   (*Correspondence*, iii. 454.)

I think the interpretation of this is clear: the bookseller who 'after I have done him many services' offered Pope 13*s*. a copy must be Gilliver, who wanted to buy the books at a wholesaler's price—which might more normally be 14*s*., two-thirds of the retail price. Gilliver's attitude may well have been reasonable for a bookseller, but it marks the beginning of the break between him and Pope. The other bookseller, 'quite a Stranger' who took 100 copies at 17*s*. is probably John Brindley of New Bond Street, who had bought a quarter share in *Sober advice from Horace* (see Griffith 347) in the preceding December; he was a bookbinder by training, held a royal appointment to the Queen and Prince of Wales, and bound for Harley and Sunderland, the two leading aristocratic collectors of the period. Many presentation copies of Pope's works from this time on are in bind-ings by him, and he became one of the regular publishers of Pope's collected volumes; he seems to have acted as the main agent in collecting subscriptions for the *Letters* of 1737, as was appropriate for a bookseller in the polite part of Town.[52] The original advertisement for *Works* II in the *Grub Street journal*, 24 April 1735, says 'printed for Lawton Gilliver . . . John Brindley . . . and sold by Robert Dodsley', but the *Daily courant* a couple of days later (26 April) puts Dodsley on a level with the others. This is Dodsley's first known publication, and we also know that Pope set him up in business in Pall Mall with £100.[53] He later was to take the place of Gilliver as Pope's publisher, his own works having pre-viously been published by Gilliver, no doubt under Pope's supervision. The next month, James, John, and Paul Knapton joined Gilliver, Brindley, and Dodsley;[54] they became the most important agents, for after Pope's death they took a third

---

[52] See *Correspondence*, iv. 65, 67.

[53] Johnson, *Lives of the English poets*, iii. 213. Pope wrote to William Duncombe on 6 May 1735: 'I beg you to accept of the New Volume of my things just printed, which will be delivered you by Mr Dodsley, the Author of the Toyshop, who has just set up a Bookseller, and I doubt not, as he has more Sense, so will have more Honesty, than most of that Profession.' (*Correspondence*, iii. 454.)

[54] So Straus, *Robert Dodsley*, p. 317, quoting the *General advertiser* of 17 May; earlier advertisements (*Daily courant* of 13 May and *Grub Street journal* of 15 May) say 'Sold by J. J. and P. Knapton . . . Lawton Gilliver . . . J. Brindley . . . R. Dodsley . . . J. and J. Brotherton . . . and W. Meadows'.

share of the copyrights that Warburton inherited, and their name appears on his editions of Pope's works.

While we can interpret Pope's letter well enough in terms of the position he found himself in with Gilliver and his attempts to extricate himself, it does not help us much in trying to determine how many copies Pope was responsible for. On the face of it, he is trying to dispose of 250 copies to be sold at a guinea. This could mean that only 500 were printed in quarto and large folio together, or it could be that Pope was so far only thinking of the quartos, which were all he had been concerned with in the Homer and *Works* I, and had not yet realized his responsibility for half the folios. If he had disposed of 100 quartos so easily to Brindley, it might seem not difficult to dispose of another 150 to Dodsley and Knapton; but Dodsley, just set up in business with Pope's £100, was scarcely in a position to pay cash. He must have sold copies on commission (as a 'publisher'), and Knapton may well have preferred a similar arrangement. In any case the small folios did not enter into Pope's reckoning, and it could well be either that he was willing to part with his share of this trade edition on Gilliver's own terms, or that it had been regarded all along as Gilliver's responsibility to pay for and sell.[55]

The original advertisement for the *Works* announced that 'This present Volume will with all convenient Speed be published in Twelves at 5s.' (Griffith, ii. 288). It was in fact published in two volumes in octavo at 3s. each bound, though even when the first volume was advertised as published in the *Grub Street journal* for 31 July the advertisement read 'In Duodecimo, and on a neat Elzevir Letter'. My former pupil, Richard Noble, has identified a duodecimo *Essay on man* in a Bodleian volume (280 n. 415) as intermediate in text between the quarto *Works* and the first octavo; it is bound with the rest of the normal first octavo edition and seems to represent a superseded first small-format version.[56] The original advertisement also complained that Lintot 'having the property of the former Volume of Poems, would never be induced to publish them compleat, but only a part of them, to which he tack'd and impos'd on the Buyer a whole additional Volume of other Men's Poems'. This is disingenuous, as it refers to the Lintot *Miscellany* that Pope had edited and divided in this way; indeed it is likely that Pope had mixed feelings about his juvenilia, for Warburton reports in the second volume of his edition of 1751 that he never intended to reprint them, but that they were reprinted because they were the copyright of others (i.e. Lintot and Tonson). However that may be, Lintot prepared to join in the publication of the edition in small octavo. He reports in a letter to Broome of 26 August 1735: 'I am again printing for Mr. Pope,—the first volume of his miscellaneous works,

---

[55] The catalogue of Gilliver's trade sale of 25 Feb. 1742 lists under lot 5 of books in quires: 43 folios in large paper, 37 in small paper, and 5 copies of vol. I in small paper. Lot 8 contains 5 copies of vol. I and 33 of vol. II in quarto. There were an unspecified number of bound copies of vol. II in all three formats. In the second sale of 16 Mar. 1742, the books in quires include '11 Pope's Works, 2 vol. large Paper' (Ward trade sales catalogues, Bodl. John Johnson).

[56] The evidence of the ornaments suggests the printer may have been Watts.

with notes, remarks, imitations, &c.,—I know not what' (*Correspondence*, iii. 489). Woodfall printed 3,000 copies for Lintot, and Pope seems to have paid separately for 75 fine-paper copies on writing demy. The third volume, of juvenilia, was printed in the same numbers by Woodfall the following year.[57] The fine-paper copies and the reference to Pope's 'notes, remarks, imitations, &c.' are sufficient indication of the attention Pope paid to these octavo editions, which are, indeed, of crucial importance to Pope's text. It is, however, too complicated a story to enter into here, and we must touch very briefly on the even more complicated story of the publication of Pope's letters.

PROSE WORKS TO 1741

After many machinations, Pope succeeded in getting Curll to buy copies of an edition of his letters which had been printed by Wright; they were delivered to him in two lots on 12 May 1735, a month after *Works* II had been completed. This publication left the way open for Pope to shout thief and have the copyright entered in the Stationers' Register by the publisher Thomas Cooper on 21 May, and deposit the nine statutory copies.[58] More to our present concern, this publication gave Pope the justification for producing an authorized edition, which appeared in two forms: as volumes V and VI of the small octavo *Works* with the imprint of J. Roberts (reprinted several times as for T. Cooper), and in the usual large formats with two sizes of folio and two qualities of quarto and with the imprint of Wright for Knapton, Gilliver, Brindley, and Dodsley. Vinton Dearing has shown that the octavo, also (largely, at least) printed by Wright, is the earlier and more complete text though its publication was held back until the quarto edition was on the market.[59] I am only concerned with the quarto edition, with its more elevated tone—it excluded trivial news and gossip—because I want to trace the history of those Scriblerus sheets that had to be withheld from the 1735 *Works*.

I think the need to find a life for these 9½ sheets may have played a part in Pope's decision to publish the letters by subscription; certainly the form of receipt he prescribed to Fortescue was for 'a Vol: of Mr Pope's Works in Prose' (*Correspondence*, iv. 10), not Letters or Correspondence. His letters to Ralph

---

[57] The entries in Woodfall's ledger are dated 15 Dec. 1735 and 30 Apr. 1736 (P.T.P., 'Pope and Woodfall', p. 377).

[58] McLaverty, 'The first printing and publication of Pope's letters', *Library*, 6th ser. ii (1980), 264–80, argues that many of the complications in this episode are attributable to Pope's special relationship with Wright and the use of a second printer, John Hughs. Cooper's role was that of publisher (as was his wife's later), even though he claimed the copy in the Register. In the same way, though Pope assigned the copyright of the 1737 edn. to Dodsley on 24 Mar. and he entered it in the Register on 17 May, the copyright is unlikely to have left Pope's hands, since it passed to Warburton on his death.

[59] Dearing, 'The 1737 editions of Alexander Pope's letters', in *Essays critical and historical dedicated to Lily B. Campbell*, ed. L. B. Wright (Berkeley and Los Angeles, 1950), pp. 185–97.

Allen, who had offered substantial help with the costs of publishing the letters, show that Pope was anxious about the financial return. On 30 July he writes: 'I'm in some doubt whether to print 300, or 500 Quarto? the first number being all I see certain; tho' there were 500 of the former volumes, of which above 100 remain on my hands' (*Correspondence*, iv. 24). The proposals Pope published in the *London gazette* on 12 February 1737 seem to say clearly that only 500 copies would be printed altogether, including quarto, large folio, and small folio, since the proprietors were considerable losers on the former poetical volumes (*Correspondence*, iv. 41). It seems possible that Pope was over-cautious in his estimate of demand, since my copy of the *Letters* in quarto is an unrecorded edition printed in half-sheets, clearly reprinted in order to complete sets. It may be that Pope's letter to Allen of 14 May 1737, where he reports sending ten copies in large folio rather than in quarto (*Correspondence*, iv. 68), is a sign that the quartos were going to be in short supply.

By the beginning of November Pope could write to Bethel and Orrery (2 and 7 November 1736) that his book was 'above half printed' and would come out by the next Lady Day. 'It makes too large a Quarto', he adds to Orrery (*Correspondence*, iv. 43). At the same time (6 November) he writes to Allen that his works 'in prose are above 3 quarters printed, & will be a book of 50 & more sheets in 4°' (*Correspondence*, iv. 41). The discrepancy between 'half printed' and '3 quarters printed' may be reconciled in part by whether Pope is including the $9\frac{1}{2}$ sheets from the 1735 *Works* in his calculations. Adding the $9\frac{1}{2}$ to the 28 sheets of letters would make $37\frac{1}{2}$ or three-quarters of a 50-sheet volume (without including the preliminary leaves), whereas 28 sheets of letters alone would not be much above half the volume.[60] What is more obscure is the reason for abandoning the prose works. Pope wrote to Orrery on 14 January 1737:

I am overruled by my friends here, as to the Miscellaneous Prose pieces, which they would omit, & make the Volume consist wholly of Letters; with a view of enlarging it hereafter with Other Letters, which may come in order, rather than to break the order by inserting things between of so different a nature. This will make the present Volume less, than I threatend you with: Both are actually printed, & are too great a bulk together.   (*Correspondence*, iv. 53.)

To take the matter of bulk first, the *Works* of 1735 had contained over 65 sheets; in comparison with this, 55 sheets is not too great in itself, but since the size of the edition was smaller it would be more difficult for it to pay its way. It would make more sense financially to make a separate volume of prose tracts which could be sold separately, and this is what Pope subsequently planned to do.

We must not forget that the copyright in the $9\frac{1}{2}$ sheets was still (at least in part) the property of Benjamin Motte. Motte took Charles Bathurst into partner-

---

[60] Before the Buckingham letters were added in *Dd⁴ and Swift's letters substituted for the *Thoughts*, the planned volume consisted of 39 sheets of letters, $2\frac{1}{2}$ of *Thoughts on various subjects*, and the $9\frac{1}{2}$ sheets from the 1735 works, making 51 sheets. $4\frac{1}{2}$ sheets of preliminaries were finally added, giving a full table of letters.

ship about 1736, and died on 12 March 1738, leaving the business in his hands. Pope seems to have had a much more satisfactory relationship with Bathurst than he did with Motte. On 24 February 1737 they signed an agreement with Pope from which I quote extracts, italicizing insertions made at the time of signing:

Whereas the said Alexander Pope Esqr doth intend to publish a Volume or Volumes of Letters and other Pieces in Prose, wherein he is willing to insert certain Peices which are the Property of Benjamin Motte and Charles Bathurst . . . [the latter] covenant and *agree* to pay the price of the Print and Paper to Alexander Pope Esqr for the same at the Delivery of the Books . . .

And further . . . [they do] agree not to sell the said Books of Letters *&c.* at a less Price than Eighteen Shillings per Book to Booksellers or one Guinea per Book to Gentlemen in Sheets for the Quarto or large Folio; or at a less Price than Eight shillings and six Pence to Booksellers and half a Guinea to Gentlemen in Sheets for the small Folio, under the Penalty of Fifty Pounds . . . In Witness whereof we have hereto set our hands and Seals this *24th* Day of *February in the year of our Lord One thousand seven hundred and thirty six.*[61]

It appears that the agreement was drawn up before Pope had decided whether to divide his volume of prose, but signed well after he had made the decision.[62] Did Motte and Bathurst realize this, or the importance of the '*&c.*' inserted after 'Books of Letters'?

Maynard Mack has shown that plans were well advanced for an expanded issue of the Scriblerus tracts; copies of preliminaries survive (Figure 58) with a title *Tracts of Martinus Scriblerus: and other miscellaneous pieces* and the joint imprints of Motte and Bathurst, and Gilliver and Clarke (the last Gilliver's apprentice whom he had taken into partnership in 1736).[63] The first two items of the contents, the *Art of sinking* and *Virgilius restauratus*, and the last, the *Key to the lock*, had been printed in 1735; the *Essay on the origine of sciences*, the *Guardians*, and the *Memoirs of P.P.* were newly printed. (The *Thoughts on various subjects* printed to follow the *Letters* of 1737 and previously printed in the first volume of *Miscellanies* (1727) are not included.) But this plan was thwarted, as we find from Pope's letter to Allen of 14 May 1737:

The Proposal I had made to myself, of publishing the Book by the tenth, was retarded by the Artifices of a Bookseller, who has some share in an Additional Part, viz. the *Tracts* which I was willing should come out in the same size at the same time, to complete the Edition; with Every Fragment, If I may so say, of my writing: tho I could not join them with my Book as not being my own Property. I mean these things written by me *in conjunction with any others.* (*Correspondence*, iv. 68.)

What does Pope mean by 'the Artifices of a Bookseller'?

The most plausible explanation seems to be that when Motte and Bathurst

---

[61] BL MS Egerton 1951, fo. 10.

[62] The proposals published in the *London gazette* on 12 Feb. 1737 say, 'Speedily to be published, In Quarto, large Folio, and small Folio, The Works of Mr. Pope in Prose; *And First, an* Authentic Edition of his *Letters*' (*Correspondence*, iv. 41).

[63] Mack, 'Two variant copies', pp. 51–2 (*Collected in himself*, pp. 141–2).

FIG. 58. Title-page of 'Tracts of Martinus Scriblerus' in enlarged *Works* II, unnumbered p. 221 (Bodl. Vet. A4 d. 142; 289 × 225)

TRACTS

OF

MARTINUS SCRIBLERUS:

And other

Miſcellaneous PIECES.

LONDON,

Printed for BENJAMIN MOTTE and CHARLES BATHURST, at the *Temple-gate*, LAWTON GILLIVER and JOHN CLARKE, in *Fleetſtreet*, 1737.

were offered their share of this volume they found it contained no letters and only fifteen sheets in quarto, a third of the size of Pope's *Letters*, and yet they were bound to sell it at the same price, a guinea to a gentleman. Pope may have offered to vary the agreement, but it seems that this was unacceptable, for a note dated 6 May 1737 is preserved in the Stationers' Register:

Mr. Simpson
Sir
If the Tracts of Scriblerus come to be enter'd in the Hall Book I desire you'll please to give notice to

Sr Yor humble servt
B. Motte

This looks like an attempt to prevent publication in spite of the agreement three months earlier; and the Scriblerus pieces were not published until after Motte's death. The other possible interpretation of Pope's letter is that Lintot was responsible for the postponement by insisting on his rights to the *Key to the lock*; this is argued by Sherburn, in relation to the 1741 *Works in prose* II, where it finally appeared, but I think it unlikely (see Appendix A below).

Publication of the *Letters* was delayed because a new copyright bill was before parliament; Pope advertised in the *London gazette* of 22 March 1737 that 'the First Part of the Works of Mr. Pope in Prose' was already printed and ready for distribution, but asked for the forbearance of subscribers 'till the Fate of the said Bill is determined' (*Correspondence*, iv. 65 n.). The bill was finally thrown out in May, whereupon Dodsley entered the book in the Stationers' Register on 17 May; Pope sent off several books the following day and it was advertised as published in the *Daily post* for 19 May. Pope had, however, made an assignment of the copyright to Dodsley on 24 March at the time when publication was delayed, and Dodsley in the Stationers' Register added the unusual formula (also used for *The second book of the epistles of Horace imitated* entered below on the same day): 'I Robert Dodsley do Claim the Sole Property in and to the above book, by Virtue of an Assignment under the hand of Alexander Pope Esqr.' This assignment was not a normal assignment of copyright, and there is no subsequent sign that Dodsley owned the copy; the assignment did, however, dot the *i*s and cross the *t*s if legal action was called for. And so it was; action was successfully taken against James Watson for a piracy in the name of Thomas Johnson later in the year.[64]

We can sum up the story of the Scriblerus tracts so far by saying that 9½ sheets were printed to follow the *Dunciad* in the *Works* of 1735 and withheld because Pope did not own the copyright and needed equal shares with Gilliver. In preparation for the *Works in prose* of 1737, he concluded the *Letters* with *Thoughts on various subects* as an introduction to the other tracts, but then decided not to issue them with the *Letters*. Instead, he printed six additional sheets for a separate publication of *Tracts of Martinus Scriblerus* of 1737, only to find that Motte and Bathurst would not accept them under the terms of their agreement. So now, instead of 9½ unusable sheets, Pope had 15, plus the two containing *Thoughts on various subjects*. His only hope lay in the production of yet another volume of *Works* in quarto and large and small folio; and this meant waiting for the correspondence with Swift which he had long hoped to publish, was already taking practical steps to obtain in 1737, but did not finally publish until 1741, after complicated stratagems which have been elucidated by Maynard Mack and Vinton Dearing.[65] These included the printing of a small octavo text which was taken to Ireland in May 1740 by Samuel Gerrard for Swift to revise.

---

[64] See *Correspondence*, iv. 87–8 and Appendix A below. The settlement with Watson is BL MS Egerton 1951, fo. 13, and Mack identifies the edition (Griffith 470, a conjectural entry) in his review of Sherburn's edition of the *Correspondence*, *PQ* xxxvi (1957), 391 (*Collected in himself*, pp. 147–8).

[65] Mack, 'The first printing of the letters of Pope and Swift', *Library*, 4th ser. xix (1938–9), 465–85 (*Collected in himself*, pp. 93–105); Dearing, 'New light on the first printing of the letters of Pope and Swift', *Library*, 4th ser. xxiv (1944), 74–80.

Later that summer, on 29 August 1740, Pope wrote to Charles Bathurst:

I shall print some things more of Scriblerus, and add to what is already done; but it will be in Quarto, and the new part of the Vol. be above two-thirds of the old. I don't care to alienate the Property, but if you have any mind to treat for the impression I will give you the refusal.   (*Correspondence*, iv. 259.)

When Pope writes it will be in quarto, I think he means that it will be in a large size rather than in octavo; the volume appeared as usual in quarto and large and small folio. That the new part would be 'above two-thirds of the old' is a precise prediction of the make-up of *Works in prose* II of 1741; the new part, consisting of the letters and the completely rewritten *Memoirs* of Scriblerus, is just over twice the total of the sheets previously printed (Table 14). That Pope did not 'care to alienate the Property' is shown by the fact that the entry of the copyright in the Stationers' Register is made directly to Pope. But there is still a problem to be solved, for Bathurst still had a claim to most, at least, of the old sheets. Was Pope's letter written in the hope that Bathurst, having been offered the whole impression, would overlook that fact?

The miscellaneous works had been published in the Pope–Swift *Miscellanies* between 1727 and 1732, and the original agreement between Motte and Pope and Swift was signed on 29 March 1727.[66] One must ask whether there is any significance in the fact that the *Works in prose* II was published within three weeks of the expiry of fourteen years from that date? If Pope were hoping that the assignment of copyright had legally expired, he was in fact wrong, for a supplementary agreement between Pope and Motte dated 1 July 1729 explicitly sets out that the rights to the first three volumes of *Miscellanies* are granted to Motte 'for Fourteen Years from the date of the publication; and we do promise, at the expiration of the said fourteen years, to renew the said grant to him or his assigns for the further term of fourteen years for the sum of five shillings'. We know that Bathurst had taken over this agreement from Motte, since it was published by his grandson Charles Bathurst Woodman from the family papers.[67]

It may be that Bathurst overlooked his legal rights in return for Pope's help in reshaping and correcting the *Miscellanies* which is repeatedly mentioned in their correspondence.[68] This bore fruit in the fourth edition in four volumes advertised in the *London evening post* of 3 July 1742 as 'now first printed in small Octavo, the same Size with the rest of Mr. Pope's Works' (Griffith 561); volume II contains all the pieces included in the 1741 *Works in prose* except for the *Guardians* and *Memoirs of Scriblerus*, to which Bathurst had no claim. Bathurst may also have had a third share of the impression in return for paying a third of the

[66] Printed by C. B. Woodman, 'Pope's arrangements with Mr. Benj. Motte', *Gentleman's magazine*, NS xliv (1855), 363, and by Rogers, *The major satires of Alexander Pope*, pp. 115–16. The original is now in the Pierpont Morgan Library.

[67] Woodman, 'Pope's arrangements with Motte', p. 364; Pope's correspondence with Bathurst is also printed, p. 585.

[68] *Correspondence*, iv. 333–4, 353, 361, 372.

TABLE 14. *Works in prose* II, 1741

Number of pages in quarto			
New:		Old:	
Letters	204	*Art of sinking + Virgil*	58
Memoirs of Scriblerus	80	*Key to lock*	18
		Memoirs of P. P.	8
		Origin of sciences	8
		Guardians	28
		Thoughts	18
TOTAL	284		138

'The new part of the Vol. [will] be above two-thirds of the old': in fact the new part is above twice the old, but forms more than two-thirds of the vol.

costs as the agreement of 1737 envisaged; his name appears in the imprint with the Knaptons and Dodsley.[69] What is certain is that Pope's entry of the copyright of *Works in prose* in no way interfered with Bathurst's normal commercial exploitation of his copyrights.

The whole episode of the prose writings is perhaps best summed up in Pope's letter to Allen of 14 July 1741.

I am in near 200 ll arrear to my Printer. That Rascal Curl has pyrated the Letters, which would have ruin'd half my Edition, but we have got an Injunction from my Lord Chancellor to prohibit his selling them for the future, tho doubtless he'l do it clandestinly. And indeed I have done with expensive Editions for ever, which are only a Complement to a few curious people at the expence of the Publisher, & to the displeasure of the Many. I was half drawn into this, by what Motte had done formerly, & for the time to come, the World shall not pay, nor make Me pay, more for my Works than they are worth. (*Correspondence*, iv. 350.)

The 'expensive Editions' which had served Pope so well when he had driven his hard bargains with Lintot have now become 'only a Complement to a few curious people at the expence of the Publisher'; and it is perhaps fair to say that the stream of piracies—of the *Dunciad*, of the *Works* of 1735, and of the prose works of 1737 and 1741—might not have been incurred if Tonson or Lintot had produced the originals. Pope may have been right in attributing them to Curll's machinations; my impression is that Curll had a healthy respect for booksellers' copyrights, however lightly he may have treated those of Pope, with whom he had long been at war.

In this context, and when considering Pope's debt of near £200 to his printer Wright, one must remember that he was quick to produce his prose works in the

---

[69] Brindley has apparently withdrawn as publisher, and the break with Gilliver has also taken place. Gilliver had a share in *Miscellanies* III, which may be represented by his place in the imprint of vols. II and III of the 1742 edn., but his stock was sold by auction on 25 Feb. and 16 Mar. 1742, and a commission of bankruptcy entered against him on 3 Dec. (see McLaverty, 'Lawton Gilliver: Pope's bookseller', p. 122). I presume Bathurst bought out his share in the copyright, for the copyrights sold at auction are all small shares from Conger 4 (for Conger 4, see *Notebook of Thomas Bennet and Henry Clements*, App. 14).

small octavo format of his *Works*. Though they may have been produced as a defence against the pirates, they must, like Lintot's duodecimo *Iliad* in 1720, have brought in a substantial return. There were thirteen octavo volumes of prose published in the five years 1737 to 1742. On a modest computation that each edition was of 1,500 copies (most of the earlier volumes printed by Woodfall were in editions of two or three thousand) and the profit 1*s.* per volume, the proceeds would be nearly £1,000, or £200 a year. I suspect that the pirates offended Pope's dignity more than his pocket.

Finally we must interpret Pope's phrase 'I was half drawn into this, by what Motte had done formerly'. Is Pope confirming my thesis that the large editions of the letters were partly occasioned by the existence of the first Scriblerus tracts printed for the 1735 *Works* and not used there? I suspect that what Motte had done was to hold Pope to the assignment of copyright which he had made in 1727. One moral of the story is that Pope was anxious that everyone else should stick to the rules, but that he himself should be free—a not uncommon desire. The other is not so clear cut; it would be a nice moral if Pope suffered financially in his attempt to wring the maximum profit from his *Works*, but it looks as though the loss on the expensive editions was well compensated for by the sale of the octavos.

POEMS 1737–1743

The pattern of Pope's new works between 1737 and 1743 (Table 15) is more complicated than that between 1731 and 1735, largely because the agreement with Gilliver for a year's rights to each new poem has lapsed. The one constant is Wright's work as Pope's printer, only interrupted in the spring of 1737 by Woodfall's printing of the two epistles of Horace, perhaps because Wright was fully employed in printing the editions of the *Letters*. The relationship draws to a close after the publication of the *New Dunciad* in March 1742; Pope wrote to Allen: 'They have pyrated my Poem—by the foolish Delays of my Printer whom I'l pay off, & imploy less for the future. It was Charity made me use him.' (*Correspondence*, iv. 392.) He continued to reprint volumes of the octavo works into 1743, but for new work—most importantly the death-bed quartos—Pope returned to Bowyer; as a result we have precise costs for these last works.

The variation in the imprints caused the author of *The satirists*, published in December 1739, to comment:

> Or shou'd I tax him with the Venal Stain;
> Praise He, or Censure, say 'tis all for Gain,
> Wou'd he one Copy gen'rous give away?
> Nor keep his various Gillivers in Play,
> T'enhance the Price of every Satire sold,
> And wring from each Competitor more gold?[70]

[70] Quoted by Guerinot, *Pamphlet attacks on Pope*, p. 243.

TABLE 15. Pope's Works, 1737–1743

Publication	title-page date	pub date	imprint	copyright	printer
Ode to Venus (IV. i)	1737	9 Mar	for Wright, sold Roberts	Wright SR 8 Mar ass 25 Feb	Wright
2nd epistle, 2nd book Horace (+FP, +reimpr.)	1737	28 Apr	for Dodsley	Dodsley SR 28 Apr	Woodfall
Letters (2⁰ + LP, 4⁰ + FP)	1737	19 May	by Wright for Knapton, Gilliver, Brindley, Dodsley	'Dodsley' SR 17 May ass 24 Mar	Wright
1st epistle, 2nd book Horace (+FP, + new edn.)	1737	19 May	for Cooper	Dodsley SR 17 May ass	Woodfall 2000 + 150 + 500
6th epistle, 1st book Horace	1737	24 Jan 1738	for Gilliver	Pope (Gilliver 1st edn.) SR 14 Jan 1738	Wright
6th satire, 2nd book Horace (by Swift and Pope)	1738	1 Mar	for Motte & Bathurst, Knaptons	Motte, Bathurst, Knaptons SR 28 Feb	
1st epistle, 1st book Horace (+ new edn.)	1737	7 Mar 1738	for Dodsley sold Cooper	Dodsley SR 6 Mar	Wright
Epistles of Horace (4⁰ + FP)	1738	27 Apr	by Wright for Knapton, Gilliver, Brindley, Dodsley		Wright
1738 (+2 reimpr, 1 new edn.)	[1738]	16 May	for T. Cooper	Pope SR 12 May	Wright
Universal prayer	1738	22 Jun	for R. Dodsley		Wright
1738, dialogue II	1738	18 Jul	for R. Dodsley	Pope SR 17 Jul	Wright
Poems & imitations of Horace (4⁰ + FP)	1738	11 Jan 1739	for Knaptons, Gilliver, Brindley, Dodsley		Wright
Works in prose II (2⁰ + LP, 4⁰ + FP)	1741	16 Apr	for Knaptons, Bathurst, Dodsley	Pope SR 15 Apr	Wright
New Dunciad (4⁰ + FP, + new edn.)	1742	20 Mar	for T. Cooper	Pope SR 18 Mar	Wright
Dunciad in 4 books (4⁰ + FP)	1743	29 Oct	for M. Cooper	'Cooper' SR 28 Oct	Bowyer 1500 + 100

Works are in folio unless otherwise indicated

ass: assigned by Pope; FP: fine paper; LP: large paper; SR: Stationers' Register

While one cannot be certain, I suspect this may be a misinterpretation of Pope's generosity. The first work in Table 15, Horace's *Ode to Venus*, is entered in the Stationers' Register by John Wright with the added note, 'I John Wright do claim the Sole Property in and to the above book by Virtue of an Assignment under the hand of Alexr. Pope Esqr. bearing date the 25 day of February 1736/7'; a similar formula is found in Dodsley's entries on 17 May, and is clearly the result of legal advice which Pope has taken in his attempt to suppress piracies. The grant of a copyright to Wright, though doubtless for a limited period, seems to me to be possibly a reward for faithful service. It is a slim work, selling for sixpence rather than the normal shilling; and naturally enough Wright had to have it distributed by a publisher, in this case James Roberts. Similarly, though Dodsley has now taken over the role of Gilliver in issuing Pope's separate poems, the *Sixth epistle of the first book of Horace* is entered in the Register over Pope's signature on 14 January 1738 as 'Speedily to be publised[!]' with a note (in the clerk's hand):

N.B. Be it rememberd that I Alexander Pope have authorised Lawton Gilliver to Print & Publish an Edition in folio and Quarto of the said Epistle, being the first Edition thereof.
A. Pope.

This, too, I would like to think of as a farewell present to Gilliver. Finally, in the list of Pope's possible generosities, he made no claim in the Register to *The universal prayer*—a very proper gesture.

What is not at all clear is whether Dodsley made a similar agreement with Pope to that which Gilliver had, allowing him limited rights to each new work in exchange for payment. The fact that he paid Woodfall's bill for the *First epistle of the second book of Horace*[71] suggests that he had some such arrangement, but (as I have argued above) his entry of the *Letters* on 17 May is probably made purely as a cover for Pope. The fact that in some cases the imprint is for T. Cooper rather than for Dodsley himself is not unusual; as I explained in Chapter 1, Dodsley in Pall Mall needed a distributor in the city and regularly used Cooper for this purpose. It was also Dodsley's practice to suppress his own name in the case of anonymous works to avoid questions of authorship, and it is probably significant that the *First epistle of the second book* was published anonymously. Though few can have doubted that Pope was the author, it contained, in John Butt's words, 'more dangerous matter' and 'brought Pope dangerously near to punishment from the House of Lords' (*Twickenham*, iv, pp. xxxvii–xxxviii). There is no precisely parallel case for Cooper's imprint to the first dialogue of *One thousand seven hundred and thirty eight*, since it bears Pope's name; but it is sufficiently controversial to warrant Dodsley's keeping out of it, especially since Pope certainly kept the copyright himself in this case.

---

[71] The *Second epistle*, though printed by Woodfall on the evidence of the ornaments, is not included in the entries from Woodfall's ledgers printed in P.T.P., 'Pope and Woodfall', pp. 377–8.

The *Sixth satire of the second book of Horace* by Swift, but completed by Pope, has a different imprint because the first part had appeared in the Pope–Swift *Miscellanies* in 1727 and that copyright belonged to Motte and Bathurst. We can only speculate why the Knaptons' names are added, but since they were publishing Pope's collections they were an obvious choice if Pope wished to conceal his part and to omit Dodsley and Cooper as part of the manœuvre.

These imitations of Horace were all printed in folio, the first and second epistles of the second book with fine-paper copies. (I know these only from the copyright deposit copies, and if fine-paper copies of the others were not deposited, they may still remain to be identified, perhaps bound up in folio collections of Pope's works.) Pope was, as usual, concerned to provide quarto copies to be bound with his works, and these appeared first as the *Epistles of Horace* published on 27 April 1738 and were subsequently reissued with reprints of the two dialogues of *One thousand seven hundred and thirty eight* as *Poems and imitations of Horace*, dated 1738 but not published until 11 January 1739. The production of these quartos is of considerable interest for the text, and their further revision for publication in octavo gives evidence about the copyright position and the publication dates of Pope's poems. The full story of the rebuilding of the second volume of the octavo *Works* as told by Maynard Mack is even more complex than that of the original *Works* II of 1735.[72]

The first thing we see from Table 16 is that the five folio Horatian poems reprinted in quarto as *Epistles of Horace* were originally published in precisely the reverse order to their final appearance. The first, Book IV ode I, was reimposed by Wright in quarto and printed off without change as an unsigned quarto gathering which could be fitted in anywhere that seemed appropriate. The two epistles of Book II, which had been printed by Woodfall, were reset by Wright and the accidentals normalized. The second epistle, which was printed first, had used italic for proper names whereas the first did not; the reprint in quarto brings the second into line with the first by removing italics (and some capitals). Most interesting textually are the epistles of Book I, since they were reimposed in quarto but also revised in type by Pope; there are some substantive changes, but many more changes to accidentals which are clearly Pope's own. John Butt's decision in the *Twickenham* edition to follow the first edition 'in typography and punctuation' is therefore questionable. I think on rather slender evidence from the running titles that sheets E–F^4 containing the sixth epistle were printed after the folio, but before the first epistle in folio; it would not have been difficult to estimate the space needed, and this agrees with the order of publication dates.

Maynard Mack has shown that the right-hand group in octavo in Table 16 corresponds to the entry for 'Printing Epistles of Horace', in 3½ sheets octavo in Woodfall's ledger, where it is entered to Lawton Gilliver under 15 June 1737, but

---

[72] Mack, 'Pope's Horatian poems: Problems of bibliography and text', *MP* xli (1943–4), 33–44 (*Collected in himself*, pp. 106–21). I differ from him in minor details. McLaverty has added a note, 'Pope's Horatian poems: A new variant state', *MP* lxxvii (1979–80), 304–6.

TABLE 16. Horatian Epistles in Folio, Quarto, and Octavo

Folio:	Bk I, Ep 1 (7 Mar 1738) '1737'	Bk I, Ep 6 (24 Jan 1738) '1737'	Bk II, Ep 1 (19 May 1737)	Bk II, Ep 2 (28 Apr 1737)	Bk IV, Ode 1 (9 Mar 1737)
				Woodfall (12 May)	
	revised by Pope	revised by Pope			
				reset by Wright	
Quarto: [*Epistles of Horace*, 27 Apr 1738]	[A]¹ B–D⁴	E–F⁴	²A–E⁴ F1–4ʳ	F4ᵛ E–K⁴ L¹	[M⁴]
Octavo:	Rev. and repr. by Woodfall by 10 Feb 1738, with 3 other poems, of which only Sat II. 6 was separately pub. (1 Mar 1738)		Rev. and repr. by Woodfall by 15 June 1737		
	The octavo reprint of *One thousand seven hundred and thirty eight* was also probably in print by late 1738 and the quartos before then, though issued 11 Jan 1739.				

'paid by Mr. Pope, June 2, 1738', and the left-hand group to a further entry of 'Epistles of Horace' in 3¾ sheets (as 4 sheets) made to Pope on 10 February 1738 and also paid on 2 June. These sheets were originally intended to be included in an enlarged second volume of the octavo *Works*, but were subsequently used to make up Volume II, Part II (Griffith 507) probably published on 4 May 1739. The immediate conclusion is that if the *First epistle of the first book* was printed in folio by Wright, reimposed and revised in quarto, and then revised again and printed in octavo by Woodfall by 10 February 1738, then it must have been printed for at least a month before its publication on 7 March 1738. It seems likely, in fact, that the 1737 dates of the epistles of book I correspond with the date of printing, and that Pope held them back for publication in the new year.[73]

There is evidence that the book trade considered the new year the best time for publication,[74] and the issue of Pope's proposals corresponds to this pattern. Thus the proposals for the *Odyssey* were dated 10 January 1725 and inserted in the papers a fortnight later. Similarly with the *Letters* of 1737, though proposals had been circulated privately as early as April 1736, Pope writes on 6 November that he intends to publish an advertisement 'in January, when the Town fills' (*Correspondence*, iv. 41); actually it appeared on 12 February. The same consideration gives point to Pope's letter about Harte's *Essay on reason* of May or June 1734— 'whether he would chuse to publish it now, or next winter, let himself judge' (*Correspondence*, iii. 408); it appeared in February. If we look at Pope's separate

[73] The *Sixth satire of the second book of Horace* offers a more complex problem; that too was reprinted by Woodfall by 10 Feb. 1738, yet not published separately until 1 Mar., but in this case it was not printed by Wright or published by Dodsley. It seems likely, though, that Pope controlled the date it was published. It may be significant that in this case there are no substantive changes between the folio and octavo, though Mack reports the addition of inverted commas in a couplet of the octavo, retained in later edns.

[74] See Carter, *History of the Oxford University Press*, p. 213.

poems from the time he controlled their publication in 1731, only the first, the *Epistle to Burlington*, appeared in December (*Sober advice from Horace*, 21 December 1734, was published by others). Out of 21 separate works, 5 appeared in January, 3 in February, and 4 in March—over half in the first three months of the year. This practice would explain not only why five of these poems were issued in the year following their imprint date (the *Second satire of the second book of Horace*, dated 1735 but published 3 October 1734, is the only contrary case), but also why the fourth part of the *Essay on man* appeared in January 1734, eight months after the third. It looks as though Pope settled down each autumn after his summer jaunts to see his work into type, but withheld the product until the new year, when Parliament reassembled and with it Society.

The fact that the first octavo entry for the *Epistles* in Woodfall's ledger is made to Gilliver in June 1737 and the second to Pope in February 1738 seems to give more precise information on the date when Gilliver was relieved of his duties as Pope's agent than we have had hitherto; it also fits conveniently my theory that he was given rights to the first edition of the *Sixth epistle of the first book of Horace* in January 1738 as a farewell present. The fact that he was specifically given the right to the first edition in folio and quarto presumably means that he had a share in the collected *Epistles of Horace* in quarto, and thus his name appears in the imprint of that and its reissue in *Poems and imitations of Horace*, '1738'.[75] The grant excluded the octavo reprint, and the copyright entries of the other Horatian poems to Dodsley and Wright were presumably based on similar exclusions.

With the publication of *One thousand seven hundred and thirty eight*, Pope's copyrights are entered directly to him in the Stationers' Register (the first dialogue by Wright, the second by Dodsley), and Dodsley probably had no interest in them except as a distributor. In Pope's complaint about a piracy of the *New Dunciad* in 1742, he makes the statement that he

applied himself to Robert Dodsley of Pall Mall . . . who thereupon employed Thomas Cooper late of Pater Noster Row . . . to print and publish the said Book or Poem for and on your Orator's behalf and in your Orator's Right and for your Orator's Sole benefit and Advantage as the Author and Proprietor thereof.[76]

That the situation cannot be as simple as this is shown by the fact that the first edition was printed by Wright and when it was pirated 'by the foolish Delays of my Printer' the second edition was printed by Bowyer and entered to Pope in his ledgers. Clearly Pope chose his printer and did not leave it to Cooper. The Bowyer account for printing 2,000 copies for £22. 18s. 3d. (including the cost of paper, charged directly to the author as is normal in distinction to bookseller's work, where the bookseller provides the paper) is in Pope's name and paid by

[75] I have assumed that Gilliver's role in producing vol. II of the octavo *Works* from 1735 was as Pope's agent since he did not own the copyright and Dodsley took his place in 1739; vol. IV containing the *Dunciad* was his own property, so it continues to appear under the imprint of Gilliver and Clarke until he sold the copyright at the end of 1740.

[76] Vincent, 'Some *Dunciad* litigation', p. 287, quoting PRO C 11 837/14.

Mrs Cooper out of the profits of the sale. It seems that the author of *Sawney and Colley*, published on 31 August 1742, is not far wrong in his criticism of Pope for 'having lately turned *Bookseller* to himself, selling all his own Pieces by Means of a *Publisher*, without giving his Bookseller any Share in them', though we may justly question the second charge of 'practising, in all respects, the lowest Craft of the Trade, such as different Editions in various Forms, with perpetual Additions and improvements, so as to render all but the *last* worth nothing; and, by that Means, fooling many People into buying them several times over'.[77] Perpetual additions and improvements there may have been, but they were made in the pursuit of excellence, not of profit.

WARBURTON AND THE DEATH-BED EDITIONS

Pope and Warburton did not meet until April 1740, but by then they had already become firm allies. In December 1738 Warburton had deserted the ranks of Pope's critics in order to defend the *Essay on man* against the attacks of the Swiss theologian De Crousaz,[78] and from that time he began to supersede Jonathan Richardson junior and Spence as Pope's chief literary associate. Pope had great hopes of this alliance between scholar and poet: 'Could you make as much a Better Man of me, as you can make a Better Author,' he told Warburton on 7 October 1743, 'I were secure of Immortality both here & hereafter' (*Correspondence*, iv. 474). The problem was how best to put the collaboration to work. If at first the most likely project was an expansion of the *Essay on man*, attention soon shifted to the *Dunciad*. Book IV, or the *New Dunciad*, was published in 1742, followed by a reworking of the poem in four books with Cibber as hero. Warburton contributed notes to this edition, and this set the pattern for a new edition of Pope's works; the *Essay on man*, *Essay on criticism*, and the *Epistles to several persons* had all been prepared for publication when the project was halted by Pope's death on 30 May 1744.

These late editions of the poems prepared by Pope and Warburton have commonly been referred to as the death-bed quartos, but it turns out that the project included the preparation of octavo editions at the same time. The publication of Pope's works after 1735 displays a regular pattern: the poems and letters first appear in quarto (and sometimes folio) before taking their place in additional volumes of the *Works* octavo. This octavo series began with the co-operation between Gilliver and Lintot after *Works* II was published in 1735; Gilliver brought out the various epistles as *Works* II octavo and Lintot split the early poems into two volumes, *Works* I and III; Gilliver then issued the *Dunciad* as *Works* IV. The letters followed the same pattern, appearing officially in quarto in

[77] Quoted by Guerinot, *Pamphlet attacks on Pope*, p. 305.
[78] For an account of Pope's relations with Warburton see Mack, *Alexander Pope: A life* (New Haven, 1985), pp. 741–5, and his introd. to the *Essay on man*, *Twickenham*, ii/1, esp. pp. xxi–xxii.

1737 and then as octavo *Works* V and VI. The pattern was changed in 1739 to accommodate the Horatian poems; they were too closely related to the epistles to form a new volume VII, but a combined volume II would have been too large, so they were issued as Vol. II, Part II. Soon the other works were adapted to this pattern; the early poems became Vol. I, Parts I and II, and the letters, with Swift's now added, became Vol. IV, Parts I, II, and III. From this perspective the death-bed quartos can be seen as another instalment in a familiar pattern.

We have already heard Pope complaining to Allen about the expense of publishing in quarto, telling him in July 1741, 'I have done with expensive Editions for ever, which are only a Complement to a few curious people at the expence of the Publisher' (*Correspondence*, iv. 350), so it may be difficult to explain why he reverted to publishing in quarto as early as March 1742. I think there were two main reasons. The first was that Pope wanted to keep faith with those who had been collecting his quartos from the time of the subscription edition of the *Iliad* in 1715. He had conscientiously considered their interests after 1735 by issuing his poems in formats that would enable them to enlarge their sets, and he was probably reluctant to break with that tradition now. The second reason is the one his letter gives to Allen: the quartos made good presents to give to influential friends. Pope now had Warburton's career to think of as well as his own, and an appropriately grand edition would impress the right people with Warburton's wit and scholarship. His desire to help Warburton is clear from a letter of 21 February 1744:

> I would also defer till then the publication of the Two Essays with your Notes in Quarto, that (if you thought it would be taken well) you might make the Compliment to any of your Friends (& particularly of the Great ones, or of those whom I find most so) of sending them as Presents from yourself.   (*Correspondence*, iv. 500.)

And he asked Bowyer in a letter two days later to remind Warburton about these presentation copies, even suggesting some names. Bowyer's ledgers do not show Warburton responding to these suggestions, but during this period Pope himself always had fine-paper copies printed and usually took pre-publication ones.

The quartos were valuable as presents, but it was the 'little' editions, as Pope calls them, the octavos, that brought him a regular income, and he was willing to devote considerable attention to them. The various *Dunciad*s (book IV and the Cibber version) were the only late works actually published in octavo before Pope's death, but, as we shall see, the other poems printed in quarto were also associated with octavo editions. It is important in this context to recognize that priority in publication does not necessarily indicate priority in printing. In the case of the letters, for example, we now know that the octavo printing preceded that of the quarto and folio in 1737 and again in 1741;[79] the same is generally true of the poems printed at this time.

---

[79] See Dearing, 'The 1737 editions of Pope's letters', pp. 185–97, and A. C. Elias, jun., 'The Pope–Swift *Letters* (1740–41): Notes on the first state of the first impression', *PBSA* lxix (1975), 323–43.

We can follow the events of this late period in some detail because Pope's alliance with Warburton involved a change of printer. William Bowyer, whose ledgers survive and give accounts for some of Pope's early works as well as these late ones,[80] had been engaged in printing Warburton's *Divine legation of Moses* almost continuously between 1737 and 1742, and Pope now turned to him as printer in preference to John Wright. It is possible that Warburton had a hand in the change—especially as Pope's account is found facing that for the *Divine legation* in the paper stock ledger—but in a letter to Allen, Pope places the responsibility squarely on Wright's shoulders: 'They have pyrated my Poem—by the foolish Delays of my Printer whom I'l pay off, & imploy less for the future. It was Charity made me use him.' (*Correspondence*, iv. 392.) Sherburn places this letter in April 1742 and suggests the pirated poem is the *New Dunciad* in quarto (Griffith 546), first printed by Wright, and published on 20 March 1742; this dating is convincing, but Pope was already employing Bowyer by that time. The printing ledgers show that Bowyer's bill for the printing of 2,000 copies of his edition of the quarto *New Dunciad* (Griffith 549) was drawn up on 25 March and paid by Mrs Cooper, probably the following day. Warburton's *Commentary* on the *Essay on man* (later the *Vindication*) was also sent to Bowyer in March (Nichols, *Anecdotes*, ii. 152), and a letter written by Pope to Warburton on 23 April confirms that their collaborative project involving Bowyer as printer was already under way (*Correspondence*, iv. 393).

The first major task Pope allocated to Bowyer in 1742, after the quarto *New Dunciad* (or book IV) was the printing of a completely revised *Dunciad* in four books, with a new hero, Colley Cibber. Pope intended to publish the revised poem in editions in quarto and octavo to fit the usual pattern of *Works*, but the position was complicated by Henry Lintot. On 15 December 1740 Lintot had finally acquired the copyright to the three-book *Dunciad variorum*, and these rights stood in the way of the publication of a revised edition. Pope's first response had been conciliatory, offering to revise the text of the poem and later agreeing to correct the press, but the conflict between Lintot's interests and his own gradually grew more apparent, until on 16 February 1743 he brought a suit against Lintot in Chancery (see Appendix A below). The dispute seems to have concerned the date of expiry of the fourteen-year term Lintot had bought from Gilliver, and whether or not Lintot could go on selling copies after the date of expiry, whenever it was. Whatever the precise legal position, the practical consequences are clear: Pope had to arrange the printing of the Cibber version of the *Dunciad* with half an eye to Lintot's edition, and without being sure when he could publish his own.

Lintot had a large edition of the three-book *Dunciad* printed in octavo by Henry Woodfall, whose accounts show 4,000 copies, plus 100 fine, printed by 4

---

[80] The Bowyer paper stock ledger is in Bodl. (MS Don. b. 4), where the printing ledgers have been deposited by the Grolier Club (MSS Dep. b. 243–4 and Dep. c. 718–23). The accounts referred to in this section are to be found in the paper stock ledger, fo. 113ʳ, and the printing ledger, iv. fos. 30ᵛ–31ʳ.

July 1741.[81] But Lintot held the edition back for nearly a year, doubtless waiting until Pope was ready to publish book IV, the *New Dunciad*. It appeared on 20 March 1742, and Lintot entered his *Dunciad* in the Stationers' Register on 1 April 1742 (finally paying Woodfall on 5 April). Lintot called his volume *Works* III. ii, a title which is a mistake for *Works* III. i but which indicates that the book was designed as a companion volume for the *New Dunciad* (the real *Works* III. ii). Lintot probably printed so large an edition in anticipation of the stir that would be caused by the *New Dunciad*, but he must also have known that the copyright period was coming to an end and that a large edition gave him the best chance of maximizing his profits. Pope obliged Lintot by providing an octavo edition of the *New Dunciad* to accompany the Woodfall octavo:[82] *Works* III. ii (Griffith 566), which appeared in summer 1742, was probably printed by Wright; there are no references to it in the Bowyer or Woodfall ledgers and, characteristically of Wright's work, it has no press figures. This arrangement left Bowyer free to concentrate on the revised, Cibber, *Dunciad*.

The octavo version of the Cibber *Dunciad* was probably started before the quarto. In a letter to Warburton of 23 April 1742 Pope says, 'I do not intend to set Bowyer, yet, upon any thing but the *first part of* the *Dunciad*' (*Correspondence*, iv. 393), and this is most likely a reference to the octavo, which, unlike the quarto, was divided into two separate parts. Subsequent letters of 1 November, 13 November, and 27 December 1742 refer to cancels (six altogether and the first specifically of a half-sheet) which are neither charged for in Bowyer's bill for the quarto nor to be found in the printed volume; they are presumably those to be found in the octavos, which are heavily cancelled. And when Pope wrote to Allen on 27 December, 'The true Edition is at last completed' (*Correspondence*, iv. 433), he was surely referring to the octavo *Dunciad in four books* because when five months later, on 21 May 1743, he told Warburton about the progress of the quarto, he said it was 'half printed' (*Correspondence*, iv. 455–6). This fits the textual evidence, which shows that both parts of the octavo *Dunciad* originally had texts which were in most details earlier than the quarto's.[83]

The *Correspondence* suggests Pope supervised both printing and distribution of these volumes himself. A letter to Bowyer of 13 November 1742 shows that the

---

[81] See P.T.P., 'Pope and Woodfall', p. 378 and Vincent, 'Some *Dunciad* litigation', p. 287.

[82] Though an advertisement for the quarto *New Dunciad* in the *London evening post* of 24 Apr. 1742 declared: 'There is no genuine Edition of this Poem, except in Quarto; nor will there be one in any other Form till it is join'd with the rest of the Author's Works in the Duodecimo Volume.' (Quoted by Griffith, ii. 446.)

[83] Pope may have been deliberately using the octavo as a means of preparing and revising his text; it would have been cheaper to cancel in octavo than in quarto. D. Vander Meulen's invaluable dissertation, 'A descriptive bibliography of Alexander Pope's *Dunciad*, 1728–1751' (unpub. Ph.D. diss., University of Wisconsin-Madison, 1981) shows that part I has 10 cancels, at least 7 of which represent a state of the text earlier than the quarto's, and that part II has 28 lines added representing a stage in the progress to the quarto text, as well as cancels based on the quarto. The Bowyer paper stock ledger shows Pope taking 11 copies of each vol. 'at different times' before publication, and this could have been connected with the process of revision.

octavo Part II was originally designed as an independent publication, available for combination with Lintot's *Works* III. i as long as his rights lasted: 'I suppose Cooper has had part of your Edition in 8⁰ of the New Dunciad & Memoirs of Scriblerus—pray tell me what number? according to which We mu[st] print or retard the Fourth Book in [thi]s edition.' (*Correspondence*, iv. 426.) Pope seems to be saying that if stocks were getting low, work would have to start on the fourth book of the revised *Dunciad*; if not, it could wait until the quarto was completed. In fact no copies of the *Works* III. ii had been sold by this point; Mrs Cooper took her first batch of 500 from the warehouse on 14 January 1743.[84] In the end so many copies were left that it proved unnecessary to print a new edition of the fourth book at all. Instead Pope gave instructions that this volume should be reissued with cancels to bring it into line with the new Cibber quarto. He wrote to Bowyer on 3 November 1743, five days after he had finally been able to publish the quarto:

Pray close the account with Mrs Cooper of the *Octavo's Second vol.* (no more of which should, I think, now be sold) and make all that remain correspond with the present Edition, ready to be re-publishd as we shall find occasion, the 2 together.
And let me know when you have vended 500 of the Quarto?

(*Correspondence*, iv. 478.)

The account was paid off as Pope requested. £42 had been paid on 29 June, and another £14. 11s. 9d. was paid on 17 November, possibly as a direct response to this letter. On 14 January 1744 the remaining bill for these octavo *Dunciads* was settled by Mrs Cooper (acting for Pope, the ledger says specifically) with a payment of £61. 2s. 3d.—it was the last account paid before Pope's death. Mrs Cooper took 500 copies of Part I on 26 March and Dodsley 25 on 5 April; Dodsley took 50 of the fine-paper copies on 17 May, perhaps for Pope himself, who was suffering from his final illness.

Well before the quarto Cibber *Dunciad* was published on 29 October 1743, work had started on the *Essay on man* and *Essay on criticism*. Pope had decided that Warburton was to be 'in some measure' the editor of the *Dunciad*, that work becoming 'a kind of Prelude or Advertisement . . . of your Commentarys on the Essays on Man, and on Criticisme' (*Correspondence*, iv. 427–8). On 28 December 1742 he told Warburton that he was giving Bowyer the *Essay on man* to print and after that the *Essay on criticism* and the 'Pastorals' (*Correspondence*, iv. 434) and on 18 January he said he would 'put Bowyer directly on the Edition of the Ethic Epistles; and I think it will be a more dignify'd Method of declaring your intention of a General Commentary, before That, than before This Poem [i.e. the

---

[84] The year could be 1744 (the entries are exceptionally confused at this point), but 1743 is more likely. Bowyer was supposed to have printed 2,000 ordinary copies and 50 fine of each octavo vol.; but there were 80 fine copies of part I and 81 of part II. The paper stock ledger gives a good illustration of the different roles of the Coopers and Dodsley. The Coopers take copies in batches of 500; Dodsley takes 25 or, in one case, 50.

Dunciad]' (*Correspondence*, iv. 439).[85] The references to the *Essay on criticism* and the 'Pastorals' make it clear that although Pope was willing to publish his two essays in a single volume in this trial quarto edition, he intended to revert to something like the usual order later: the 'Pastorals' were to go with the *Essay on criticism* and the *Rape of the lock*, while the *Epistles to several persons*, *Epistle to Arbuthnot*, and '2 or 3 of the best of Horace' were to go with the *Essay on man* (*Correspondence*, iv. 480, 491); 'Mr. Pope thinks that his Works will be comprehended in two volumes of 60 sheets each', Warburton told Bowyer on 9 March 1744 (Nichols, *Anecdotes*, ii. 164). In the mean time the two *Essays* were to be published together 'to try the Taste of the Town' (*Correspondence*, iv. 480). Some tact was called for because Lintot had a recognized claim to the *Essay on criticism*, but Pope thought he could see his way round that (see Appendix A below). Bowyer printed 1,500 ordinary quartos and 100 fine paper—the same as for the quarto *Dunciad*—and 250 were taken by Mrs Cooper on 20 January 1744 ready for publication.

In a remarkable letter of 23 February 1744 (further discussed in Appendix A) Pope gave Bowyer instructions about the details of publication. First he told him to enter the book in the Register, giving him a wording which drew attention to Warburton's notes; then he explained how he was to deal with Lintot—by offering him a proportion of the edition of the *Essay on criticism* and writing down (for use in evidence) anything he might say about retaliation—finally he dealt with the octavo printing:

Let Wright immediatly send you in what he has done, gathered, of the *little* Essay on Man. As I remember one or two Sheets at the beginning were first done by you, so that you must put them & his together. he has finishd it, but possibly a Title leaf may be wanting to the whole which pray Supply, & have it ready gatherd. (*Correspondence*, iv. 501–2.)

This is the octavo corresponding to the '*little Edition*' of the *Dunciad*, but in this case most of the printing had been done by Wright. A fine piece of detective work by Richard Noble has identified these sheets as eventually appearing in a volume issued in 1748 (Foxon P873); the first two sheets, B–C^8, were printed by Bowyer before the quarto was printed, but sheets D–L^8 were printed by Wright using the quarto as copy. It was completed with prelims and M^4 (which reprinted the last leaf of the *Essay* and added the *Universal prayer*); this last sheet is apparently the one entered in the Bowyer printing ledgers (iv, fos. 74^v, 114^r) on 23 August 1748.[86] Once again the printing of the octavo straddles the printing of the quarto; the text of the *Essay on man* was much more settled than that of the new *Dunciad*,

[85] There has been a dispute over the meaning of 'Ethic Epistles' in this letter, but in view of the proximity of the letter of 28 Dec. and the fact that the Bowyer ledgers (iv, fo. 21^r) list Warburton's *Commentary* on the *Essay on man* as commentary on 'Ethic Epistles', there can be little doubt the *Essay on man* is meant here.

[86] K. I. D. Maslen discussed these editions, prior to Noble's discovery, in 'New editions of Pope's *Essay on man* 1745–48', *PBSA* lxii (1968), 177–88.

and it may have been thought safe to switch to printing directly in quarto soon after starting the octavo printing.

The same problem of whether to print first in quarto or octavo occurred with the *Epistles to several persons* (the epistles to Cobham, a Lady, Bathurst, and Burlington). Warburton was at work on his commentry by 15 November 1743 (*Correspondence*, iv. 480) and on 27 January 1744 Pope told him:

> I have gone over all your Papers on the 2 Epistles, to my Satisfaction, and I agree with you to make shorter work with those to the Lady,˙& to Lord Burlington (tho I have re-placed most of the omitted lines in the former) I wish next for your Remarks, on that to Dr Arbuthnot . . . These I propose to print next together.   (*Correspondence*, iv. 495.)

Pope decided to print first in octavo (again possibly to give himself opportunities for second thoughts) but on 3 March 1744 he wrote to Bowyer to tell him he had changed his mind:

> On Second thoughts, let the Proof of the Epistle to Lord Cobham, I, be done in the *Quarto*, not the *Octavo*, size: contrive the Capitals & evry thing exactly to correspond with that Edition. The first proof send me, the Number of the whole but *1000*, and the Royal *over* & above.   (*Correspondence*, iv. 504.)

By this time Pope was seriously ill and could spare less energy for writing and publication; to print in quarto straightaway would be more economical of effort. These late instructions to Bowyer are among the most interesting in Pope's correspondence for they show he had quite distinct conceptions of the quarto and octavo editions: they had different stylings and Bowyer is given instructions to depart from his copy and bring the 'Capitals & evry thing' into line with the other quartos.

Pope had got as far as reviewing the *Epistle to a lady* in proof, for that seems the correct interpretation of the letter to Warburton saying, 'I have just run over the Second Epistle from Bowyer. I wish you could add a Note at the very End of it' (*Correspondence*, iv. 516), and he had probably also seen the proof of the *Epistle to Burlington*, when he died, leaving Warburton with the responsibility of completing the edition.

At Pope's death the state of his publishing activities was, to say the least, untidy. We have no ledgers from Wright's warehouse but we do know he had 1,500 copies of Bolingbroke's *The idea of a patriot king* which Pope had printed without the author's knowledge, because Wright approached Bolingbroke and asked what he was to do with them.[87] In Bowyer's warehouse there were: 1,483 ordinary (and 28 fine) copies of the octavo *Dunciad* Part I and probably 981 (and 29) of Part II, but these had at least been paid for and they could be sold by Mrs Cooper; in addition there were 955 (and 80) copies of the quarto *Dunciad*; 1,224 (and 95) of the quarto *Essay on man and Essay on criticism*; and all the copies of the quarto *Epistles to several persons* (1,000 and 100). Pope owed £136. 8s. 0d. for the *Dunciad*; £62. 6s. 9d. and £33. 6s. 3d. for the essays; and £40. 10s. 8d. for the

---

[87] G. Barber, 'Bolingbroke, Pope, and the *Patriot king*', *Library*, 5th ser. xix (1964), 67–89.

epistles, making a grand total of £272. 11s. 8d., which was only slightly reduced by payments of £31. 8s. by Mrs Cooper; the rest was left to the executors.

In his will Pope left all his manuscripts to Bolingbroke 'either to be preserved or destroyed', but he bequeathed to Warburton 'the Property of all such of my Works already printed, as he hath written, or shall write Commentaries or Notes upon, and which I have not otherwise disposed of, or alienated; and all the Profits which shall arise after my Death from such Editions as he shall publish without future Alterations'. The books in the warehouse became the property of Martha Blount, who inherited the residue of the estate. Warburton was to have some trouble relating to these books later but he had immediate problems with Lintot over the *Essay on criticism* and with the additions Pope had made to the *Epistle to a lady*. The *Correspondence* shows that the portrait of Atossa, which was later to be construed as an attack on the Duchess of Marlborough, had been added to the poem by Pope himself; but that was not widely known and Warburton did not have the protection a published edition would have afforded him: he could have been accused of altering the poem and consequently have lost his rights to the profits. In a letter to Bowyer of 20 June 1744, he tried to deal with this problem, but first he had to allay Lintot's anxieties about the effect of Pope's will on his copyrights:

You will oblige me with telling me that beast Lintot's steps. I would do him all reason while he acts with decency and justice, and shall never print any part of his property with my Notes and Commentary without his leave; but if he acts like a rogue, I have but one word with him, the Chancery and Mr. Murray. This *inter nos*.—If the executors inquire of you, and when they do, about the state of Mr. Pope's Works in your hands yet unfinished (that is to say, of the Epistles), I then desire you would let Mr. Murray have a copy of all those Epistles; and you may tell him I desired you would do so: but say nothing till then. Pray preserve all the Press Copy, to the least scrap.   (Nichols, *Anecdotes*, ii. 165.)

Pope's policy towards Lintot was to be maintained: there was to be co-operation, but with the threat of Murray and the law lurking in the background.[88] As for the *Epistle to a lady*, all Warburton could do was ask Bowyer to keep the copy, hoping it included the new lines, and finish off the volume. But in the event he was not able to avert the controversy threatened by the Atossa lines; Bolingbroke discovered them and interpreted them as an attack on the Duchess of Marlborough, and the volume was suppressed.[89]

This account of Pope's publishing ventures with Warburton may close with a glance at Warburton's subsequent progress. He bought the quartos in the warehouse from the executors in 1748 (*Twickenham*, iii/2, p. xiii n.); the *Epistles to several persons* volume was suppressed but the others were sold to Knapton to be issued with cancel title-pages. Warburton took on the responsibility of producing

---

[88] Warburton was at this time a supporter of perpetual copyright (see his *A letter from an author to a Member of Parliament*, 1747), though he later changed his mind.

[89] The liveliest account of these difficulties is to be found in Brady's 'The history and structure of Pope's *To a lady*'.

an authorized edition of Pope but did not publish it until 1751. It appeared not as a quarto but as an octavo in nine volumes. We can see now that that was not a betrayal of Pope's intentions but a realization of them: Warburton produced the edition for a general readership that would always have been the result of the grand edition in quarto. The enterprise was a profitable one to Warburton (just as his friendship with Pope was profitable in leading to an alliance with Ralph Allen through marriage to his niece), but one episode in his financial career may illustrate the problems for an amateur in dealing with the trade. In 1755 Knapton, thought to be the richest of the booksellers, went bankrupt, with Warburton a major creditor. Knapton had to sell off the remaining volumes that had been bought from the executors and the one-third share in Pope's *Works* he had bought from Warburton. Warburton took half the volumes and Draper and Millar the other half; these two booksellers were also willing to buy the one-third of the copyright, but they wanted half in all and persuaded Warburton to part with one-sixth in order to make up their half. Warburton takes up the story in a letter to Knapton:

I agreed as I thought to sell them in proportion as they bought of you & they understood me, in proportion as I had sold to you. I was extreme vexed: but my honour was concerned so I have got 250 l instead of 425 l. All the satisfaction I have in this ugly affair is that you have got a better price. For they say they would never have given what they have to you but for the reason above, their understanding me in a sense I never thought of.[90]

Warburton must have sold one-third of the copy to Knapton for £500 in 1748, while Millar and Draper paid Knapton £850 for it in 1755—a good indication of the rise in the market value of Pope's *Works*. Warburton, stating his intention ambiguously, only succeeded in getting the price he sold at in 1748 rather than the current value. 'I have sold what I proposed this Morning', he says in the same letter to Knapton, 'But have bit my selfe.' Reviewing the advantages that came to him from his relationship with Pope, it is difficult to feel any considerable sympathy.

---

[90] BL MS Egerton 1954, fo. 5. I have placed this letter in 1755, rather than the customary 1748, on the basis of its connection with Warburton's letter to Dodsley of 26 Dec. 1755 in the Edinburgh University Library, MS La. II. 153. The letter is transcribed by D. W. Nichol in 'Pope, Warburton and the Knaptons: Problems of literary legacy (with a book-trade correspondence)' (unpub. Ph.D. diss., University of Edinburgh, 1984), a study which carries the story of Pope and Warburton up to 1756.

4

POPE'S TEXT: THE EARLY WORKS

THE *ILIAD* PROOFS

My account of Pope's relations with the trade has shown how closely he was concerned with the production of his works; I now turn to the effects this detailed concern with book production and typography had on his text. I hope to show two things: first, that Pope was a pioneer in evolving our modern style by abandoning the capitalization of substantives and the italicizing of proper nouns, and second, that he revised the accidentals of his works—that is, the spelling, use of capitals and italic, and punctuation—at least as thoroughly as he revised the words of his text. The assumption of modern English textual criticism, and indeed the basis of the distinction between substantive and accidental readings proposed by Greg,[1] has been that while a scribe or a compositor will do his best to copy an author's words correctly, he will feel free to impose his own conventions on the accidentals, though doubtless preserving some of the author's own. A first edition, on this argument, is as close as we can normally get to an author's manuscript, and so to his own habits of spelling and punctuation, and we should for this reason follow its accidentals; it is, however, merely the most expedient course, for we cannot say whether any particular comma, for example, is the author's or the compositor's. On the same assumption, a revised edition of a work will contain the author's substantive revisions to the text, but the accidentals will be further removed from the author's by a second compositorial intervention as well as by chance errors.[2] Pope's works contradict this assumption in every way. His manuscripts do not show his final intentions for the accidentals; he expected his compositors to follow his punctuation and other conventions (unless he had given specific instructions, as he probably did for the *Iliad* and *Odyssey*), which

---

[1] W. W. Greg, 'The rationale of copy-text', in *Collected papers*, ed. J. C. Maxwell (Oxford, 1966), pp. 374–91; the paper was read before the English Institute on 8 Sept. 1949 and printed in *SB* iii (1950–1), 19–36.

[2] The best guides to the vast literature on this topic are three articles by G. T. Tanselle: 'Greg's theory of copy-text and the editing of American literature', *SB* xxviii (1975), 167–229; 'Recent editorial discussion and the central questions of editing', *SB* xxxiv (1981), 23–65; and 'Historicism and critical editing', *SB* xxxix (1986), 1–46.

he would then adjust in proof; and when he produced a revised edition (and the majority of editions were revised) his changes to accidentals would far out-number his changes to substantive readings.

Having made such a bold assertion, my first concern is to show the process as directly as I can in Pope's proof-correcting. We are fortunate that Simonne Le Gal brought to light a volume in the Bibliothèque de l'Arsenal in Paris which contains the proofs of the first two volumes of the *Iliad* corrected in Pope's own hand.[3] These are the only proofs corrected by Pope which seem to have survived, though I still hope that those for the *Dunciad* which were preserved by Daniel Prince may yet surface in Oxford where the latter took over the operation of the Clarendon Press.[4] The *Iliad*, like Pope's other subscription works over the next twenty years, was first printed for subscribers in quarto with twenty lines to a page, and then reimposed in folio with twenty-two lines to a page for public sale. The Arsenal proofs show both stages: the second volume consists of proofs of the quarto, and the first, proofs of the folio after the type had been reimposed. By the time of the folio proof-reading the text of the poem had been well worked over in preparing the subscribers' copies and we find only one or two typographical changes being made; more changes are to be found in the prose sections. R. M. Schmitz has counted corrections at 126 places in this volume, of which the printer followed 107;[5] he attributes all of them to Pope, but there must always be doubt about the conventional proof-reader's signs and I have become sceptical of Pope's wholesale involvement. One result of this second revision between quarto and folio was that the public and not the subscribers received the more polished text; the readers of the *Twickenham* are in the same plight as the subscribers.

There is much more substantial revision to the text in the earlier stage of proof represented by the quarto proofs of the second volume; and it is clearly by Pope himself.[6] In Figure 59 we find, apart from the rewriting in two lines, five commas added in proof as well as the hyphen in 'Heav'n Gates' in the last line but one, while 'Pale' in the second line has its intitial capital changed to lower case. (The opening parenthesis in line 465 was not added as directed.) It is very difficult to find a typical page, for the amount of correction varies enormously; this may in part reflect the varying amount of polish which Pope gave to different parts of his manuscript. Two sheets of book VII contain no changes at all except for Pope's revision of lines 135–8, written between the lines (Figure 60)—in someone else's hand, I think.[7] Pope was guilty of a howler here, and as the change is made in the

[3] Le Gal, 'En marge de l'exposition du "livre anglais"; l'Homère de Pope', *Bulletin du bibliophile et du bibliothécaire*, NS i (1952), 49–54. See also Callan, 'Pope's *Iliad*: A new document'.

[4] See above pp. 109–10 and Nichols, *Anecdotes*, iii. 705.

[5] R. M. Schmitz, 'The "Arsenal" proof sheets of Pope's *Iliad*: A third report', *MLN* lxxiv (1959), 486–9.

[6] Mack in App. D to the *Twickenham* Homer (x. 469) writes of the corrections to the quarto: 'Most substantive verbal changes seem to be in Pope's hand; most spelling and punctuation changes are probably, though not demonstrably, his; instructions to the printer are in most cases (perhaps in all) *not* his.'

[7] This revision is discussed at length by Callan. 'Pope's *Iliad*', pp. 118–21.

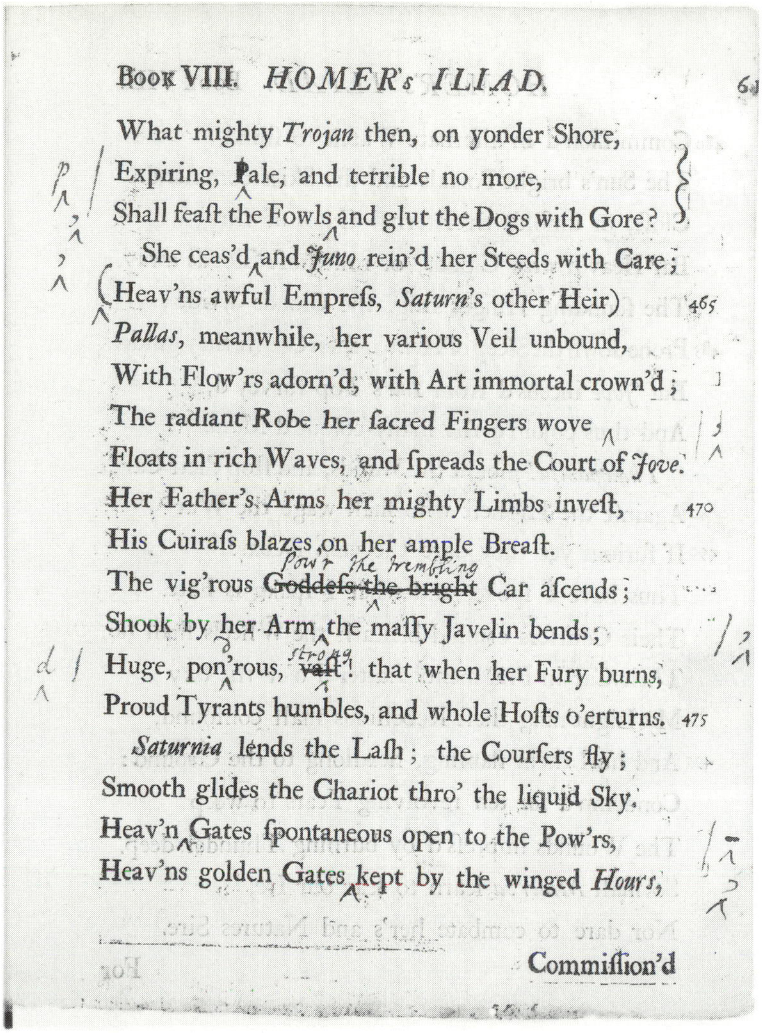
Book VIII. *HOMER's ILIAD.* 613

What mighty *Trojan* then, on yonder Shore,
Expiring, Pale, and terrible no more,
Shall feast the Fowls and glut the Dogs with Gore?
She ceas'd and *Juno* rein'd her Steeds with Care;
(Heav'ns awful Empress, *Saturn*'s other Heir)
Pallas, meanwhile, her various Veil unbound,
With Flow'rs adorn'd, with Art immortal crown'd;
The radiant Robe her sacred Fingers wove
Floats in rich Waves, and spreads the Court of *Jove*.
Her Father's Arms her mighty Limbs invest,
His Cuirass blazes on her ample Breast.
The vig'rous ~~Goddess the bright~~ Car ascends;
Shook by her Arm the massy Javelin bends;
Huge, pon'rous, ~~vast~~! that when her Fury burns,
Proud Tyrants humbles, and whole Hosts o'erturns.
Saturnia lends the Lash; the Coursers fly;
Smooth glides the Chariot thro' the liquid Sky.
Heav'n Gates spontaneous open to the Pow'rs,
Heav'ns golden Gates kept by the winged *Hours*,

Commission'd

FIG. 59. Pope's corrections in *Iliad* proofs, p. 613 (Arsenal 4° B 1572; approx. 255 × 185)

quarto by a cancel leaf 3R3, these sheets (3R and 3S) may well be part of the edition as printed, not proofs. It is also possible that in some cases the sheets we have are revises, not first proofs, though this is clearly not the case when Pope has—rather endearingly—left blanks for epithets to be decided in proof (Figure 61, last line). Though there may be some still unresolved problem about the status of all the proofs, they are immaterial to our present purpose of seeing the sort of changes Pope made to the accidentals of his text. I shall concentrate on the analysis of capitals and italics, but the illustrations will also tell a tale of Pope's revisions to punctuation.

Figure 61 shows Pope capitalizing the adjectives 'Hospitable' in the second line and 'Orient' in line 215, and it is in capitalization that these proofs are most interesting in relation to Pope's manuscript. At this period in English printing it

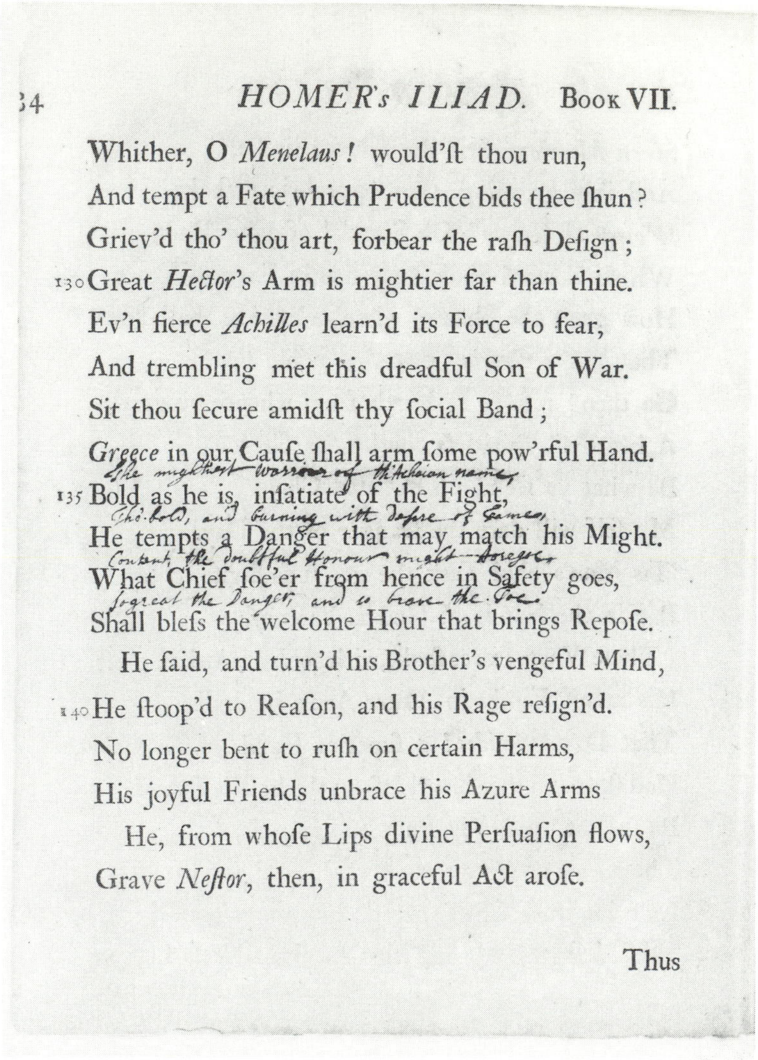

FIG. 60. Corrections by an unidentified hand in the *Iliad* proofs, p. 534 (Arsenal 4° B 1572; approx. 255 × 185)

HOMER's ILIAD. BOOK VII.

Whither, O *Menelaus!* would'st thou run,

And tempt a Fate which Prudence bids thee shun?

Griev'd tho' thou art, forbear the rash Design;

130 Great *Hector's* Arm is mightier far than thine.

Ev'n fierce *Achilles* learn'd its Force to fear,

And trembling met this dreadful Son of War.

Sit thou secure amidst thy social Band;

Greece in our Cause shall arm some pow'rful Hand.

135 Bold as he is, insatiate of the Fight,

He tempts a Danger that may match his Might.

What Chief soe'er from hence in Safety goes,

Shall bless the welcome Hour that brings Repose.

He said, and turn'd his Brother's vengeful Mind,

140 He stoop'd to Reason, and his Rage resign'd.

No longer bent to rush on certain Harms,

His joyful Friends unbrace his Azure Arms

He, from whose Lips divine Persuasion flows,

Grave *Nestor*, then, in graceful Act arose.

Thus

was normal practice to capitalize every substantive, but Pope's rough drafts are not consistent in this respect; his fair copies get much closer to the conventions of print. Pope seems to have differed from the routine capitalization of printers in two important respects. When a noun was in a subordinate position, as in an adverbial phrase, he sensibly withheld the capital (we shall see examples in a moment). His other idiosyncrasy was the capitalization of certain adjectives, and that is what we see in Figure 61.

But what was the nature of the manuscript the printer used? We know that Pope's friend from Binfield, Thomas Dancastle, transcribed the last four volumes of the *Iliad* for the printer, but we have no clear evidence about the first two volumes that concern us here. It is clear from their correspondence that Dan-

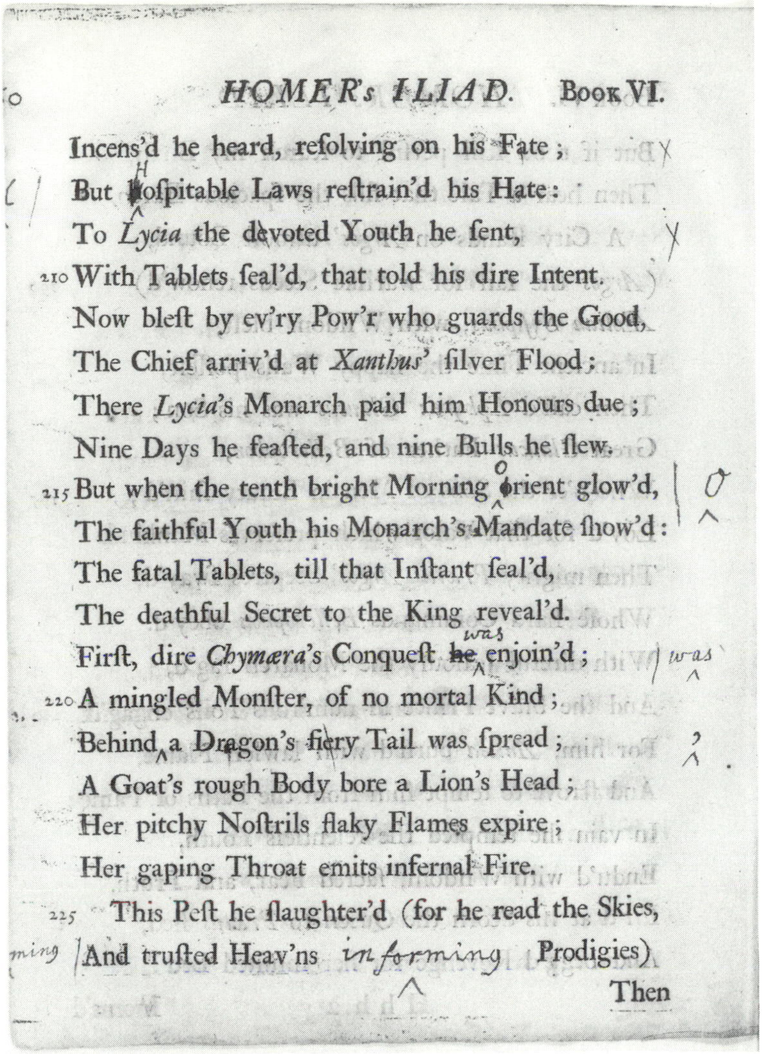

HOMER's ILIAD. Book VI.

Incens'd he heard, reſolving on his Fate;
But Hoſpitable Laws reſtrain'd his Hate:
To *Lycia* the devoted Youth he ſent,
With Tablets ſeal'd, that told his dire Intent.
Now bleſt by ev'ry Pow'r who guards the Good,
The Chief arriv'd at *Xanthus'* ſilver Flood;
There *Lycia*'s Monarch paid him Honours due;
Nine Days he feaſted, and nine Bulls he ſlew.
But when the tenth bright Morning orient glow'd,
The faithful Youth his Monarch's Mandate ſhow'd:
The fatal Tablets, till that Inſtant ſeal'd,
The deathful Secret to the King reveal'd.
Firſt, dire *Chymæra*'s Conqueſt he enjoin'd;
A mingled Monſter, of no mortal Kind;
Behind a Dragon's fiery Tail was ſpread;
A Goat's rough Body bore a Lion's Head;
Her pitchy Noſtrils flaky Flames expire;
Her gaping Throat emits infernal Fire.
This Peſt he ſlaughter'd (for he read the Skies,
And truſted Heav'ns informing Prodigies)

Then

FIG. 61. Blank left for insertion of epithet in *Iliad* proofs, p. 460, line 226 (Arsenal 4° B 1572; approx. 255 × 185)

castle worked from Pope's 'foul papers', which were always incomplete in capitalization; Dancastle's own letters do not suggest that he would have been very good at normalizing them.[8] We have no direct reference to Dancastle's work before volume III, but we do have a fair copy of the opening to book VIII in Pope's print hand (*Twickenham*, vii, Plates 7 and 9). It has an endorsement 'Mr.

[8] See *Correspondence*, i. 519 and ii. 19 for references to Dancastle's copying, and ii. 8 for his eccentric capitals. Callan, 'Pope's *Iliad*', p. 114n., says, 'Several leaves of the transcription are preserved in MS Add. 4809: Pope has used the backs for his version of the *Odyssey*'. I can find only two leaves not in Pope's hand (or Fenton's revised by Pope): fo. 173, in an unidentified hand, which is reproduced in *Twickenham*, x, Plates 15 and 16, and bears the text of *Odyssey* XVII. 582–637; and fo. 119^{v}, which bears a passage from *Odyssey* I.

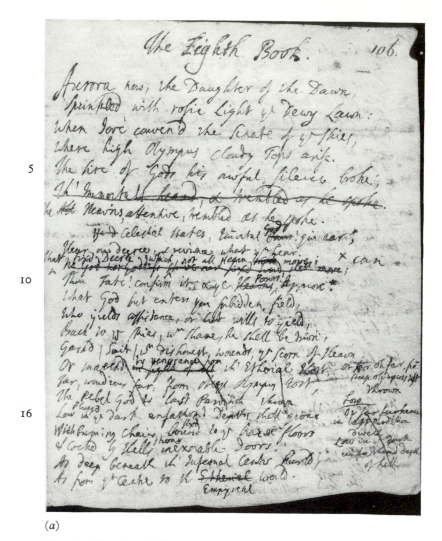

(a)

FIG. 62. Opening of *Iliad*, book VIII in draft (BL Add. MS 4807, fo. 106^r; 190 × 149) and fair copy (Beinecke; approx. 241 × 190)

Pope's own hand|writing to send to the|Press' but it bears no printer's marks, so it was probably not used. Nevertheless, a comparison of Pope's draft and fair copy for this passage (Figure 62) is of some interest, to show the sort of changes Pope made, as well as his inconsistency. The two styles of handwriting coexist throughout Pope's life. Pope's first draft is more heavily capitalized in the opening of the book than elsewhere, but still fails to be consistent in capitalizing nouns such as 'silence' in line 5, 'decree' in line 8, 'field' in line 11, and 'assistance' in line 12. In the fair copy these are all capitalized. So, surprisingly, is 'Ear' in the phrase at the end of line 7 'give Ear': this is just the sort of case where Pope norm-

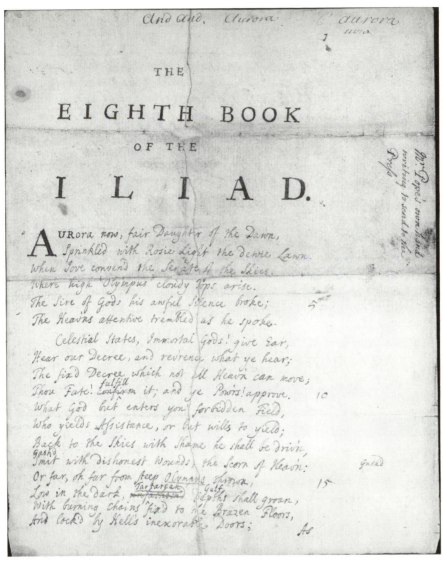

(b)

ally made an exception. It is printed in lower case in the proofs, which is confirmation that this fair copy was not used by the printer.

The fair copy capitalizes more adjectives than the first draft, but observe the change in the second line where the draft capitalizes 'Dewy' but not 'rosie' while the fair copy does the opposite. The printed text makes both lower case. It also standardizes the inconsistent spelling of the draft to a '-y' ending where the fair copy standardized with '-ie'. This is surely a splendid example of the dangers of over-simplification in textual matters. We happily talk of an author's preferred spelling, yet here is Pope varying his spelling of an ending within a single line and then from draft to fair copy. The evidence of the concordance is that he normally spelt dewy 'y' and rosie 'ie' or 'y';[9] the change to 'ie' for dewy must have been made to match what he had just written, and the standardization to 'y' in print is the right one. It simplifies matters enormously to assume that an author's manuscript will show the accidentals he wanted, but when we have two manuscripts we find the accidentals vary widely. How can we be sure when we have only one that it shows what the author wished to preserve?

Some of the capitalized adjectives are clearly deliberate: 'Immortal Gods' in line 7 is capitalized at every stage, and in line 17 the capitalization of 'Brazen Floors' introduced by the fair copy is also found in print. Similarly in the first two lines on the next page (Figure 63) with 'Infernal Centre' and 'Ætherial World'. But below in line 25 'our Golden everlasting Chain' and in line 41 'Human State' we have capitals for adjectives which are not found in the proof. Elsewhere capitals like these are removed in the process of proof-reading.

One thing, I think, is clear from our look at the manuscripts: even when Pope made a fair copy, he did not get every detail into its final form. When his foul papers were transcribed by Dancastle the sort of revision we see in proof would be all the more necessary. In fact there is clear evidence that at some stage there was a very heavy-handed attempt at standardizing the capitalization, which made revision in proof absolutely vital. In the passage from book V illustrated in Figure 64, 'fear' had been capitalized in line 1021. There can be no doubt that someone thought it was a noun instead of a verb. Pope changes it, and capitalizes 'Immortal' at the same time. There are quite a number of similar cases. Possibly these errors are due to Dancastle; I hope they are not the responsibility of Bowyer's compositor, though certainly the corrector (perhaps Bowyer himself) is at fault in not spotting it. This nicely makes the point that the Homer differs from Pope's other poems in being based on Pope's rough drafts; the accidentals had to be subjected to some process of normalization and then finally checked by Pope.

An example of Pope's refusal to capitalize a noun when it is part of an adverbial phrase can be found in Figure 65. Originally 'chace' was capitalized in line 408, but Pope corrects the routine capitalization to bring it in line with his own preferences. There are other similar changes in proof, as well as examples where the capital is left uncorrected.

Pope was often tempted to use italic to point the stress of a line. We probably find a relic of the practice in line 97 in Figure 66, as well as Pope's rejection of the temptation. The marginal note, somewhat cropped, 'in the common letter' is probably not in Pope's hand; he was often content to underline italic as a sign to change it back to roman. (Note on this page the removal of the capital for 'Gain' in line 101; another example of a verb being capitalized under the misapprehension that it was a noun.) Except in his earliest manuscripts Pope was never consistent in underlining proper names and their adjectives as a sign that they should be italicized, and the result is that when the compositor fails to make the change

---

[9] *A concordance to the poems of Alexander Pope*, compiled by E. G. Bedford and R. J. Dilligan (Detroit, 1974).

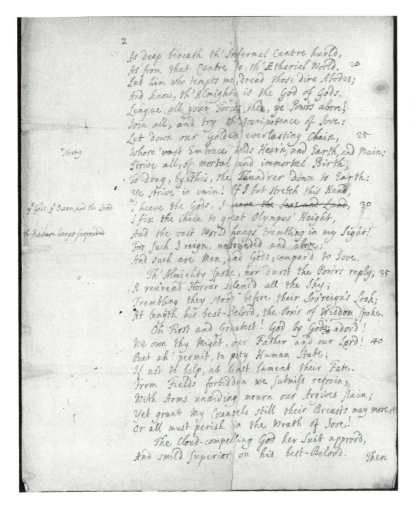

FIG. 63. Fair copy of opening of *Iliad*, book VIII, p. 2 (Beinecke; approx. 241 × 190)

on his own initiative, as with Athenian in line 987 of Figure 67, Pope has to correct him.

I have concentrated on capitals and italics, which will concern us later; let me remind you again of the changes to punctuation we have seen—three in the last three lines of Figure 67. Pope might leave the compositor to print proper names in italic and he might expect, or even ask for, some normalization of his capitals and his spelling; but he would change the compositor's work when he disagreed with the result. There is one proviso which continually returns; Pope, like the rest of us, often missed things in proof that he would have corrected if he had noticed them. Despite this I think the proofs clearly show that the first edition of the *Iliad* is much closer to what Pope wanted than his manuscripts. The proofs mark a further stage of revision of the accidentals as well as the substantives of his text.

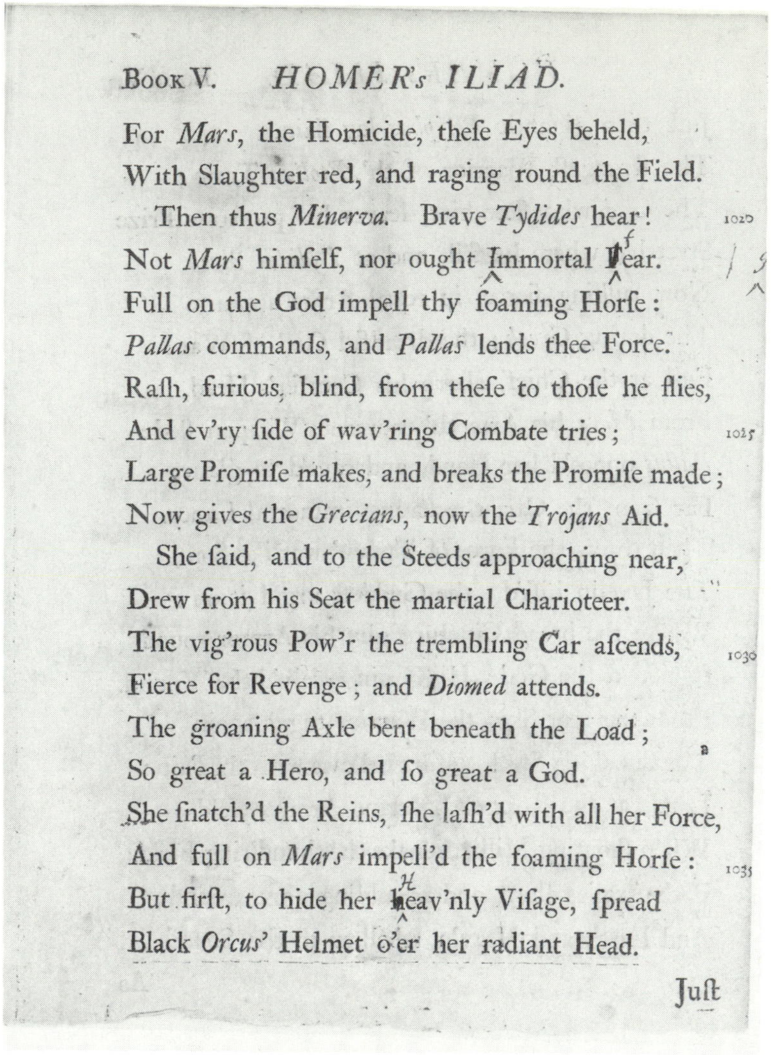

FIG. 64. Change in capitalization of
verb in *Iliad* proofs, p. 389, line 1021
(Arsenal 4° B 1572; approx. 255 × 185)

BOOK V.   *HOMER's ILIAD.*

For *Mars*, the Homicide, thefe Eyes beheld,
With Slaughter red, and raging round the Field.
    Then thus *Minerva.*   Brave *Tydides* hear!   1020
Not *Mars* himfelf, nor ought Immortal Fear.
Full on the God impell thy foaming Horfe :
Pallas commands, and *Pallas* lends thee Force.
Rafh, furious, blind, from thefe to thofe he flies,
And ev'ry fide of wav'ring Combate tries ;   1025
Large Promife makes, and breaks the Promife made ;
Now gives the *Grecians*, now the *Trojans* Aid.
    She faid, and to the Steeds approaching near,
Drew from his Seat the martial Charioteer.
The vig'rous Pow'r the trembling Car afcends,   1030
Fierce for Revenge ; and *Diomed* attends.
The groaning Axle bent beneath the Load ;
So great a Hero, and fo great a God.
She fnatch'd the Reins, fhe lafh'd with all her Force,
And full on *Mars* impell'd the foaming Horfe :   1035
But firft, to hide her heav'nly Vifage, fpread
Black *Orcus'* Helmet o'er her radiant Head.

                                        Juft

AN ESSAY ON CRITICISM

The Bodleian manuscript of the *Essay on criticism* (MS Eng. Poet c. 1) seems to
be the only surviving Pope manuscript that served as printer's copy, except for
parts of the *Odyssey*, and it makes it possible for us to gain some further insight
into the changes Pope made in proof. From the use of one pair of head-line skel-
etons for all the gatherings, it seems that both formes of each sheet were set and
imposed, and perfected proofs then read by Pope; when corrections had been
made (and in at least one case, a revise corrected) the sheet was printed off and
proofs of the next prepared.

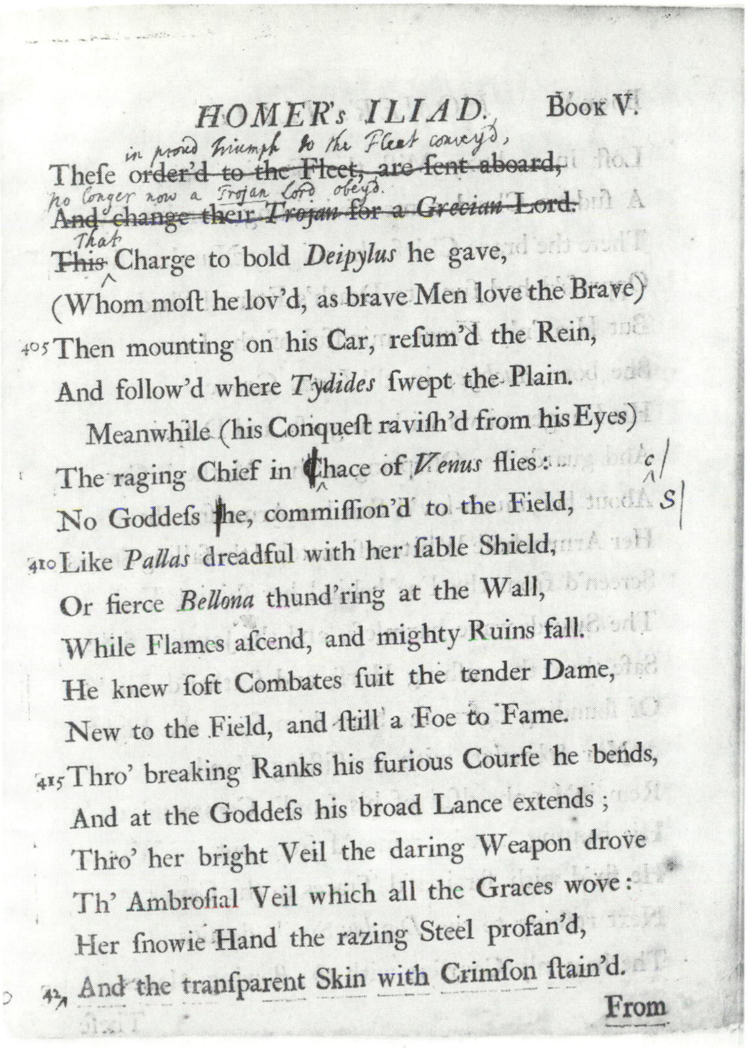

HOMER's ILIAD. Book V.

These ~~order'd to the Fleet, are ſent aboard,~~ in proud Triumph to the Fleet convey'd,
~~And change their *Trojan* for a *Grecian* Lord.~~ no longer now a Trojan Lord obey'd.
~~This~~ That Charge to bold *Deipylus* he gave,
(Whom moſt he lov'd, as brave Men love the Brave)
Then mounting on his Car, reſum'd the Rein,
And follow'd where *Tydides* ſwept the Plain.
Meanwhile (his Conqueſt raviſh'd from his Eyes)
The raging Chief in Chace of *Venus* flies:
No Goddeſs ſhe, commiſſion'd to the Field,
Like *Pallas* dreadful with her ſable Shield,
Or fierce *Bellona* thund'ring at the Wall,
While Flames aſcend, and mighty Ruins fall.
He knew ſoft Combates ſuit the tender Dame,
New to the Field, and ſtill a Foe to Fame.
Thro' breaking Ranks his furious Courſe he bends,
And at the Goddeſs his broad Lance extends;
Thro' her bright Veil the daring Weapon drove
Th' Ambroſial Veil which all the Graces wove:
Her ſnowie Hand the razing Steel profan'd,
And the tranſparent Skin with Crimſon ſtain'd.
From

FIG. 65. Change in capitalization of noun in *Iliad* proofs, p. 358, line 408 (Arsenal 4° B 1572; approx. 255 × 185)

Looking at the heading in Pope's pen and ink imitation of type (Figure 68), one can only echo John Butt's rhetorical question 'Who can doubt that a man capable of such calligraphy would insist on comparable workmanship from his printer?'[10] Having done that, I must go on to disagree with what has been written of the relationship between the manuscript and the first edition.[11] Previous writers have

[10] Butt, 'Pope's poetical manuscripts', Warton Lecture on English poetry, *Proceedings of the British Academy*, xl (1954), 28. R. M. Schmitz has edited a facsimile of the manuscript, *Pope's Essay on criticism 1709* (St Louis, 1962).

[11] P. Simpson, *Proof-reading in the sixteenth, seventeenth, and eighteenth centuries* (Oxford, 1935), pp. 99–104; Schmitz, *Pope's Essay on criticism 1709*, pp. 14–16.

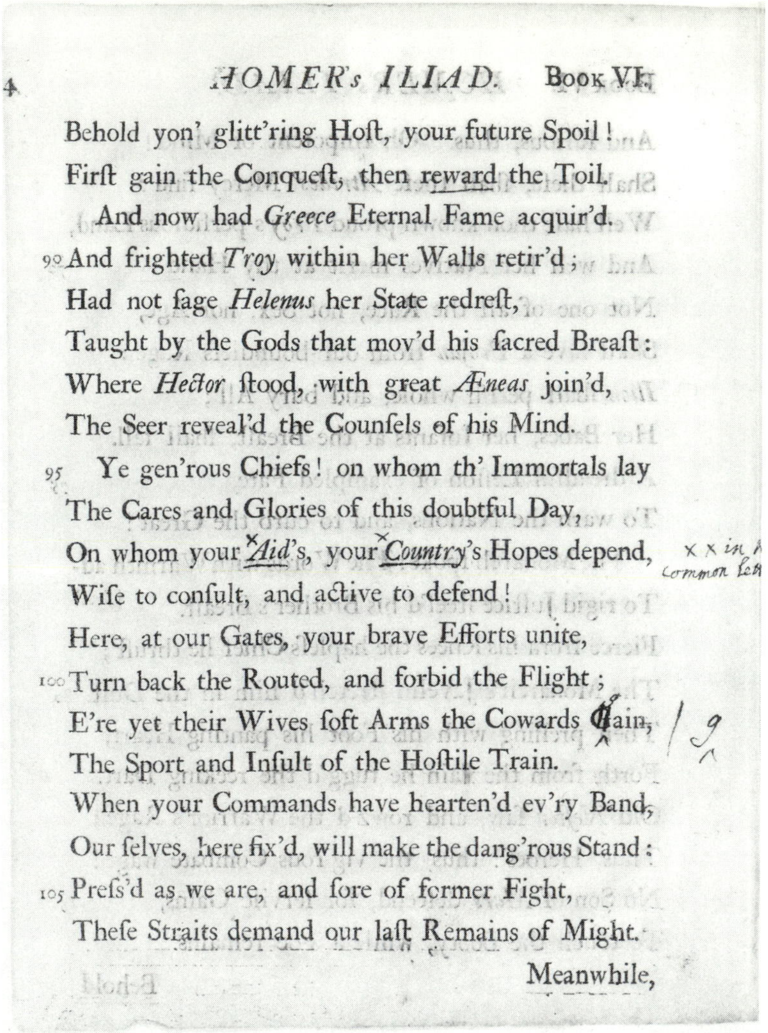

FIG. 66. Removal of italic in *Iliad*
proofs, p. 454, line 97 (Arsenal 4° B
1572; approx. 255 × 185)

4   *HOMER's ILIAD.*   Book VI.

Behold yon' glitt'ring Hoft, your future Spoil !
Firft gain the Conqueft, then reward the Toil.
    And now had *Greece* Eternal Fame acquir'd,
90 And frighted *Troy* within her Walls retir'd ;
Had not fage *Helenus* her State redreft,
Taught by the Gods that mov'd his facred Breaft :
Where *Hector* ftood, with great *Æneas* join'd,
The Seer reveal'd the Counfels of his Mind.
95   Ye gen'rous Chiefs ! on whom th' Immortals lay
The Cares and Glories of this doubtful Day,
On whom your *Aid's,* your *Country's* Hopes depend,
Wife to confult, and active to defend !
Here, at our Gates, your brave Efforts unite,
100 Turn back the Routed, and forbid the Flight ;
E're yet their Wives foft Arms the Cowards Chain,
The Sport and Infult of the Hoftile Train.
When your Commands have hearten'd ev'ry Band,
Our felves, here fix'd, will make the dang'rous Stand :
105 Prefs'd as we are, and fore of former Fight,
Thefe Straits demand our laft Remains of Might.
                                    Meanwhile,

noted that Pope writes 'end the first page here', and the printer does so; they have
not realized that Pope is also responsible for marking '3d page', '4th', '5', and
'6th'.[12] This brings us to two alternative markings by the printer for the beginning of sheet B, 'B. prima', or page 9 of the printed edition (Figure 69). The
second mark on page 5 of the manuscript corresponds to the beginning of the new

[12] The Morgan MS of the first book of the *Essay on man* (a fair copy, though later superseded by the
Houghton MS) is marked off in the same way for 20 pages, some of 14 and some of 16 lines; this would
suggest that Pope was thinking of a quarto format more heavily leaded than his normal 20 lines to a page
(as in the *Works* of 1717 and 1735), or conceivably a smaller size. The Washington University MS of
Windsor Forest is marked less conspicuously as '2 +', '3 +', etc., in 24-line sections, a normal folio page.
Facsimiles of the Morgan and Houghton MSS have been made available by Mack, *The last and greatest
art* (Newark, London, Toronto, 1984), pp. 190–418, and of *Windsor Forest* by Schmitz, *Pope's Windsor
Forest 1712* (St Louis, 1952).

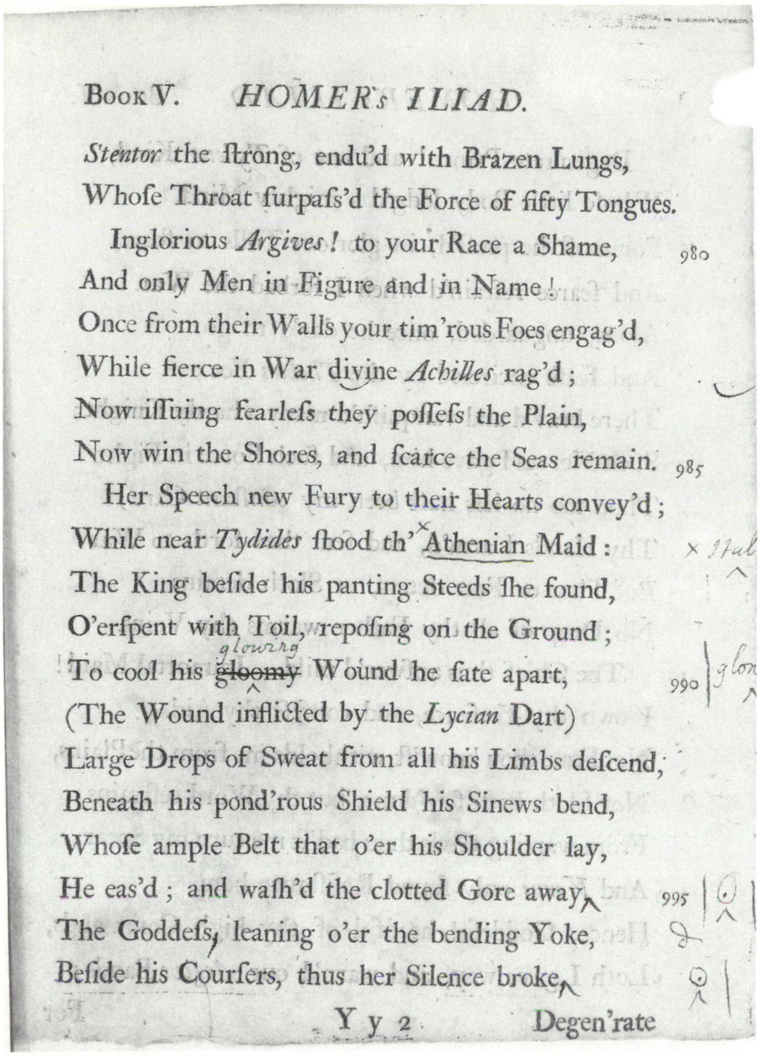

Book V. *HOMER's ILIAD.*

Stentor the ſtrong, endu'd with Brazen Lungs,
Whoſe Throat ſurpaſs'd the Force of fifty Tongues.
Inglorious *Argives!* to your Race a Shame,
And only Men in Figure and in Name!
Once from their Walls your tim'rous Foes engag'd,
While fierce in War divine *Achilles* rag'd;
Now iſſuing fearleſs they poſſeſs the Plain,
Now win the Shores, and ſcarce the Seas remain.
Her Speech new Fury to their Hearts convey'd;
While near *Tydides* ſtood th' Athenian Maid:
The King beſide his panting Steeds ſhe found,
O'erſpent with Toil, repoſing on the Ground;
To cool his ~~gloomy~~ glowing Wound he ſate apart,
(The Wound inflicted by the *Lycian* Dart)
Large Drops of Sweat from all his Limbs deſcend,
Beneath his pond'rous Shield his Sinews bend,
Whoſe ample Belt that o'er his Shoulder lay,
He eas'd; and waſh'd the clotted Gore away,
The Goddeſs, leaning o'er the bending Yoke,
Beſide his Courſers, thus her Silence broke,
Y y 2 Degen'rate

Fig. 67. Addition of italic in *Iliad* proofs, p. 387, line 987 (Arsenal 4° B 1572; approx. 255 × 185)

sheet in the edition, so it is the earlier mark that has been superseded. The later mark corresponds to Pope's marking of pages of 20 lines each (19 on pages 4 and 8, which contain triplets), as opposed to the printer who originally set only 18 lines a page; and made the earlier mark. The printer, John Watts, had cast off the poem and calculated that it would almost fill six sheets at 18 lines a page; under Pope's instructions it finished with 6 lines on the third page of F (Figure 70) and it only ends in this comparatively satisfactory manner because the printer cast off again when he began sheet E and, as a result, he set E with 18 lines to a page and F with 16 lines. Otherwise the poem would have ended with 4 lines on page 42. Pope, then, was responsible for overriding the printer's decision of how many lines a page he would set, a pretty formidable interference and one from whose

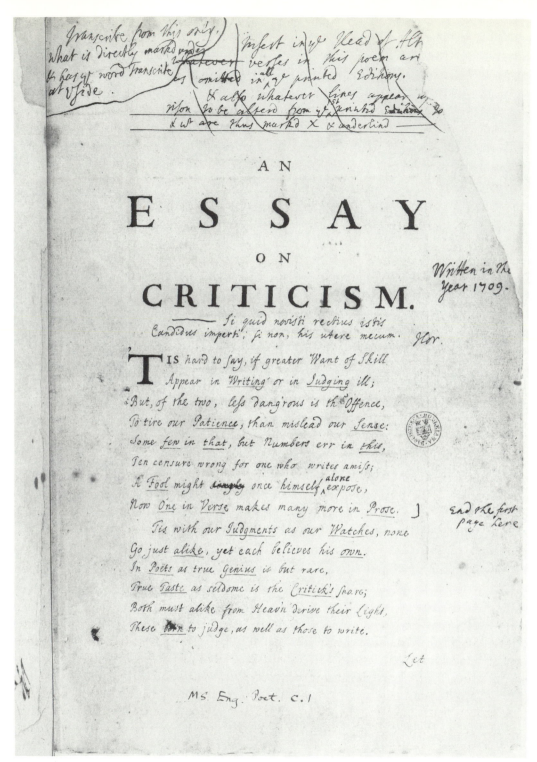

FIG. 68. Autograph manuscript of *Essay on criticism*, p. 1 (Bodl. MS Eng. Poet. c. 1; 323 × 206)

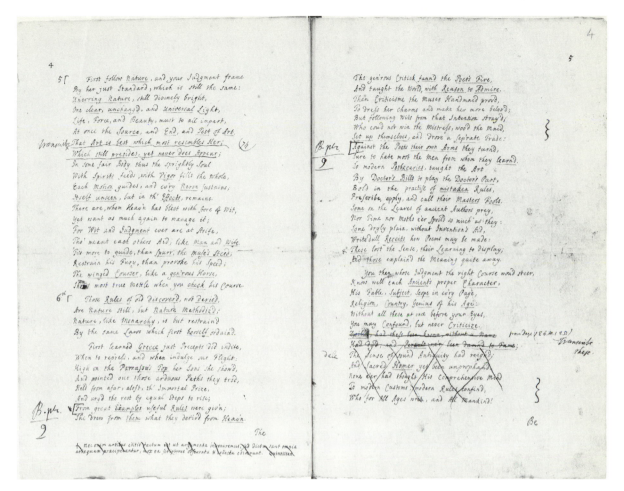

FIG. 69. Alternative markings for beginning of sheet B in *Essay on criticism* manuscript, pp. 4–5
(Bodl. MS Eng. Poet. c. 1; 323 × 410)

consequences he had to be rescued. We can draw one further conclusion from the double marking of the beginning of sheet B; since the printer had to set 9 extra lines in sheet A, and since they have been changed in proof, Pope must have had a second, revise proof pulled for him to correct.

It is instructive to take one passage from the manuscript and trace its changes through successive editions; it will, I think, show Pope's continuing concern with the accidental details of his text as well as with its substantives. I have cut up photographed pages and reassembled them and I have adjusted the scale so that they can be easily compared with the manuscript and with one another. Here as elsewhere my comments are based on the passage in question; a study of whole

42   *An* E S S A Y

Who durſt aſſert the *juſter Ancient Cauſe,*
And here *reſtor'd* Wit's *Fundamental Laws.*

Such was *Roſcomon*——not more *learn'd* than *good,*
With Manners gen'rous as his Noble Blood;
To him the Wit of *Greece* and *Rome* was known,
And ev'ry Author's *Merit,* but his own.

Such late was *Walſh,* —— the Muſes Judge and Friend,
Who juſtly knew to blame or to commend;
To Failings *mild,* but *zealous* for Deſert;
The *cleareſt Head,* and the *ſincereſt Heart.*
This humble Praiſe, lamented *Shade!* receive,
This Praiſe at leaſt a grateful Muſe may give!
The Muſe, whoſe early Voice you taught to Sing,
Preſcrib'd her Heights, and prun'd her tender Wing,
(Her Guide now loſt) no more attempts to *riſe,*
But in low Numbers ſhort Excurſions tries:

            Con-

on C R I T I C I S M.   43

Content, if hence th' Unlearn'd their Wants may view,
The Learn'd reflect on what before they knew:
Careleſs of *Cenſure,* nor too fond of *Fame,*
Still pleas'd to *praiſe,* yet not afraid to *blame,*
Averſe alike to *Flatter,* or *Offend,*
Not *free* from Faults, nor yet too vain to *mend.*

F I N I S.

FIG. 70. Final pages of first edition of *Essay on criticism* (1711), pp. 42–3 (Bodl. Don. e. 70; 219 × 323)

poems would no doubt clarify certain practices as well as revealing inconsistencies.[13]

The one substantive change between manuscript and first edition in the passage illustrated in Figure 71 is at 5 where 'positive persisting Fops' is substituted for 'positive abandon'd Fops'. As for punctuation, an apostrophe is added to 'Judges' at 2, and to 'Tis' in the following line 3 and at 8; one cannot tell whether this is the work of Pope or his printer, though the use of italic for the apostrophe 's' in '*Judge's*', where normal printing practice was to use roman, suggests this

---

[13] Schmitz records 'some 241 changes in punctuation, type face and spelling, drawn from a comparison between the manuscript and the first edition in print. These changes, forever subject to recount, may be listed as: Punctuation 30; Capital to small letter 11; Small letter to capital 47; Italic to roman type 81; Roman to italic type 26; Italic to roman capitals 6; Spelling 40.' He says that although we may safely assign responsibility to the printer for normalizing Pope's spelling and mending his indifference to apostrophes, 'there are still so many changes in type face and punctuation that the only reasonable explanation seems to be Pope's own polishing down to the last detail' (*Pope's Essay on criticism 1709*, p. 15).

change was made in proof.[14] This is, in any case, the sort of normalization of which any author is glad. The addition of the comma at the end of the line at 4 I think is the work of Pope. It is difficult to know who is responsible for the normalizing of spelling; it seems likely that the printer removed at least some of Pope's idiosyncratic spellings. Here he was probably responsible for the change from '*Moralls*' at 1—the doubling of consonants is Pope's most common idiosyncrasy at this stage—and for normalizing 'ne'r' to 'ne'er' at 11 and 14; the division of 'Goodbreeding' at 12 into two words may equally be Pope's; we shall see it undergoing further changes. The adding of capitals to all common nouns was by now a regular practice, and at 9 the capitalization of 'mischief' is routine. The capitalization of nouns in the adverbial phrases at 6 'with pleasure' and 13 'on no pretence' is more interesting. We have already seen Pope removing capitals from such occurrences in the *Iliad* proofs, and in his later manuscripts it is noticeable that Pope, while maintaining a high level of capitalization, does not extend it to such phrases and John Wright follows his copy. When, as here, other printers normalize, I think it is against Pope's wishes.[15] I think that Pope's intentions were distorted here, but he makes no change in the next couple of editions.

The most interesting changes, and the most certainly by Pope, are the changes of emphasis. Pope leaves a blank line in the printed text before the first line shown here, and closes up the blank in the manuscript below; I have created a blank in the printed text here to keep the two in step, but the actual arrangement can be seen in Figure 72. To mark the change of subject he stresses 'Morals' (1) by changing it to caps. and small caps., as he had earlier in the poem changed '*Nature*' (p. 7), '*Rules*' (p. 8), '*Ancient*' (p. 9), and '*Licenced*' (p. 11)—though I wouldn't claim that he was consistent in this: he read proof one sheet at a time. The other changes, between roman and italic, are more concerned with the balance of a phrase. The first example at 6 may be linked with the capitalization of 'pleasure' in the phrase 'with pleasure *own*'; the manuscript clearly suggests the stress should fall on the verb, and Pope perhaps changed his view of the line after the printer had capitalized 'Pleasure'. The italic at 7 'on the *last*' seems unnecessary and is removed. At 10 the printed version '*taught* as if you taught them *not*' is a clear improvement, as is the change at 11 of '*Things*' to roman in the line 'And Things *ne'er known* propos'd as Things *forgot*'.

Looking at this passage as a whole one may not be able to allot each and every change confidently to either Pope or his printer; uncertain, marginal cases remain. Nor may one be certain that all the changes are improvements. What is clear is that the changes are purposive and not just the result of a compositor's rules or a compositor's whim. This becomes even clearer when we compare the

[14] Line 12 of the poem shows the correct usage, '*Critick*'s'; line 45 with the incorrect '*Wit's*' is a change from the roman of the manuscript. Lines 100 and 109 are incorrect, but follow the underlining of Pope's manuscript.

[15] A good example is the new edn. of the *First satire of the second book of Horace* (Foxon P889), where the printer changes 'sleep a wink' to 'sleep a Wink' (p. 7) and 'It stands on record' to 'It stands on Record' (p. 19).

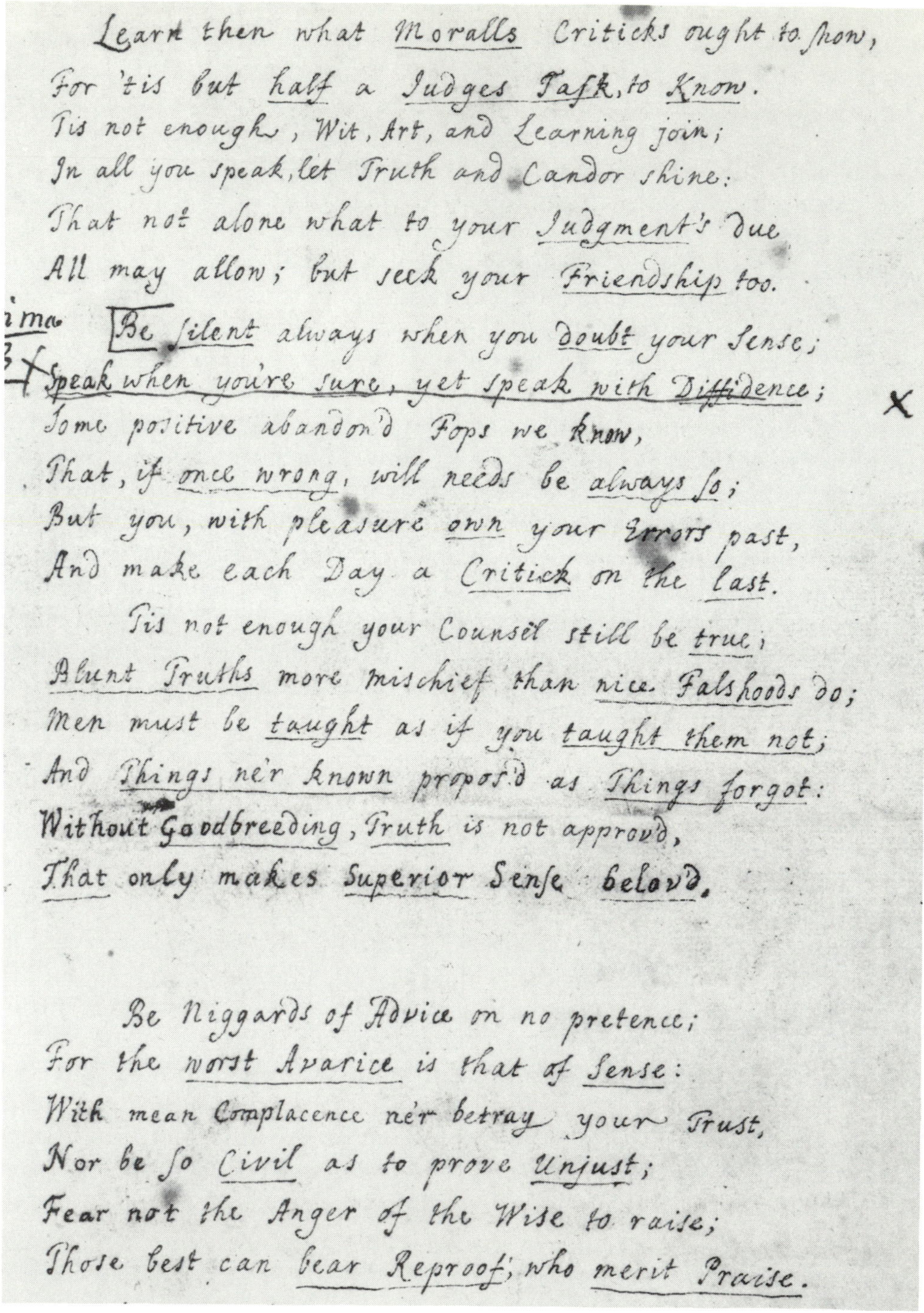

Learn then what Moralls Criticks ought to show,   1
For 'tis but half a Judges Task, to Know.   2
Tis not enough, Wit, Art, and Learning join;   3
In all you speak, let Truth and Candor shine:
That not alone what to your Judgment's due   4
All may allow; but seek your Friendship too.

ima Be silent always when you doubt your Sense;
Speak when you're sure, yet speak with Diffidence;   X
Some positive abandon'd Fops we know,   5
That, if once wrong, will needs be always so;
But you, with pleasure own your Errors past,   6
And make each Day a Critick on the last.   7

Tis not enough your Counsel still be true;   8
Blunt Truths more mischief than nice Falshoods do;   9
Men must be taught as if you taught them not;   10
And Things ne'r known propos'd as Things forgot: 11
Without Goodbreeding, Truth is not approv'd,   12
That only makes Superior Sense belov'd.

Be Niggards of Advice on no pretence;   13
For the worst Avarice is that of Sense:
With mean Complacence ne'r betray your Trust,   14
Nor be so Civil as to prove Unjust;
Fear not the Anger of the Wise to raise;
Those best can bear Reproof, who merit Praise.

(a)

FIG. 71. Manuscript of *Essay on criticism*, p. 21 (Bodl. MS Eng. Poet. c. 1; 323 × 206) and first edition (1711), pp. 32–3 (Bodl. Don. e. 70; 219 × 161)

Learn then what MORALS Criticks ought to fhow,
For 'tis but *half* a *Judge's Task*, to *Know*.
'Tis not enough, Wit, Art, and Learning join;
In all you fpeak, let Truth and Candor fhine:
That not alone what to your *Judgment's* due,
All may allow; but feek your *Friendfhip* too.

Be *filent* always when you d ubt your Senfe;
Speak when you're *fure*, yet fpeak with *Diffidence*;
Some pofitive perfifting Fops we know,
That, if *once wrong*, will needs be *always fo*;
But you, with Pleafure own your Errors paft,
And make each Day a *Critick* on the laft.
'Tis not enough your Counfel ftill be *true*,
Blunt Truths more Mifchief than *nice Falfhoods* do;
Men muft be *taught* as if you taught them *not*;
And Things *ne'er known* propos'd as Things *forgot*:
Without *Good Breeding*, *Truth* is not approv'd,
That only makes *Superior* Senfe *belov'd*.

Be Niggards of Advice on no Pretence;
For the *worft Avarice* is that of *Senfe*:
With mean Complacence ne'er betray your Truft,
Nor be fo *Civil* as to prove *Unjuft*;
Fear not the Anger of the Wife to raife;
Thofe beft can *bear Reproof*, who *merit Praife*.

(*b*)

Learn then what MORALS Criticks ought to fhow,   1

For 'tis but *half a Judge's Task,* to *Know.*

'Tis not enough, Wit, Art, and Learning join ;

In all you fpeak, let Truth and Candor fhine :

That not alone what to your *Judgment*'s due,

All may allow ; but feek your *Friendfhip* too.

Be *filent* always when you d ubt your Senfe;

Speak when you're *fure,* yet fpeak with *Diffidence* ;   2

Some pofitive perfifting Fops we know,   3

That, if *once wrong,* will needs be *always fo* ;

But you, with Pleafure own your Errors paft,

And make each Day a *Critick* on the laft.   4

'Tis not enough your Counfel ftill be *true,*   5

Blunt Truths more Mifchief than *nice Falfhoods* do;

Men muft be *taught* as if you taught them *not* ;   6

And Things *ne'er known* propos'd as Things *forgot* :   7

Without *Good Breeding,* *Truth* is not approv'd,   8

That only makes *Superior* Senfe belov'd.

Be Niggards of Advice on no Pretence ;

For the *worft Avarice* is that of *Senfe* :   9

With mean Complacence ne'er betray your Truft,

Nor be fo *Civil* as to prove *Unjuft* ;   10

Fear not the Anger of the Wife to raife ;

Thofe beft can *bear Reproof,* who *merit Praife.*

(*a*)

FIG. 72. First edition of *Essay on criticism* (1711), pp. 32–3 (Bodl. Don. e. 70;
219 × 161) and second edition (1713), pp. 28–9 (Bodl. G. Pamph. 1283(30);
176 × 105)

Learn then what *Morals* Criticks ought to ſhow,
For 'tis but *half* a *Judge's Task*, to *Know.*

'Tis not enough, Wit, Art, and Learning join;
In all you ſpeak, let Truth and Candor ſhine:
That not alone what to your *Judgment's* due,
All may allow; but ſeek your *Friendſhip* too.

Be *ſilent* always when you *doubt* your Senſe;
And ſpeak, tho' *ſure,* with *ſeeming Diffidence:*
Some poſitive, perſiſting Fops we know,
That, if *once wrong,* will needs be *always ſo;*
But you, with Pleaſure own your Errors paſt,
And make, each Day, a *Critick* on the laſt.

'Tis not enough your Counſel ſtill be *true;*
Blunt Truths more Miſchief than *nice Falſhoods* do;
Men muſt be *taught* as if you taught them not;
And things *ne'er known* propos'd as Things *forgot.*
Without *Good Breeding,* Truth is not approv'd;
That only makes *Superior* Senſe *belov'd.*

Be Niggards of Advice on no Pretence;
For the *worſt Avarice* is that of *Senſe.*
With mean Complacence ne'er betray your Truſt,
Nor be ſo *Civil* as to prove *Unjuſt:*
Fear not the Anger of the Wiſe to raiſe;
Thoſe beſt can *bear Reproof,* who *merit Praiſe.*

(b)

first edition with the second, an octavo of 1712, post-dated 1713 (Figure 72).
Here the one substantive change is as 2, one of those which Pope noted for correction on the last page of the manuscript in reply to John Dennis's criticisms of
the poem; the logical contradiction between '*sure*' and '*Diffidence*' is reconciled by
the change to '*seeming Diffidence*' with appropriate changes to punctuation and
italic. Punctuation is freely adjusted, always in the direction of making it heavier;
the addition of a comma at 3 adds force to the alliteration in 'positive, persisting
Fops' introduced in the first edition, and the added commas at 4 are an attempt to
ensure the correct reading of the line, which is that every day one should make a
critique of the preceding. The change at 5 to a semicolon is justifiable in itself,
but leads to an unsatisfactory series of three lines ending in semicolons (which
was revised in 1717); the period at 7 is a clear improvement. At 8 the heavier
semicolon is again substituted, at 9 Pope substitutes a period for a colon and at 10
colon for semicolon. The only change of capitalization is at 7, where the first
occurrence of 'things' is now lower case; I think it results from the need to make
the line fit the octavo measure, which is perceptibly tight—notice the spacing two
lines higher.

As for stress, it seems that at 1 Pope had second thoughts about stressing
'*Morals*' so heavily; the white line already sufficiently points the change of theme.
Nevertheless, the capitals reappear in the death-bed edition, and it may be that
shortage of space in the line was again the reason for the change. At 6 Pope
further reduces the italic by printing 'not' in roman; the force of the line is not
changed. Similarly at 8, the italicization of 'Truth' is seen to be unnecessary. One
might, indeed, regard the heavy use of italic throughout as something of an in-
trusion, a continual nudging of the reader, and it vanishes in the next edition.

The third edition of 1713, you may remember, is Pope's first use of the Elzevier style with its classical associations; the classics, of course, had no italics, and
relied on the perspicuity of their style and the inflexions of the language without
recourse to such adventitious aids to sense. At this stage of his career, as Figure
73 shows, Pope abandons the heavy use of italic for stress, though he keeps it for
particular purposes: '*Morals*' (1) in the first line is still stressed, and so is '*That*' at
3; italics are also kept in their traditional use for proper nouns like '*Appius*' below
at 4. The result is certainly a much more attractive page. The spacing between
verse paragraphs follows the precedent of Addison's *Campaign*, and Pope later
abandons it. The only other revision on this page is the change at 2 to lower-case
'good Breeding'; the *Iliad* manuscripts show Pope in similar difficulties over how
one should capitalize such compounds.

One other change in this edition, though not on this page, is significant: the use
of capitals and small caps. for proper names to confer honour (Figure 74). It was,
of course, a 'classical' device. Joseph Moxon speaks of the compositor setting
proper names in capitals to dignify them, with 'a *Space* between every *Letter* . . .
to make it shew more Graceful and Stately. For *Capitals* express Dignity wher-

ever they are *Set*, and Space and Distance also implies stateliness.'[16] Possibly the practice originated with the authorized version of the Bible, which uses caps. and small caps. for 'LORD' wherever it refers to the deity. In Pope's later poems we find patrons like Cobham and St John thus distinguished, and Homer had already been dignified in this way in the second edition of the *Essay*. In Figure 74 the presence of an italic exclamation mark after 'VIDA' (1) strongly suggests the change was made in proof, for Watts followed the usual printer's convention of putting italic punctuation marks after an italic word, and it is hard to explain the italic mark in any other way than that it first followed *Vida* in italic.

With Pope's *Works* of 1717 (reduced to half-scale on the right-hand side of Figure 73) we come to a more historic departure, to which I shall have to return: the abandonment of that use of capitals for each noun which had become the norm of English printing in the last twenty years. As with italic in the third edition, there is no absolute prohibition, and this passage is lighter than most. Capitals are kept for 'Morals' in the first line, as well as for 'Critics' (1) though 'Morals' loses its italic in accordance with a policy of further reduction ('That' at 4 is also made roman). In fact all trace of italic for stress is removed in this edition: capitals now have that function. 'Critics' (as a generic term) is also capitalized at 5, but so is 'Critic' in the sense of French *critique* at 2, probably with the different intention of calling attention to a foreign word, though it actually encourages confusion with the 'Critics'. Throughout the volume 'critic' loses its final 'k', but whether this change was made with Pope's agreement remains uncertain; he does seem to have abandoned it in his later manuscripts. The other changes all lie in the six lines (3): a comma inserted in the first, and the semicolon of the third changed into a comma to good effect. The next two lines are neatly improved by the use of 'unknown' and 'dis-approv'd'. In this edition proper names are only printed in italic; the honorific caps. and small caps. we saw in the third edition are laid aside for the time being.

It is worth looking briefly at two later forms of the *Essay* before we return to the question of capitalizing nouns (Figure 75). The first is the octavo series of pocket editions of the *Works*, which was published between 1736 and 1743. In these volumes Pope allowed the use of italic, both for stress and for proper names, which he had generally abandoned for his collected works in larger formats; accordingly we have '*Morals*' emphasized once more in the first line, and at line 578 '*good-breeding*', now hyphenated, is given some stress. At the same time the hyphen is removed from 'disapprov'd' in the same line, and in the next, 579, 'superiour' is spelt with a second 'u' for the first time, a highly atypical spelling for Pope or printer. The other changes are all to punctuation: line 566 loses the terminal comma it gained in the first edition; in 573 we lose again the commas

---

[16] Moxon, *Mechanick exercises on the whole art of printing*, ed. H. Davis and H. Carter, 2nd edn. (Oxford, 1962; first printed in parts, 1683), p. 216.

Learn then what *Morals* Criticks ought to fhow,
For 'tis but *half* a *Judge's Task*, to *Know*.
'Tis not enough, Wit, Art, and Learning join;
In all you fpeak, let Truth and Candor fhine:
That not alone what to your *Judgment's* due,
All may allow; but feek your *Friendfhip* too.

Be *filent* always when you *doubt* your Senfe;
And fpeak, tho' *fure*, with *feeming Diffidence*:
Some pofitive, perfifting Fops we know,
That, if *once wrong*, will needs be *always fo*;
But you, with Pleafure own your Errors paft,
And make, each Day, a *Critick* on the laft.

'Tis not enough your Counfel ftill be *true*;
Blunt Truths more Mifchief than *nice Falfhoods* do;
Men muft be *taught* as if you taught them not;
And things *ne'er known* propos'd as Things *forgot*.
Without *Good Breeding*, Truth is not approv'd;
That only makes *Superior* Senfe *belov'd*.

Be Niggards of Advice on no Pretence;
For the *worft Avarice* is that of *Senfe*.
With mean Complacence ne'er betray your Truft,
Nor be fo *Civil* as to prove *Unjuft*:
Fear not the Anger of the Wife to raife;
Thofe beft can *bear Reproof*, who *merit Praife*.

'Twere well, might Criticks ftill this Freedom take;
But *Appius* reddens at each Word you fpeak,

(a)

Learn then what *Morals* Criticks ought to fhow,   1
For 'tis but half a Judge's Task, to Know.
'Tis not enough, Wit, Art, and Learning join;
In all you fpeak, let Truth and Candor fhine:
That not alone what to your Judgment's due,
All may allow; but feek your Friendfhip too.

Be filent always when you doubt your Senfe;
And fpeak, tho' fure, with feeming Diffidence:
Some pofitive, perfifting Fops we know,
That, if once wrong, will needs be always fo;
But you, with Pleafure own your Errors paft,
And make, each Day, a Critick on the laft.

'Tis not enough your Counfel ftill be true;
Blunt Truths more Mifchief than nice Falfhoods do;
Men muft be taught as if you taught them not;
And things ne'er known propos'd as Things forgot.
Without good Breeding, Truth is not approv'd;   2
That only makes Superior Senfe belov'd.   3

Be Niggards of Advice on no Pretence;
For the worft Avarice is that of Senfe.

With mean Complacence ne'er betray your Truft,
Nor be fo Civil as to prove Unjuft:
Fear not the Anger of the Wife to raife;
Thofe beft can bear Reproof, who merit Praife.

'Twere well, might Criticks ftill this Freedom take;
But *Appius* reddens at each Word you fpeak,   4

(b)

round 'each day' that were added in the second edition; and in 583 a period takes the place of a colon.

With the so-called death-bed edition of 1743–4 we come face to face with the problem of Warburton's influence in the last years of Pope's life. I shall be prepared to argue that in matters of accidentals Warburton abandoned his own preference for a complex and varied type-page in favour of Pope's classic simplicity, but his influence on the substantives and the punctuation of Pope's text is much more sinister. R. H. Griffith has shown that a couplet (*Twickenham*, lines 152–3) which was transposed in this edition was moved back to its original place by

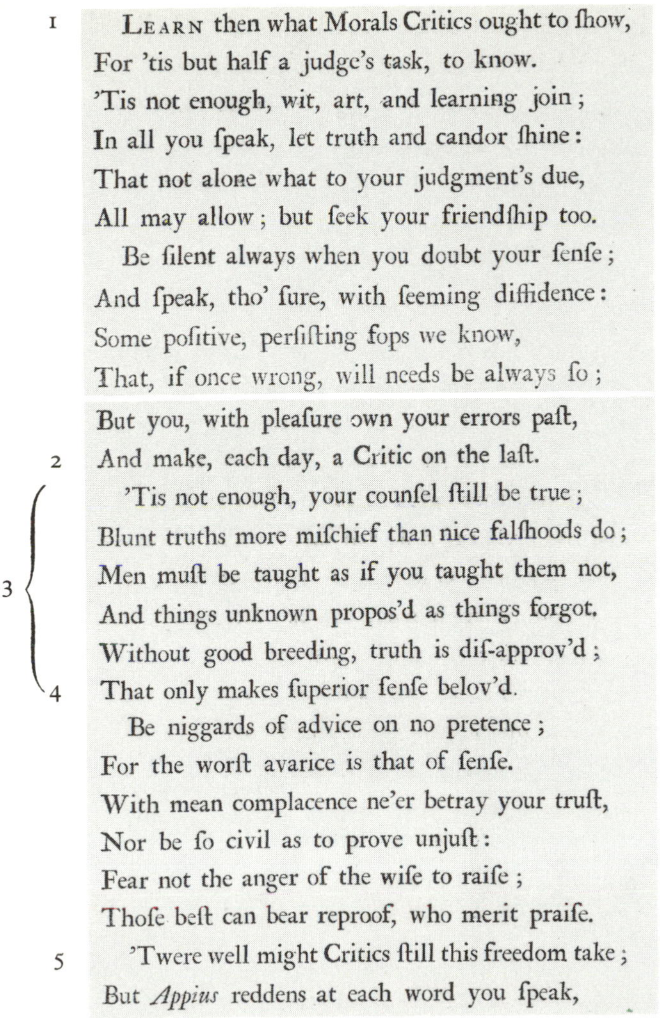

1 LEARN then what Morals Critics ought to ſhow,
For 'tis but half a judge's task, to know.
'Tis not enough, wit, art, and learning join;
In all you ſpeak, let truth and candor ſhine:
That not alone what to your judgment's due,
All may allow; but ſeek your friendſhip too.

Be ſilent always when you doubt your ſenſe;
And ſpeak, tho' ſure, with ſeeming diffidence:
Some poſitive, perſiſting fops we know,
That, if once wrong, will needs be always ſo;

But you, with pleaſure own your errors paſt,
2 And make, each day, a Critic on the laſt.

'Tis not enough, your counſel ſtill be true;
Blunt truths more miſchief than nice falſhoods do;
3 Men muſt be taught as if you taught them not,
And things unknown propos'd as things forgot.
Without good breeding, truth is diſ-approv'd;
4 That only makes ſuperior ſenſe belov'd.

Be niggards of advice on no pretence;
For the worſt avarice is that of ſenſe.
With mean complacence ne'er betray your truſt,
Nor be ſo civil as to prove unjuſt:
Fear not the anger of the wiſe to raiſe;
Thoſe beſt can bear reproof, who merit praiſe.

5 'Twere well might Critics ſtill this freedom take;
But *Appius* reddens at each word you ſpeak,

(*c*)

FIG. 73. Three editions of *Essay on criticism*: (*a*) second (1713), pp. 28–9 (Bodl. G. Pamph. 1283(30); 176 × 105); (*b*) third (1713), pp. 28–9 (Bodl. 2799 f. 438; 160 × 87); (*c*) quarto *Works* (1717), pp. 105–6 (Bodl. Vet. A4 d. 140; 290 × 218)

means of a cancel in Warburton's edition of 1764, twenty years later.[17] The implication must be that Warburton was responsible for moving the couplet in the first place as part of his attempt to rationalize Pope's poems; and if we admit that, we cannot be sure where Warburton is revising and where Pope. There is no

[17] Griffith, 'Early Warburton? or late Warburton?', *Studies in English* (University of Texas), xx (1940), 123–31. See also F. W. Bateson's argument that Warburton was similarly responsible for the transposition of passages in the death-bed edition of the *Ethic epistles* (*Twickenham*, iii/2, pp. 5–12), and R. W. Rogers, 'Notes on Pope's collaboration with Warburton in preparing a final edition of the *Essay on man*', *PQ* xxvi (1947), 358–66, though his account must be modified in the light of the manuscript evidence discussed in the next chapter.

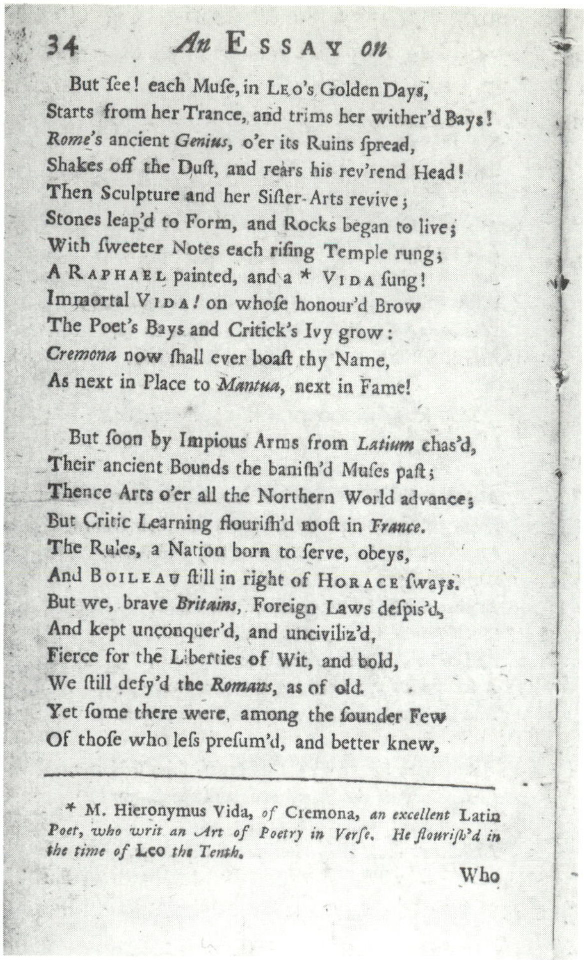
34 *An* ESSAY *on*

But ſee! each Muſe, in LEO's Golden Days,
Starts from her Trance, and trims her wither'd Bays!
Rome's ancient *Genius*, o'er its Ruins ſpread,
Shakes off the Duſt, and rears his rev'rend Head!
Then Sculpture and her Siſter-Arts revive;
Stones leap'd to Form, and Rocks began to live;
With ſweeter Notes each riſing Temple rung;
A RAPHAEL painted, and a * VIDA ſung!
Immortal VIDA! on whoſe honour'd Brow
The Poet's Bays and Critick's Ivy grow:
Cremona now ſhall ever boaſt thy Name,
As next in Place to *Mantua*, next in Fame!

But ſoon by Impious Arms from *Latium* chas'd,
Their ancient Bounds the baniſh'd Muſes paſt;
Thence Arts o'er all the Northern World advance;
But Critic Learning flouriſh'd moſt in *France*.
The Rules, a Nation born to ſerve, obeys,
And BOILEAU ſtill in right of HORACE ſways.
But we, brave *Britains*, Foreign Laws deſpis'd,
And kept unconquer'd, and unciviliz'd,
Fierce for the Liberties of Wit, and bold,
We ſtill defy'd the *Romans*, as of old.
Yet ſome there were, among the ſounder Few
Of thoſe who leſs preſum'd, and better knew,

* M. Hieronymus Vida, *of* Cremona, *an excellent* Latin *Poet, who writ an Art of Poetry in Verſe. He flouriſh'd in the time of* Leo *the Tenth.*

Who

FIG. 74. Use of caps. and small caps. in third edition of *Essay on criticism* (1713), p. 34 (Bodl. 2799 f. 438; 160 × 87)

doubt that Pope in some sense approved the result, but whether he was at this time in full possession of his poetic faculties, or whether he was subject to undue influence, remain open questions.

Typographically the death-bed editions revert to Pope's preferred style without the use of italic (except for such uses as to distinguish colloquial phrases). If we compare this edition of the *Essay on criticism* with the octavo *Works* (Figure 75), we first note the substantive revisions in lines 562 and 564 and must wonder whether they are due to Warburton's influence. In line 562 'wit, art and learning' become 'taste, judgment, learning' and below 'what to your judgment's due' becomes 'what to your sense is due'. This certainly suggests the imposition of Warburton's terminology on Pope. The revisions for emphasis are fairly straightforward; 'MORALS' in the first line reverts to caps. and small caps., and 'Judge' in the next line is capitalized as such generic figures normally are in

Pope's later works; so is the virtue 'Good Breeding' in line 576. This is all right and proper, but I must confess I am still not satisfied with the capital that 'Critic' in line 571 has had since 1717. The one spelling change is 'superior' for 'superiour' again in line 577, while the general reduction of italic extends even to 'Appius' in line 585.

If we try to see the changes in accidentals through these editions in perspective, I think two features stand out. In punctuation, most changes are clear improvements while a few show indecision in the face of a real difficulty; the line

And make each day a Critic on the last

is a good example. Unpunctuated, one is tempted to read it as if one was to make each day into a critic; but if one marks off 'each day' with commas one breaks the rhythm of the line. It would help if one could change the spelling to the French form, 'critique', as Pope did in the *Dunciad* between the editions of 1728 and 1729.[18] The typographical changes—first the abandonment of profuse italic, then of initial capitals for common nouns, and finally the abandonment of italic for proper nouns—are more fundamental. I hope to show that these are indeed Pope's own typographical decisions; if you will accept that for the moment, I think it is quite clear that the *Twickenham* edition, in following the orthodoxy of taking the accidentals of the first edition, leaves us with a text which, though historically interesting, entirely misrepresents Pope's later intentions and ignores his detailed revision of punctuation. Unfortunately, the alternative policy, as represented by Herbert Davis's reprinting of the death-bed edition, leaves us with Warburton's interference and the unresolved inconsistencies of that edition.

This inconsistency can be seen in the treatment of 'Nature' on the last page of sheet B and the first of C (Figure 76). At the beginning of the paragraph on page 8 '*Nature*' should be in caps. and small caps., not italic, while the second occurrence on page 9 should by the analogy of previous editions be in ordinary lower case with a capital—there is no reason for stressing it on its second occurrence.[19] Similarly in this first sheet italic is still used for proper names (Figure 77), though it is removed in the rest of the poem: an editor must surely standardize here. Later in the poem (Figure 78), Dryden's name appears in caps. and small caps. (so does Leo later on); one suspects Pope would have wanted a similar honour granted to writers like Horace and Homer, who had been treated this way in the earlier editions—in fact a feature of Pope's later revisions is the restoration of caps. and small caps. which had been removed in the *Works* of 1717. The change has not been properly worked out in this case.

This page also shows '*Extremes*' italicized as the keyword of a new paragraph. This is probably to link the text with the progress of Warburton's commentary below, but again the convention is not systematically executed. Perhaps it is

---

[18] See Fig. 81. *OED* records the occurrence in the *Essay on criticism* as the first in this sense with the '-ick' spelling, and the *Dunciad* occurrence as early evidence of the change to the French spelling.
[19] Warburton's edn. of 1751 corrects the first, but preserves caps. and small caps. for the second.

LEARN then what Morals Critics ought to fhow,
For 'tis but half a judge's task, to know.
'Tis not enough, wit, art, and learning join;
In all you fpeak, let truth and candor fhine:
That not alone what to your judgment's due,
All may allow; but feek your friendfhip too.

Be filent always when you doubt your fenfe;
And fpeak, tho' fure, with feeming diffidence:
Some pofitive, perfifting fops we know,
That, if once wrong, will needs be always fo;

But you, with pleafure own your errors paft,
And make, each day, a Critic on the laft.

'Tis not enough, your counfel ftill be true;
Blunt truths more mifchief than nice falfhoods do;
Men muft be taught as if you taught them not,
And things unknown propos'd as things forgot.
Without good breeding, truth is dif-approv'd;
That only makes fuperior fenfe belov'd.

Be niggards of advice on no pretence;
For the worft avarice is that of fenfe.
With mean complacence ne'er betray your truft,
Nor be fo civil as to prove unjuft:
Fear not the anger of the wife to raife;
Thofe beft can bear reproof, who merit praife.

'Twere well might Critics ftill this freedom take;
But *Appius* reddens at each word you fpeak,

(a)

LEARN then what *Morals* Critics ought to fhow,
For 'tis but half a judge's tafk, to know.
'Tis not enough, wit, art, and learning join;
In all you fpeak, let truth and candour fhine: 565
That not alone what to your judgment's due
All may allow; but feek your friendfhip too.

Be filent always when you doubt your fenfe;
And fpeak, tho' fure, with feeming diffidence:

Some pofitive, perfifting fops we know,   570
That, if once wrong, will needs be always fo;
But you, with pleafure own your errors paft,
And make each day a Critic on the laft.
   'Tis not enough, your counfel ftill be true;
Blunt truths more mifchief than nice falfhoods do;   575
Men muft be taught as if you taught them not,
And things unknown propos'd as things forgot.
Without *good-breeding*, truth is difapprov'd;
That only makes fuperiour fenfe belov'd.
   Be niggards of advice on no pretence;   580
For the worft avarice is that of fenfe.
With mean complacence ne'er betray your truft,
Nor be fo civil as to prove unjuft.
Fear not the anger of the wife to raife;
Thofe beft can bear reproof, who merit praife.   585
   'Twere well might Critics ftill this freedom take;
But *Appius* reddens at each word you fpeak,

(b)

enough to say that consistent typography has not been achieved in this edition, and it looks as though an editor ought to make some changes to avoid inconsistencies. In fact, so far as I am aware, no editor has made any record of the changes in accidentals, even though it is clear once one studies them that Pope was responsible.

CAPITALIZATION OF NOUNS

Let us return to the systematic capitalization of common nouns which Pope abandoned in the *Works* of 1717. The origin of this capitalization is obscure, and it is not limited to England: the trend is also found in Holland and to a lesser

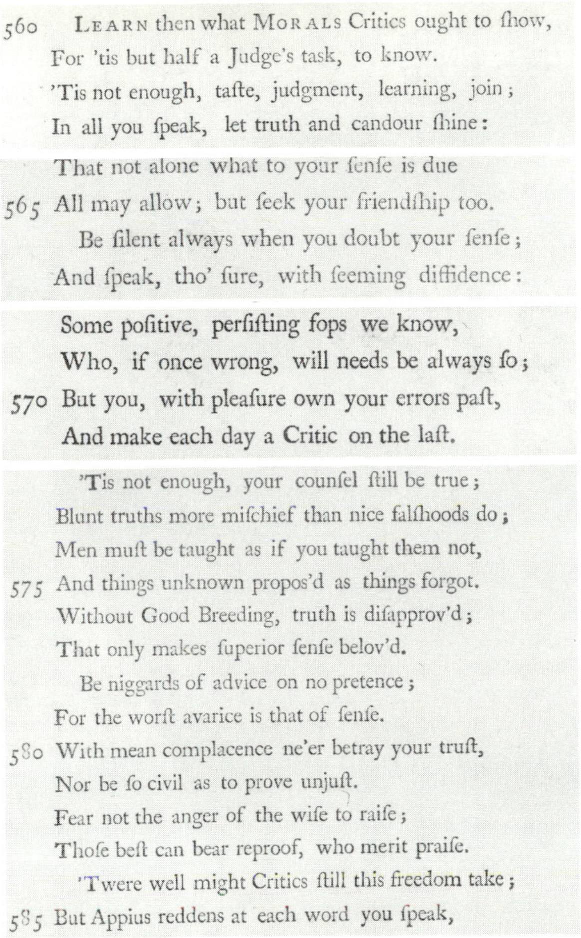

560 LEARN then what MORALS Critics ought to fhow,
    For 'tis but half a Judge's task, to know.

    'Tis not enough, tafte, judgment, learning, join;
    In all you fpeak, let truth and candour fhine:

    That not alone what to your fenfe is due
565 All may allow; but feek your friendfhip too.

    Be filent always when you doubt your fenfe;
    And fpeak, tho' fure, with feeming diffidence:

    Some pofitive, perfifting fops we know,
    Who, if once wrong, will needs be always fo;
570 But you, with pleafure own your errors paft,
    And make each day a Critic on the laft.

    'Tis not enough, your counfel ftill be true;
    Blunt truths more mifchief than nice falfhoods do;
    Men muft be taught as if you taught them not,
575 And things unknown propos'd as things forgot.
    Without Good Breeding, truth is difapprov'd;
    That only makes fuperior fenfe belov'd.

    Be niggards of advice on no pretence;
    For the worft avarice is that of fenfe.
580 With mean complacence ne'er betray your truft,
    Nor be fo civil as to prove unjuft.
    Fear not the anger of the wife to raife;
    Thofe beft can bear reproof, who merit praife.

    'Twere well might Critics ftill this freedom take;
585 But Appius reddens at each word you fpeak,

(c)

FIG. 75. Three later editions of *Essay on criticism*: (*a*) quarto *Works* (1717), pp. 105–6 (Bodl. Vet. A4 d. 140; 290 × 218); (*b*) octavo *Works* I (1736), pp. 127–8 (Bodl. Radcliffe. f. 243; 170 × 101); (*c*) death-bed quarto edition (1743), pp. 47–50 (Bodl. CC 51(3) Art; 255 × 191)

degree in France, and in Latin as well as the vernacular.[20] English grammarians seem to have been opposed to the practice, despite their printers. Thus Guy Miège in *The English grammar* (1688), p. 126, says that capitals are proper at the beginning of 'any Noun that has an Emphasis with it, or that is predominant', though his printer capitalizes all nouns and some adjectives. (He also opposes the use of italic for frequent emphasis 'as if the Reader had not Sense to apprehend it, without so visible and palpable a Distinction'.) John Jones in *Practical phonography* (1701), pp. 140–1, gives the traditional rules for the use of capitals and concludes:

[20] I am grateful to the late Wytze Hellinga for information on Dutch practice; for the French, see R. Laufer, *Introduction à la textologie* (Paris, 1972), p. 71, referring to *Le Diable boiteux* of 1707.

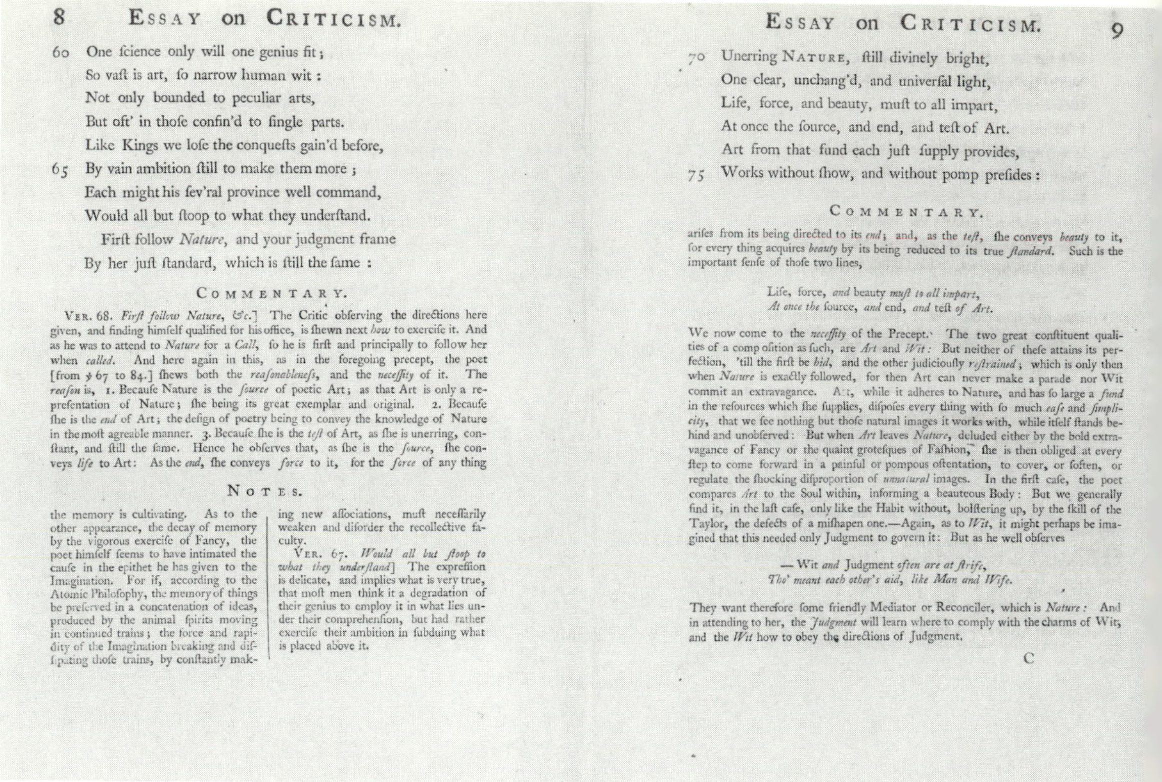

8   ESSAY on CRITICISM.

60 One science only will one genius fit;
  So vast is art, so narrow human wit:
  Not only bounded to peculiar arts,
  But oft' in those confin'd to single parts.
  Like Kings we lose the conquests gain'd before,
65 By vain ambition still to make them more;
  Each might his sev'ral province well command,
  Would all but stoop to what they understand.
    First follow *Nature*, and your judgment frame
  By her just standard, which is still the same:

COMMENTARY.

VER. 68. *First follow Nature, &c.*] The Critic observing the directions here given, and finding himself qualified for his office, is shewn next *how* to exercise it. And as he was to attend to *Nature* for a *Call*, so he is first and principally to follow her when *called*. And here again in this, as in the foregoing precept, the poet [from ℣ 67 to 84.] shews both the *reasonableness*, and the *necessity* of it. The *reason* is, 1. Because Nature is the *source* of poetic Art; as that Art is only a representation of Nature; she being its great exemplar and original. 2. Because she is the *end* of Art; the design of poetry being to convey the knowledge of Nature in the most agreable manner. 3. Because she is the *test* of Art, as she is unerring, constant, and still the same. Hence he observes that, as she is the *source*, she conveys *life* to Art: As the *end*, she conveys *force* to it, for the *force* of any thing

NOTES.

the memory is cultivating. As to the other appearance, the decay of memory by the vigorous exercise of Fancy, the poet himself seems to have intimated the cause in the epithet he has given to the Imagination. For if, according to the Atomic Philosophy, the memory of things be preserved in a concatenation of ideas, produced by the animal spirits moving in continued trains; the force and rapidity of the Imagination breaking and dissipating those trains, by constantly making new associations, must necessarily weaken and disorder the recollective faculty.
VER. 67. *Would all but stoop to what they understand*] The expression is delicate, and implies what is very true, that most men think it a degradation of their genius to live in what lies under their comprehension, but had rather exercise their ambition in subduing what is placed above it.

ESSAY on CRITICISM.   9

70 Unerring NATURE, still divinely bright,
  One clear, unchang'd, and univerfal light,
  Life, force, and beauty, must to all impart,
  At once the source, and end, and test of Art.
  Art from that fund each just supply provides,
75 Works without show, and without pomp presides:

COMMENTARY.

arises from its being directed to its *end*; and, as the *test*, she conveys *beauty* to it, for every thing acquires *beauty* by its being reduced to its true *standard*. Such is the important sense of those two lines,

Life, force, *and* beauty *must to all impart,*
At once the source, *and end, and test of Art.*

We now come to the *necessity* of the Precept. The two great constituent qualities of a composition as such, are *Art* and *Wit*: But neither of these attains its perfection, 'till the first be *hid*, and the other judiciously *restrained*; which is only then when *Nature* is exactly followed, for then Art can never make a parade nor Wit commit an extravagance. Art, while it adheres to Nature, and has so large a fund in the resources which she supplies, disposes every thing with so much *ease* and *simplicity*, that we see nothing but those natural images it works with, while itself stands behind and unobserved: But when *Art* leaves *Nature*, deluded either by the bold extravagance of Fancy or the quaint grotesques of Fashion, she is then obliged at every step to come forward in a painful or pompous ostentation, to cover, or soften, or regulate the shocking disproportion of *unnatural* images. In the first case, the poet compares *Art* to the Soul within, informing a beauteous Body: But we generally find it, in the last case, only like the Habit without, bolstering up, by the skill of the Taylor, the defects of a misshapen one.—Again, as to *Wit*, it might perhaps be imagined that this needed only Judgment to govern it: But as he well observes

— *Wit and Judgment often are at strife,*
Tho' meant each other's aid, like Man and Wife.

They want therefore some friendly Mediator or Reconciler, which is *Nature*: And in attending to her, the *Judgment* will learn where to comply with the charms of Wit, and the *Wit* how to obey the directions of Judgment.

C

FIG. 76. Inconsistent use of italic and caps. and small caps. in death-bed quarto *Essay on criticism* (1743), pp. 8–9, lines 68 and 70 (Bodl. CC 51(3) Art; 255 × 380)

Note, That in Print, they generally put *great* or *capital Letters*, in the Beginning of the common Names of Things, to adorn it; but that is not yet become customary in Writing, tho' it daily gains ground . . . But 'tis unsufferable to write *capital Letters* in the Beginning of *Verbs*, *Adjectives*, &c. [So much for Milton!]

Thomas Dyche in his *Guide to the English tongue* (1707), p. 118, says, "'Tis esteem'd Ornamental to begin any Substantive in the Sentence with a Capital, if it bear any considerable stress of the Author's Sense upon it, to make it the more Remarkable and Conspicuous', but adds in a note in smaller type, "'Tis grown Customary in Printing to begin every Substantive with a Capital, but in my Opinion 'tis unnecesary, and hinders that remarkable Distinction intended by the Capitals'.

  I believe it is to Pope and Gay that we owe the abandonment of this practice, though to prove it would call for a more systematic search than I have yet been able to make. My present evidence is set out in Table 17, where I have italicized those editions where systematic capitalization has been abandoned. It is not only Pope's *Works* which abandon capitalization in 1717 but also the collection *Poems*

ESSAY on CRITICISM. 5

Some are bewilder'd in the maze of ſchools,
And ſome made coxcombs Nature meant but fools.
In ſearch of wit theſe loſe their common ſenſe,
And then turn Critics in their own defence :
Each burns alike, who can, or cannot write,
Or with a Rival's, or an Eunuch's ſpite.
All fools have ſtill an itching to deride,
And fain would be upon the laughing ſide.
If *Mævius* ſcribble in *Apollo*'s ſpight,
There are, who judge ſtill worſe than he can write.
Some have at firſt for Wits, then Poets paſt,
Turn'd Critics next, and prov'd plain fools at laſt.

COMMENTARY.

wit which is not regulated by *ſenſe*; or *falſe politeneſs* which is ſolely regulated by the *faſhion*. Both theſe ſorts, who have their Judgments thus doubly depraved, the poet obſerves are naturally turned to cenſure and reprehenſion; only with this difference, that the *Dunce* always affects to be on the *reaſoning*, and the *Fool* on the *laughing* ſide.—And thus, at the ſame time, our author proves the truth of his inductory obſervation, *that the number of bad Critics is vaſtly ſuperior to that of bad Poets.*

VER. 28. *In* ſearch of Wit *theſe loſe their* common ſenſe] This obſervation is extremely juſt. *Search of wit* is not only the *occaſion*, but the *efficient cauſe* of *loſs of common ſenſe*. For *wit* conſiſting in chuſing out, and ſetting together, ſuch ideas from whoſe likeneſſes pleaſant pictures are made in the fancy; the *Judgment* thro' an habitual *ſearch* of *Wit*, loſes by degrees its faculty of ſeeing the true relations of things; in which conſiſts the exercise of *common ſenſe*.

VER. 36. *Some have at firſt for Wits, &c.*] The poet having thus enumerated, in this account of the nature of Judgment and its ſeveral depravations, the ſeveral ſort of *bad Critics*, and ranked them into two general Claſſes; as the firſt ſort, namely thoſe ſpoiled by *falſe learning*, are but few in compariſon of the other, and likewiſe come leſs within his main view (which is *poetical Criticiſm*) but keep groveling at the bottom amongſt *words* and *letters*, he thought it here ſufficient juſt to have mentioned them, propoſing to do them right elſewhere. But thoſe ſpoiled by *falſe taſte* are in-

FIG. 77. Use of italic for proper names in death-bed quarto *Essay on criticism* (1743), p. 5 (Bodl. CC 51(3) Art; 255 × 191)

34 ESSAY on CRITICISM.

But when loud ſurges laſh the ſounding ſhoar,
The hoarſe, rough verſe ſhould like the torrent roar.
When Ajax ſtrives, ſome rock's vaſt weight to throw,
The line too labours, and the words move ſlow;
Not ſo, when ſwift Camilla ſcours the plain,
Flies o'er th' unbending corn, and ſkims along the main.
Hear how Timotheus' vary'd lays ſurprize,
And bid alternate paſſions fall and riſe!
While, at each change, the ſon of Libyan Jove
Now burns with glory, and then melts with love:
Now his fierce eyes with ſparkling fury glow,
Now ſighs ſteal out, and tears begin to flow:
Perſians and Greeks like turns of nature found,
And the World's victor ſtood ſubdu'd by Sound!
The pow'r of Muſic all our hearts allow,
And what Timotheus was, is DRYDEN now.
Avoid *Extremes*; and ſhun the fault of ſuch,
Who ſtill are pleas'd too little or too much.

COMMENTARY.

VER. 384. *Avoid extremes, &c.*] Our author is now come to the *laſt cauſe of wrong Judgment*, PARTIALITY; the parent of the immediately preceding cauſe, a *bounded capacity*: Nothing ſo much narrowing and contracting the mind as *prejudices* entertained for or againſt things or perſons. This, therefore, as the main root of all the foregoing, he proſecutes at large from ℣ 383 to 473.

Firſt he, *previouſly* expoſes that capricious turn of miud, which, by running men

NOTES.

VER. 374. *Hear how* Timotheus, *&c.*] *See* Alexander's Feaſt, *or* the Power of Muſic; *an Ode by Mr.* Dryden.

FIG. 78. Use of caps. and small caps. for proper names in death-bed quarto *Essay on criticism* (1743), p. 34 (Bodl. CC 51(3) Art; 255 × 191)

on several occasions, better known by Ault's title *Pope's own miscellany*; Pope imposes his new convention on his contributors. Since he is in the middle of the translation of the *Iliad*, he does not change its typography, but the duodecimo reprint, of which the first volume was already printed by September 1719, shows the new style, which is also used for the *Odyssey*. It is probably only a coincidence that Thomas Johnson wrote in the preface to his Dutch piracy begun in 1718: 'it has been thought necessary for its greater beauty to leave out that prodigious number of Capitals, which disfigures the page, by an abuse introduced, thro' want of taste, into English books more than any other.'[21]

All these works were printed by Bowyer, but I must stress that the change is not due to him; all the other works printed by him at this time that I have examined follow the traditional style. Reprints of Pope's earlier separate works are brought into line in 1719. When he came to edit the posthumous works of Parnell

[21] Quoted by R. H. Griffith, 'A piracy of Pope's *Iliad*', *SP* xxviii (1931), 740.

TABLE 17. Works Printed without Initial Capitals (italicized)

	Pope	Gay	Others
1714		Shepherd's week, 1st & 2nd (Watts)	
1715	Iliad I (Bowyer)	What d'ye call it, 1st & 2nd	
1716	Iliad II (Bowyer)	Trivia (25 Jan, Bowyer) *Trivia*, 2nd (9 June) (Bowyer) *What d'ye call it*, 3rd (? 9 June) [? *Shepherd's week*, 3rd] (Watts)	
1717	Iliad III (Bowyer) *Works* (Bowyer) *Poems on several occasions* (Bowyer)	Three hours after marriage Two epistles	
1718	Iliad IV (Bowyer) Rape of the lock, 5th		
1719	*Essay on criticism*, 6th *Ode for musick*, 3rd *Eloisa to Abelard*, 2nd (Bowyer) *Windsor Forest*, 4th (Bowyer)		
1720	Iliad V, VI (Bowyer) *Iliad* 12⁰ (Bowyer) *Miscellaneous Poems* (Bowyer)	*Poems on several occasions* (Watts)	
1721	Parnell, Poems, ed. Pope	Panegyrical epistle (Bowyer)	*Addison, Works*, ed. Tickell (Watts)
1722		*An epistle to her grace* (Watts)	
1723	Sheffield, Works, ed. Pope		*Fenton, Mariamne* (Watts)
1724		*Captives* (Watts)	
1725	*Shakespeare*, ed. Pope (Watts) *Odyssey* I–III (Watts)		*Milton* ed. Fenton (Watts)

and Sheffield Pope did not impose his style, but the Shakespeare and the *Odyssey*, both printed by Watts, follow the new convention. The change is clear-cut and systematic.

The role of Gay is more difficult to assess, but the evidence is, if possible, more convincing still. In June 1716 the second edition of *Trivia* was printed by Bowyer for Lintot, and at about the same time the third edition of *The what d'ye call it* was produced for Lintot by an unidentified printer. The third edition of the *The shepherd's week* (Foxon G73) printed by John Watts for Tonson, but published by Rebecca Burleigh, has a false date 'MDCCXIV'; but it is certainly no later than 1717, and I suspect it was also printed in 1716.[22] We therefore have three reprints, produced by three different printers, for two booksellers. Now reprints normally follow copy precisely, but these three all abandon capitalization. There is no common factor in their production but Gay, so one must assume that the change was made on his instructions. We know from his poem in Lintot's miscellany of 1712 that he had some interest in typography, but it cannot be coincidence that he made this departure from convention at the time when Pope must have been about to put his *Works* of 1717 to press. Either Pope and Gay agreed to make this change, or else Pope persuaded Gay to try a small-scale experiment before he committed himself to using the new style in his *Works*. The fact that Gay's next two works revert to the old style may suggest the latter, though (as I show in Appendix B below) Gay's experiment was more radical in also abandoning italic than the convention adopted by Pope in 1717.

The publication of Gay's subscription quarto of 1720, *Poems on several occasions*, is of interest because as far as I can see it is the first work printed by John Watts to use the new style with the exception of the earlier experiment with *The shepherd's week*. The work of Watts is significant because of his importance as a printer and because his work for Tonson is of such literary interest. In fact the Tonsons' concern with the text of Milton is relevant here. The early editions of *Paradise lost* form a normal ancestral series and follow quite carefully the idiosyncrasies of the first and second editions printed in Milton's lifetime, though inevitably some corruption and modernization creeps in; only the illustrated folio of 1695 shows marks of editorial intervention. After Tonson had set up Watts as his printer, there was a determined effort in 1705 to purify the text of the poem by collating it with the second edition of 1674, and restoring many of Milton's archaic spellings and his own idiosyncratic capitalization, which includes capitalizing some verbs. After the younger Tonson took over the business in 1718 a second attempt was made to purify the text, collating it with the first edition.[23] Then, for the illustrated quarto of 1720, Thomas Tickell for the first time undertook a discreet modification of Milton's capitalization, apparently restricting the use of capitals to words which bore a stress in the reading.

Having done that, Tickell went on to edit Addison's works in four quarto volumes, and in the first two sheets (B–C) of the poems in the first volume he tried to reduce Addison's heavy capitalization on the same principles (often capitalizing adjective rather than noun, as in 'Prophetick numbers', 'Human gore'). After these first two sheets, he abandoned the attempt, and from then on capitals

---

[22] Vinton Dearing has privately suggested that the date should be 'MDCCXVI' and that the last two letters are transposed. He discusses the complexities of the text in his edn. of Gay's *Poetry and prose* (Oxford, 1974), ii. 511–12.

[23] See a letter of John Hughes to Tonson dated 17 Aug. 1719 (BL Add. MS 28275, fo. 61), excerpted in *TLS*, 22 Feb. 1934, p. 126. The edition of 1711 is noteworthy only for its undistinguished and unnecessary emendations, many of which are preserved by 1719; the early editions were used only eclectically. I am indebted to A. G. Lilly for the use of his unpublished work on the text.

are reserved for words like 'Saint', 'Muse', 'Harmony'; as a result Addison appears in modernized dress.[24]

The other author whose work appears in the third column of Table 17 is Elijah Fenton, who worked with Pope on his edition of Shakespeare and, of course, collaborated with Pope and Broome on the *Odyssey* translation. His poem 'Florelio' had been printed with Pope's *Eloisa to Abelard* in 1719 (dated 1720) in Pope's decapitalized style; unlike Broome he took this for his own, and his drafts for the *Odyssey*, which will be discussed shortly, show fewer capitals than Pope's. It is therefore not surprising to find his play *Mariamne* printed in 1723 without capitals, and in 1725 he edited Milton again for Tonson, this time completely modernizing spelling and other accidentals. His edition of Waller for Tonson in 1729 follows the same principles. It is worth pointing out that the younger Tonson does not seem to have imposed this style on his authors willy-nilly, while Watts himself seems to have had a definite preference for old-fashioned capitalization; his *Musical miscellany* of 1729–30, for example, follows the traditional style.

Although the evidence shows that Pope abandoned capitalization of nouns in print, his later manuscripts, particularly in his formal print hand, keep close to the pattern he had established early in life, though unstressed nouns in subordinate positions are often not capitalized. Two aspects of this deserve consideration. The first is theoretical; textual criticism since Greg has tended to assume that an author's manuscripts will normally be more informative about what accidentals an author preferred than printed texts which have—so the argument goes—been modified by the printer. With Pope the situation is quite different; his later manuscripts show a system of accidentals which he rejected for his printed works. Nor does the further doctrine of textual criticism apply, that the accidentals of the first edition will normally be closer to what the author wished than later editions which have been corrupted in the printing house. By contrast, since Pope's manuscripts do not represent his intentions for his printed text, he was led to a continual refinement of accidentals in subsequent editions of his works.

The second and practical result of the fact that Pope's manuscripts do not represent the way he wished his works to be printed, is that either his printers had to try to standardize his copy or else to follow its inconsistencies and submit to subsequent correction. I think the first course is seen in Watts's printing of the *Odyssey* and the second in Pope's long association with John Wright.

A page of copy for the opening of the fourth book of the *Odyssey* in Fenton's hand (Figure 79) shows the nature of the problem. I have already remarked on Fenton's ready adoption of the uncapitalized style, and this is found in his manuscript for the *Odyssey*. Only 'Son' in line 7, 'Nymph' in line 12, and 'regal Line' in line 18 are capitalized in the manuscript but not in the first edition, and they are all cases where an honorific capital might be appropriate. By contrast, Pope's

---

[24] Tickell's own manuscripts and printed poems are less heavily capitalized from this period, though he seems not to have achieved a consistent practice.

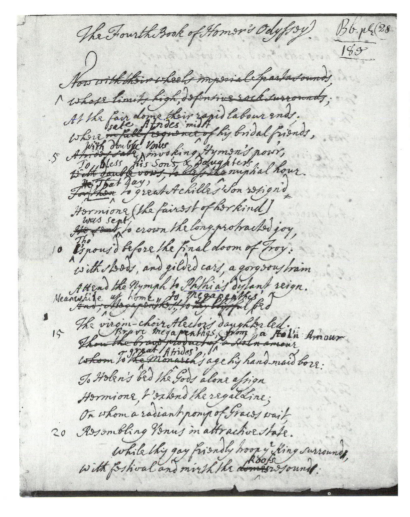

FIG. 79. Printer's copy for opening of *Odyssey*, book IV in Fenton's hand, with Pope's corrections (BL Add. MS 4809 fo. 28ʳ; 246 × 183)

revisions in his print hand capitalize most nouns: 'Vows' in line 5, 'Sons & Daughters' in line 6, 'Amour' in line 15, 'Roofs' in the last line. (For the absence of capitals in unstressed nouns, see 'That day' in line 7 and 'at home' in line 13.) Proper nouns are not underlined for italics (except '*Phthia*'s' in line 12) but they are printed in italics in the first edition. Since we can see from the notation in the top right-hand corner that this manuscript served as printer's copy, we must assume that John Watts was expected to italicize proper names as a matter of routine; and I think it follows that he was also told not to follow copy in capitalizing nouns. No doubt there must have been some agreement as to what words were to be granted an honorific capital—'Gods', 'Graces', and 'King' in this passage—and some adjustments made in proof.

Another page of copy, for the seventeenth book of the *Odyssey* (Figure 80), shows Pope rewriting in his informal hand a passage he had already once revised.

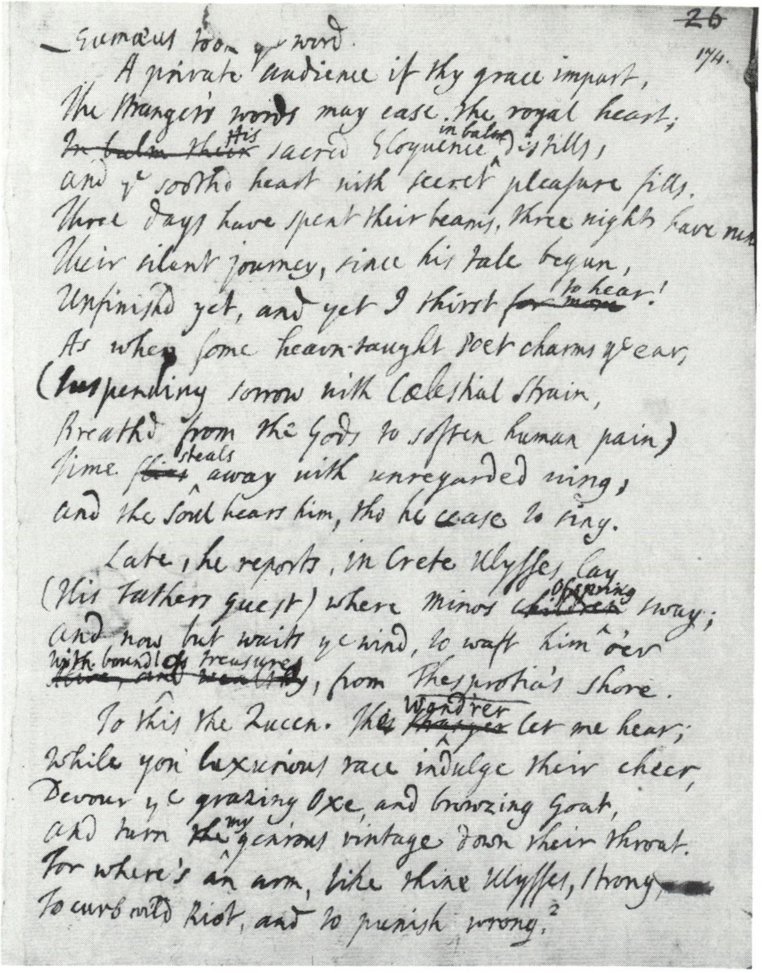

FIG. 80. Printer's copy for *Odyssey*, book XVII in Pope's informal hand (BL Add. MS 4809 fo. 174^r; 230 × 175)

Here (as in his letters) Pope's capitalization is much more fitful, but in the first twelve lines we see 'Strangers' and 'Eloquence' in lines 3 and 4, 'Poet' in line 9 and 'Cœlestial Strain' in 10, 'Gods' in 11 and 'Soul' in 13. Of these, only 'Poet' and 'Gods' are capitalized in the first edition—which seems a good decision. These manuscripts were written after Pope had abandoned the use of capitals in his printed works, but they show that his old habits of capitalization survived and had to be modified by the printer.

THE *DUNCIAD* AND ITALIC

A simple example of the way Pope's old habits might mislead his printer is shown in Jonathan Richardson's marking of a textual change in the first edition of the

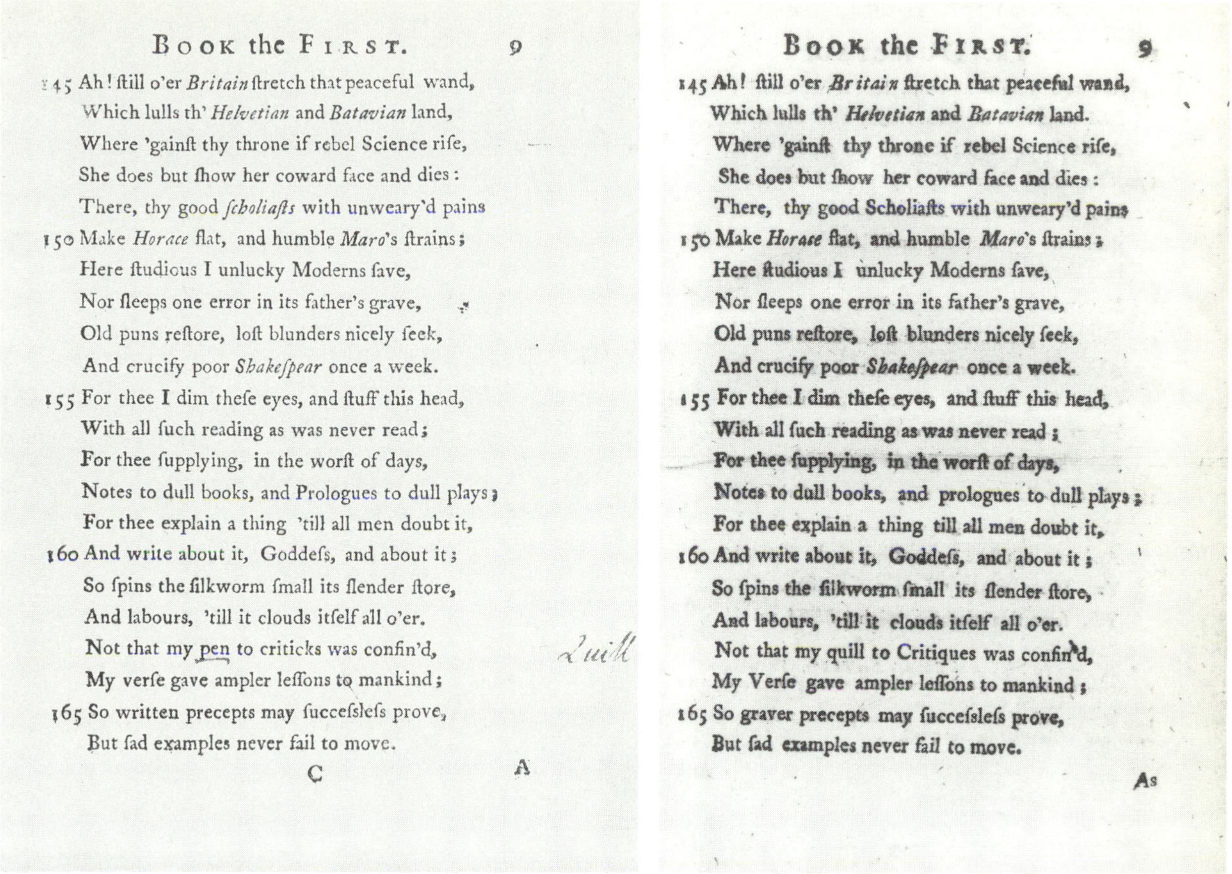

BOOK the FIRST.   9

145 Ah! still o'er *Britain* stretch that peaceful wand,
    Which lulls th' *Helvetian* and *Batavian* land,
    Where 'gainst thy throne if rebel Science rise,
    She does but show her coward face and dies:
    There, thy good *scholiasts* with unweary'd pains
150 Make *Horace* flat, and humble *Maro's* strains;
    Here studious I unlucky Moderns save,
    Nor sleeps one error in its father's grave,
    Old puns restore, lost blunders nicely seek,
    And crucify poor *Shakespear* once a week.
155 For thee I dim these eyes, and stuff this head,
    With all such reading as was never read;
    For thee supplying, in the worst of days,
    Notes to dull books, and Prologues to dull plays;
    For thee explain a thing 'till all men doubt it,
160 And write about it, Goddess, and about it;
    So spins the silkworm small its slender store,
    And labours, 'till it clouds itself all o'er.
    Not that my pen to criticks was confin'd,
    My verse gave ampler lessons to mankind;
165 So written precepts may successless prove,
    But sad examples never fail to move.

             C   A

BOOK the FIRST.   9

145 Ah! still o'er *Britain* stretch that peaceful wand,
    Which lulls th' *Helvetian* and *Batavian* land.
    Where 'gainst thy throne if rebel Science rise,
    She does but show her coward face and dies:
    There, thy good Scholiasts with unweary'd pains
150 Make *Horace* flat, and humble *Maro's* strains;
    Here studious I unlucky Moderns save,
    Nor sleeps one error in its father's grave,
    Old puns restore, lost blunders nicely seek,
    And crucify poor *Shakespear* once a week.
155 For thee I dim these eyes, and stuff this head,
    With all such reading as was never read;
    For thee supplying, in the worst of days,
    Notes to dull books, and prologues to dull plays;
    For thee explain a thing till all men doubt it,
160 And write about it, Goddess, and about it;
    So spins the silkworm small its slender store,
    And labours, 'till it clouds itself all o'er.
    Not that my quill to Critiques was confin'd,
    My Verse gave ampler lessons to mankind;
165 So graver precepts may successless prove,
    But sad examples never fail to move.

                                As

FIG. 81. *Dunciad*, first edition, octavo (1728), p. 9 (Huntington 106517; approx. 210 × 130); and second edition, p. 9 (Bodl. Vet. A4 f. 632; 149 × 93)

Dunciad, 1728 (Figure 81).[25] It would not have been surprising if the second edition had capitalized 'Quill' and given it a false emphasis, but in fact its accidentals are carefully revised, beginning with a period for a comma at the end of line 146 (which is changed to a colon in the quarto *Works* of 1735, and to a semicolon in the following octavos). The word '*scholiasts*' of line 149, which was presumably in italic because of its Greek origins, becomes roman with a capital. Below in line 158 the capital for 'Prologues' had the air of having crept in by accident—perhaps such an accident as I invented for 'Quill'; it is duly removed. In the following line (159) the apostrophe for ''till' has been removed—the spacing round 'thing' sug-

[25] The annotations were copied from Pope's copy by Jonathan Richardson, who wrote on the title-page, 'All here added or changed is from a Copy of Mr Pope's of this Same Edition' (Huntington Library, 106517); see Rogers, *Major satires of Pope*, App. B, pp. 120–3.

FIG. 82. Removal of catchwords and italic from text of *Dunciad variorum* (1729), p. 15 (Bodl. CC 76(1) Art; 249 × 188)

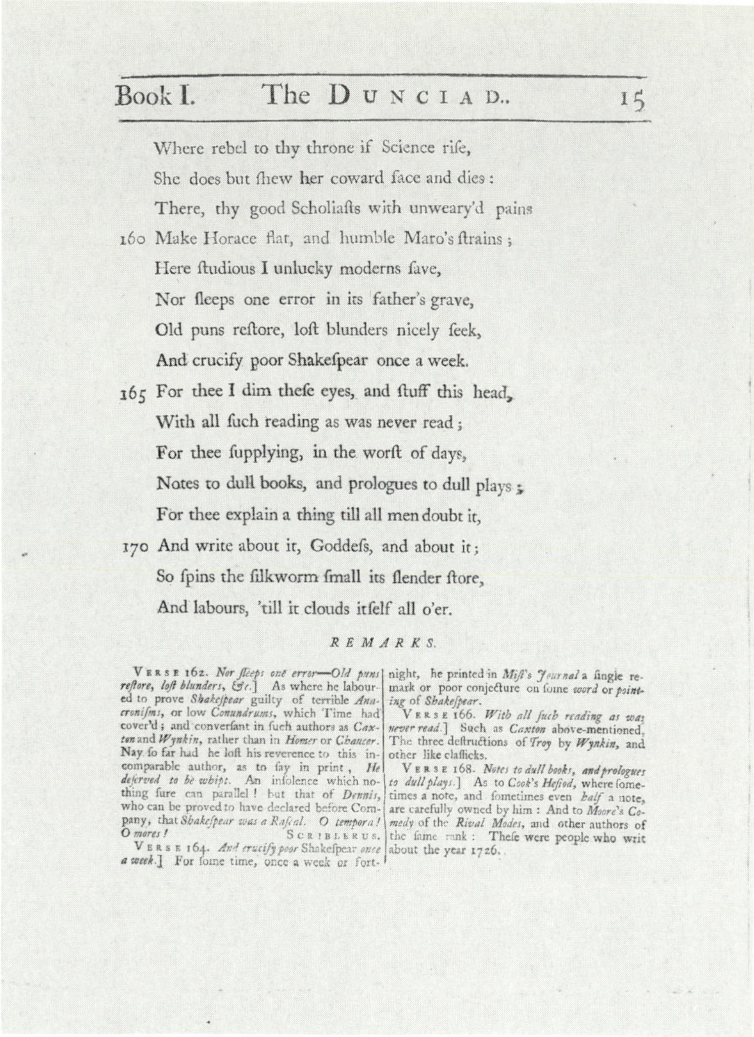

Book I.   The D U N C I A D.   15

Where rebel to thy throne if Science rife,
She does but fhew her coward face and dies:
There, thy good Scholiafts with unweary'd pains
160 Make Horace flat, and humble Maro's ftrains;
Here ftudious I unlucky moderns fave,
Nor fleeps one error in its father's grave,
Old puns reftore, loft blunders nicely feek,
And crucify poor Shakefpear once a week.
165 For thee I dim thefe eyes, and ftuff this head,
With all fuch reading as was never read;
For thee fupplying, in the worft of days,
Notes to dull books, and prologues to dull plays;
For thee explain a thing till all men doubt it,
170 And write about it, Goddefs, and about it;
So fpins the filkworm fmall its flender ftore,
And labours, 'till it clouds itfelf all o'er.

REMARKS.

VERSE 162. *Nor fleeps out error—Old puns reftore, loft blunders, &c.*] As where he laboured to prove *Shakefpear* guilty of terrible *Anacronifms*, or low *Conundrums*, which Time had cover'd; and converfant in fuch authors as *Caxton* and *Wynkin*, rather than in *Homer* or *Chaucer*. Nay fo far had he loft his reverence to this incomparable author, as to fay in print, *He deferved to be whipt.* An infolence which nothing fure can parallel! but that of *Dennis*, who can be proved to have declared before Company, that *Shakefpear was a Rafcal. O tempora! O mores!*   SCRIBLERUS.
VERSE 164. *And crucify poor Shakefpear once a week.*] For fome time, once a weak or fort-

night, he printed in *Mift's Journal* a fingle remark or poor conjecture on fome word or pointing of *Shakefpear.*
VERSE 166. *With all fuch reading as was never read.*] Such as *Caxton* above-mentioned. The three deftructions of *Troy* by *Wynkin,* and other like claffics.
VERSE 168. *Notes to dull books, and prologues to dull plays.*] As to *Cook's Hefiod,* where fometimes a note, and fometimes even *half* a note, are carefully owned by him: And to *Moore's Comedy of the Rival Modes,* and other authors of the fame rank: Thefe were people who writ about the year 1726.

gests it was done in proof; this seems to represent Pope's preference, but the apostrophe is kept in line 162, and this inconsistency is found in the subsequent editions. In the couplet 163–4 the change of 'pen' to 'quill' is made, and at the same time not only is the antithesis of 'Critiques' and 'Verse' pointed by means of capitals, but Pope also avoids confusion by adopting the French spelling—a change which would give some authority to a bold editor for removing the ambiguity we found in a line of the *Essay on criticism* by making the same change.

In my next chapter I shall have a good deal to say about Pope's abandonment of italic type; this started in the *Dunciad variorum* of 1729. Catchwords vanished at the same time. This is the first work printed for Pope by his own printer John Wright. In Figure 82 we see that in addition to the removal of italic, the first line

APPENDIX. 185

Id. March 29. A Letter about Therſites, accuſing the Author of Diſaffection to the Government, by James Moore Smyth.

Miſt's Weekly Journal, March 30. An Eſſay on the Arts of a Poet's ſinking in reputation, or a ſupplement to the Art of ſinking in Poetry [ſuppoſed by Mr. Theobald.]

Daily Journal, April 3. A Letter under the name of Philo-ditto, by James Moore Smyth.

Flying Poſt, April 4. A Letter againſt Gulliver and Mr. P. [by Mr. Oldmixon.]

Daily Journal, April 5. An Auction of Goods at Twickenham, by James Moore Smyth.

Flying Poſt, April 6. A Fragment of a Treatiſe upon Swift and Pope, by Mr. Oldmixon.

The Senator, April 9. On the ſame, by Edward Roome.

Daily Journal, April 8. Advertiſement, by James Moore Smyth.

Flying Poſt, April 13, Verſes againſt Dr. Swift, and againſt Mr. P——'s Homer, by J. Oldmixon.

Daily Journal, April 23, Letter about a Tranſlation of the cha-racter of Therſites in Homer, by Tho. Cook, &c.

Miſt's Weekly Journal, April 27. A Letter of Lewis Theobald.

Daily Journal, May 11. A Letter againſt Mr. P. at large, Anon. John Dennis.

All theſe were afterwards reprinted in a Pamphlet entitled, A collection of all the Verſes, Eſſays, Letters and Advertiſements occaſion'd by Pope and Swift's Miſcellanies, Prefaced by Conca-nen, Anonymous, 8° and printed for A. Moore, 1728, price 1 s. Others of an elder date, having lain as waſte paper many years, were upon the publication of the Dunciad brought out, and their Authors betrayed by the mercenary Bookſellers (in hope of ſome poſſibility of vending a few) by advertiſing them in this man-ner —— " The Confederates, a Farce, by Capt. Breval, (for which " he is put into the Dunciad.) An Epilogue to Powel's Puppet-" ſhow, by Col. Ducket, (for which he is put into the Dunciad.)

FIG. 83. Removal of catchwords and italic from apparatus of *Dunciad* in *Works* II (1735), p. 185 (Bodl. Vet. A4 d. 142; 289 × 225)

96   The DUNCIAD.   Book I.

O thou, of busines the directing soul,
To human heads like byass to the bowl,
Which as more pond'rous makes their aim more true,
Obliquely wadling to the mark in view.   150
O ever gracious to perplex'd mankind!
Who spread a healing mist before the mind,
And, lest we err by Wit's wild, dancing light,
Secure us kindly in our native night.
Ah! still o'er *Britain* stretch that peaceful wand,   155
Which lulls th' *Helvetian* and *Batavian* lands;
Where rebel to thy throne if Science rise,
She does but shew her coward face and dies:
There, thy good Scholiasts with unweary'd pains
Make *Horace* flat, and humble *Maro's* strains;   160
Here studious I unlucky moderns save,
Nor sleeps one error in its father's grave,

REMARKS.

V. 162. *Nor sleeps one error —— Old puns restore,
lost blunders,* &c.] As where he laboured to prove
Shakespear guilty of terrible *Anachronisms,* or low *Conundrums,* which Time had cover'd; and conversant in
such authors as *Caxton* and *Wynkin,* rather than in *Homer* or *Chaucer.* Nay, so far had he lost his reverence
to this incomparable author, as to say in print, *He de-
serv'd to be whipt.* An insolence which nothing sure
can parallel! but that of *Dennis,* who can be proved
to have declared before company, that *Shakespear was
a Rascal. O tempora! O mores!*
SCRIBLERUS.

Book I.   The DUNCIAD.   97

Old puns restore, lost blunders nicely seek,
And crucify poor *Shakespear* once a week.
For thee I dim these eyes, and stuff this head,   165
With all such reading as was never read;

REMARKS.

V. 164. *And crucify poor* Shakespear *once a week.*]
For some time, once a week or fortnight, he printed
in *Miss's Journal* a single remark or poor conjecture
on some *word* or *pointing* of *Shakespear,* either in his
own name, or in letters to himself as from others without name. He since published an edition of *Shakespear,*
with alterations of the Text, upon bare conjectures either of his own, or any others who sent them to him,
to which Mr. *M.* alludes in these Verses of his excellent
Poem on Verbal Criticism:

*He with low industry goes gleaning on,
From good, from bad, from mean, neglecting none:
His brother Bookworm so, on shelf or stall,
Will feed alike on Woolston and on Paul——
Such the grave bird in northern seas is found,
(Whose name a Dutchman only knows to sound)
Where're the king of fish moves on before,
This humble friend attends from shore to shore;
With eye still earnest, and with bill declin'd,
He picks up what his patron drops behind;
With such choice cates his palate to regale,
And is the careful Tibbald of a whale.*

V. 166. *With all such reading as was never read.*]
Such as *Caxton* above-mention'd, the three destructions
of *Troy* by *Wynkin,* and other like classicks.
G

FIG. 84. Reintroduction of italic in text and apparatus of *Dunciad* in octavo trade edition (1735), pp. 96–7 (Bodl. 12 θ 937; 175 × 208)

(formerly 147) is revised, and in line 161 (formerly 151) the 'unlucky Moderns' lose their capital. I am not going to pursue the textual history of this passage further, except to note that it had its punctuation revised at least twice (in the *Works* of 1735 and its octavo reprints) before it was dropped with the change in the *Dunciad*'s hero in 1743. Instead, a word on the typography and inconsistencies of the later editions. Although in 1729 the text is purged of italic, the complicated apparatus of both footnotes and appendices remains unreformed and to some extent inconsistent—no doubt partly (as Scriblerus remarks on the variant spellings *Satire* and *Satyr*) 'from the different Orthography of the various Annotators'.[26] The edition printed in the winter of 1732–3, which was to form part of

26 'M. Scriblerus lectori', 1729 quarto, p. 81 (*Twickenham*, v. 196).

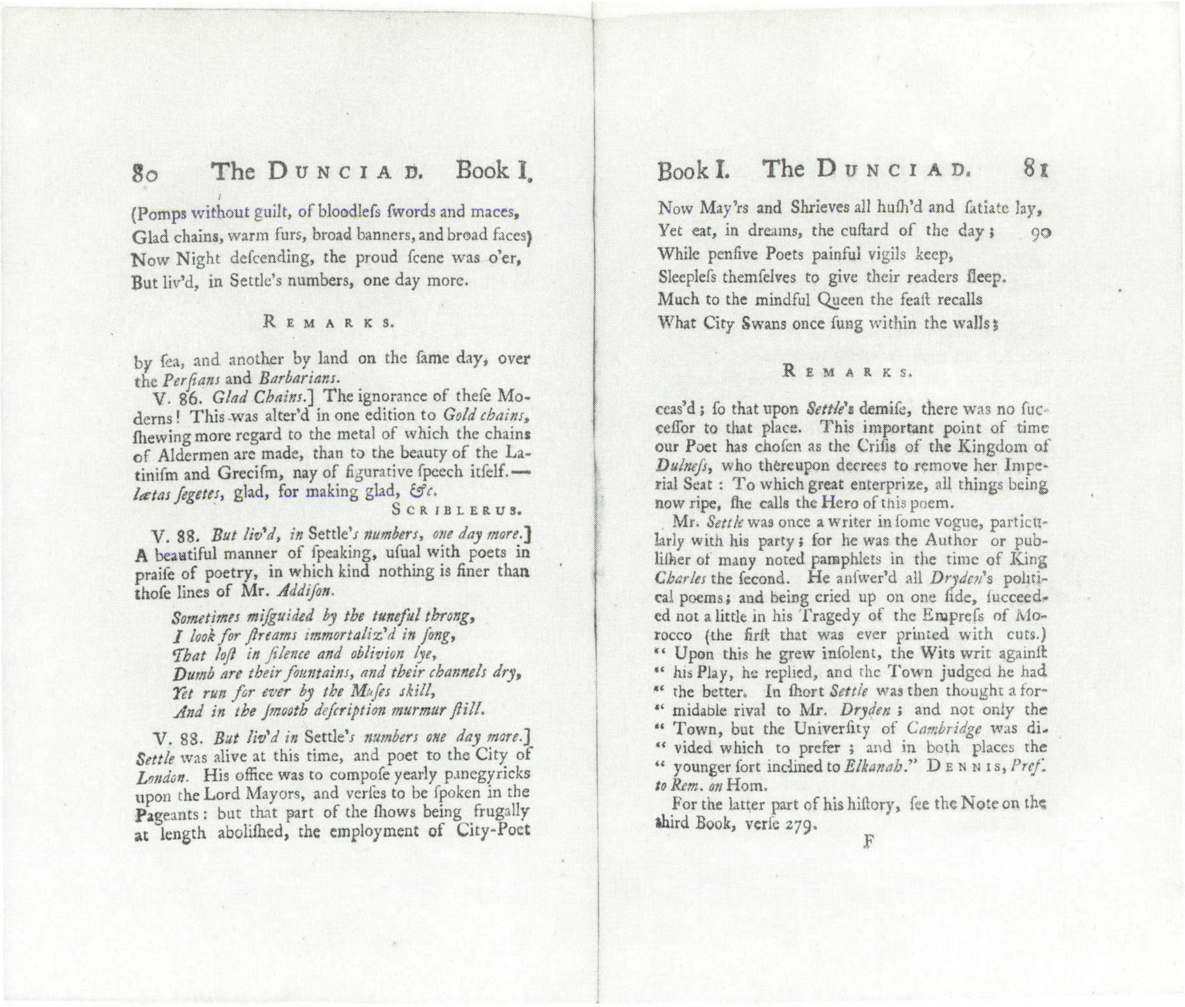

FIG. 85. Failure to italicize 'Settle' in *Dunciad* octavo (1735), pp. 80–1 (Bodl. 12 θ 937; 175 × 208)

the 1735 *Works*, made a determined attempt to remove italic from this peripheral material. The effect, as can be seen in Figure 83, is surprisingly modern. Even the abbreviation 'Id.' in the first line is in roman. I can think of nothing to parallel this until the 1760s.

The octavos which formed part of the cheap edition of the *Works* between 1735 and 1742 (Figure 84) reintroduce italic for proper names in the text and use it more widely in the notes, as do their companion volumes. But, as happened on other occasions, the change was not made to the poem in the first sheet of text, E (Figure 85), so 'Settle' is in roman (line 88), though the lemma to the note suggests in its inverted style that the name should be in italic. This inconsistency was preserved through all the following octavos.

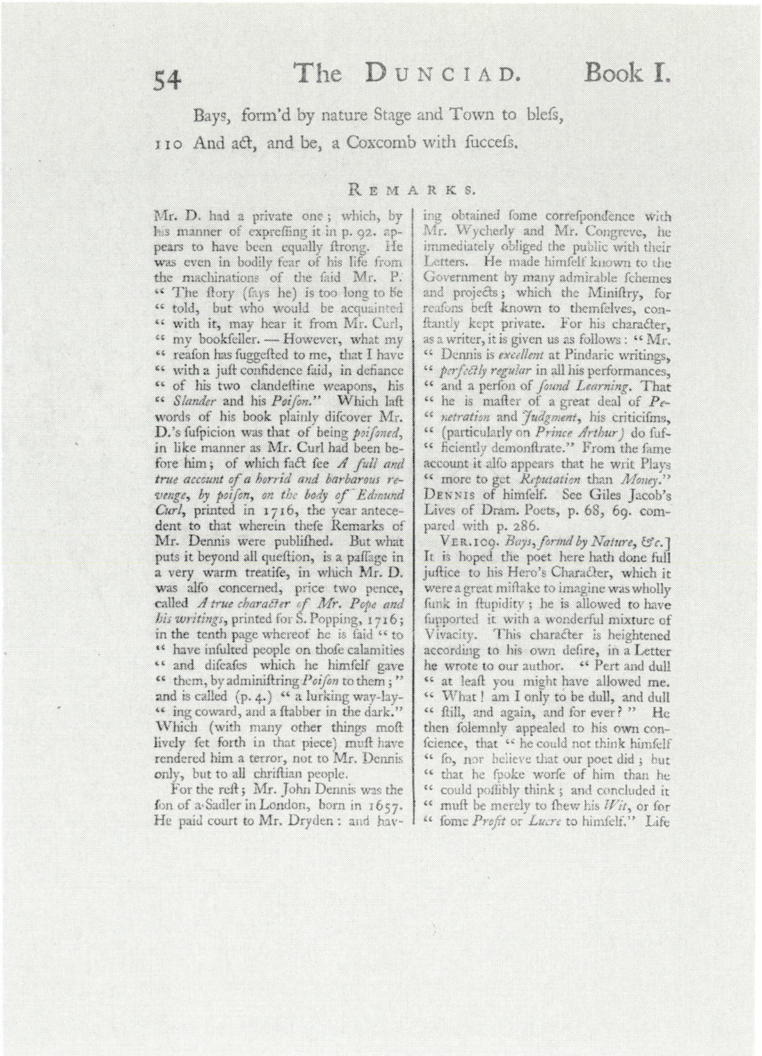

FIG. 86. Retention of some italic from octavos in apparatus of death-bed quarto *Dunciad* (1743), p. 54 (Bodl. CC 51(1) Art; 255 × 191)

54   The DUNCIAD.   Book I.

Bays, form'd by nature Stage and Town to blefs,
110 And act, and be, a Coxcomb with fuccefs.

REMARKS.

Mr. D. had a private one; which, by his manner of expreffing it in p. 92. appears to have been equally ftrong. He was even in bodily fear of his life from the machinations of the faid Mr. P. " The ftory (fays he) is too long to be " told, but who would be acquainted " with it, may hear it from Mr. Curl, " my bookfeller.—However, what my " reafon has fuggefted to me, that I have " with a juft confidence faid, in defiance " of his two clandeftine weapons, his " *Slander* and his *Poifon.*" Which laft words of his book plainly difcover Mr. D.'s fufpicion was that of being *poifoned*, in like manner as Mr. Curl had been before him; of which fact fee *A full and true account of a horrid and barbarous revenge, by poifon, on the body of Edmund Curl*, printed in 1716, the year antecedent to that wherein thefe Remarks of Mr. Dennis were publifhed. But what puts it beyond all queftion, is a paffage in a very warm treatife, in which Mr. D. was alfo concerned, price two pence, called *A true character of Mr. Pope and his writings*, printed for S. Popping, 1716; in the tenth page whereof he is faid " to " have infulted people on thofe calamities " and difeafes which he himfelf gave " them, by adminiftring *Poifon* to them; " and is called (p. 4.) " a lurking way-lay- " ing coward, and a ftabber in the dark." Which (with many other things moft lively fet forth in that piece) muft have rendered him a terror, not to Mr. Dennis only, but to all chriftian people.

For the reft; Mr. John Dennis was the fon of a Sadler in London, born in 1657. He paid court to Mr. Dryden: and hav-

ing obtained fome correfpondence with Mr. Wycherly and Mr. Congreve, he immediately obliged the public with their Letters. He made himfelf known to the Government by many admirable fchemes and projects; which the Miniftry, for reafons beft known to themfelves, conftantly kept private. For his character, as a writer, it is given us as follows: " Mr. " Dennis is *excellent* at Pindaric writings, " *perfectly regular* in all his performances, " and a perfon of *found Learning*. That " he is mafter of a great deal of *Pe-* " *netration* and *Judgment*, his criticifms, " (particularly on *Prince Arthur*) do fuf- " ficiently demonftrate." From the fame account it alfo appears that he writ Plays " more to get *Reputation* than *Money.*" DENNIS of himfelf. See Giles Jacob's Lives of Dram. Poets, p. 68, 69. compared with p. 286.

VER.109. *Bays, form'd by Nature, &c.*] It is hoped the poet here hath done full juftice to his Hero's Character, which it were a great miftake to imagine was wholly funk in ftupidity; he is allowed to have fupported it with a wonderful mixture of Vivacity. This character is heightened according to his own defire, in a Letter he wrote to our author. " Pert and dull " at leaft you might have allowed me. " What! am I only to be dull, and dull " ftill, and again, and for ever?" He then folemnly appealed to his own confcience, that " he could not think himfelf " fo, nor believe that our poet did; but " that he fpoke worfe of him than he " could poffibly think; and concluded it " muft be merely to fhew his *Wit*, or for " fome *Profit* or *Lucre* to himfelf." Life

The death-bed edition of 1743 (Figure 86) as usual abandons italic, with two exceptions. In the footnotes, though italic is pretty consistently removed from proper names, and sometimes from the titles of books (for example, 'Lives of Dram. Poets', but not the Curll pamphlets), the italicized phrases in the notes are reprinted unchanged from the octavo. The most blatant exception, though, is that the new book IV (misprinted VI in Figure 87) has some passages where italic is actually added. On this page only 'Mad *Mathesis*' in line 31 was italicized in the first edition, no doubt because of her Greek origins. One can, of course, see that these are all personifications, but why should they have italics when the mighty mother Dulness in line 30 has none? The notes on this passage also italicize

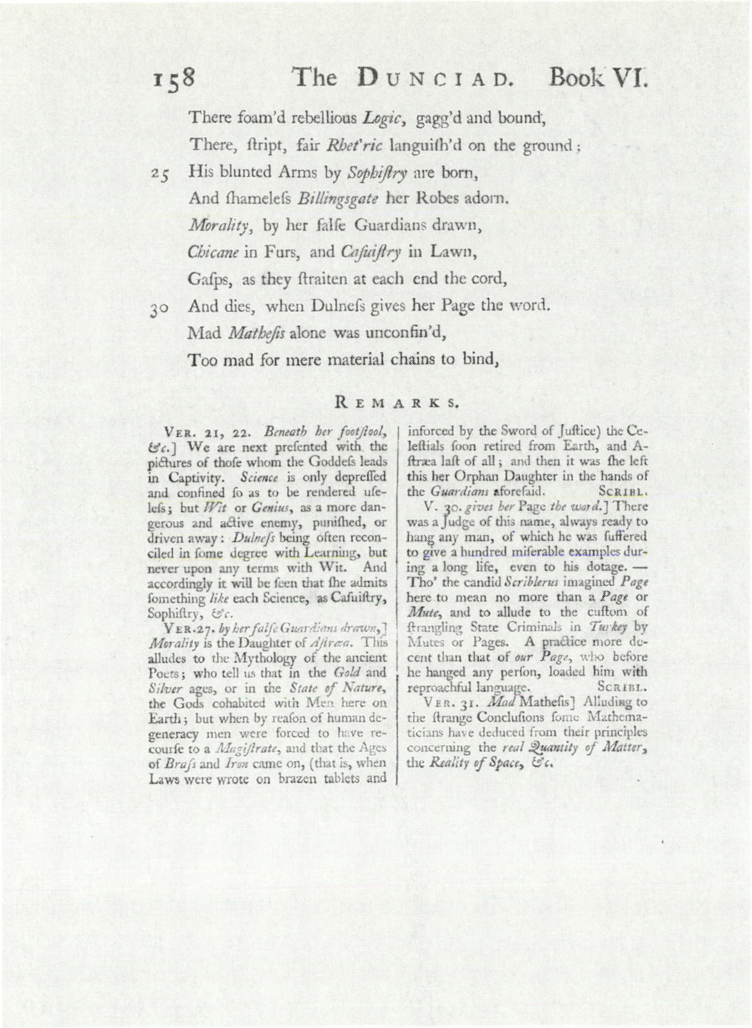
158 The DUNCIAD. Book VI.

There foam'd rebellious *Logic*, gagg'd and bound,
There, ſtript, fair *Rhet'ric* languiſh'd on the ground;
His blunted Arms by *Sophiſtry* are born,
And ſhameleſs *Billingsgate* her Robes adorn.
Morality, by her falſe Guardians drawn,
Chicane in Furs, and *Caſuiſtry* in Lawn,
Gaſps, as they ſtraiten at each end the cord,
And dies, when Dulneſs gives her Page the word.
Mad *Matheſis* alone was unconfin'd,
Too mad for mere material chains to bind,

REMARKS.

VER. 21, 22. *Beneath her footſtool, &c.*] We are next preſented with the pictures of thoſe whom the Goddeſs leads in Captivity. *Science* is only depreſſed and confined ſo as to be rendered uſeleſs; but *Wit* or *Genius*, as a more dangerous and active enemy, puniſhed, or driven away: *Dulneſs* being often reconciled in ſome degree with Learning, but never upon any terms with Wit. And accordingly it will be ſeen that ſhe admits ſomething *like* each Science, as Caſuiſtry, Sophiſtry, *&c.*

VER. 27. *by her falſe Guardians drawn,*] *Morality* is the Daughter of *Aſtræa*. This alludes to the Mythology of the ancient Poets; who tell us that in the *Gold* and *Silver* ages, or in the *State of Nature*, the Gods cohabited with Men here on Earth; but when by reaſon of human degeneracy men were forced to have recourſe to a *Magiſtrate*, and that the Ages of *Braſs* and *Iron* came on, (that is, when Laws were wrote on brazen tablets and inforced by the Sword of Juſtice) the Celeſtials ſoon retired from Earth, and Aſtræa laſt of all; and then it was ſhe left this her Orphan Daughter in the hands of the *Guardians* aforeſaid. SCRIBL.

V. 30. *gives her* Page *the word.*] There was a Judge of this name, always ready to hang any man, of which he was ſuffered to give a hundred miſerable examples during a long life, even to his dotage. — Tho' the candid *Scriblerus* imagined *Page* here to mean no more than a *Page* or *Mute*, and to allude to the cuſtom of ſtrangling State Criminals in *Turkey* by Mutes or Pages. A practice more decent than that of *our Page*, who before he hanged any perſon, loaded him with reproachful language. SCRIBL.

VER. 31. *Mad* Matheſis] Alluding to the ſtrange Concluſions ſome Mathematicians have deduced from their principles concerning the *real Quantity of Matter*, the *Reality of Space*, *&c.*

FIG. 87. Addition of italic in text of book IV of death-bed quarto *Dunciad* (1743), p. 158 (Bodl. CC 51(1) Art; 255 × 191)

Astræa, *Scriblerus*, and *Turkey*. This inconsistency is preserved in Warburton's edition of 1751, and it may be that since he had such a large share in preparing the *New Dunciad*, he was the Dunce responsible in the first place.

But the point I want to make is that the death-bed edition of the *Dunciad*, like that of the *Essay on criticism* we looked at earlier, does not solve all our editor's problems since it presents him with these inconsistencies that Pope would not have wished to have preserved.

5

POPE'S TEXT:
THE LATER WORKS

In the last chapter I talked of the way Pope abandoned the routine capitalization of nouns in his 1717 *Works*, and how he went on to abandon italics in the *Dunciad* of 1729. I cannot pretend that Pope's usage of capitals and italics in his poems of the thirties is always easy to interpret—to some extent it reflects his subjective judgement of what seems right in a given passage, and that will vary from time to time. We have already seen from one of the *Odyssey* manuscripts (Figure 79) that in his formal hand Pope continued to use a heavy capitalization that did not reflect his intentions for the printed page, and so the printer had to make the necessary changes. John Wright, who printed almost all Pope's later work, took on no such responsibility and followed the capitalization of his manuscript copy, leaving Pope to reduce it in proof or in subsequent editions; later, Pope sometimes seems to feel he had gone too far in reducing them and restores some capitals.

Italicization is a more complex problem, and there seem to be three conflicting factors at work. Pope's desire to classicize or romanize can be seen not only in the changes made in successive editions but also in the successive manuscripts of the *Essay on man*. At the same time he was always tempted to italicize to make a point or mark an antithesis. Apart from these opposing tendencies, there seems a clear decision to use italics in trade editions, whether these are the original folio publications of his poems or the collected works in octavo, and to avoid italics as far as possible in the large formats intended for a select circle; and I think that this decision must be Pope's. Perhaps he felt that the vulgar needed help in reading his work correctly, or at least that they should have italics for proper names as they would expect; and if these italics, why not others?

ETHIC EPISTLES

Comparison of Pope's later manuscripts with the printed editions is complicated by the fact that the printer's copy was probably in all cases a lost scribal transcript, though there is no reason to believe that Warton's account of Dodsley's transcribing the *Epilogue to the satires* is typical of the long-worked Epistles.

After Dodsley had made a fair copy, 'Every line was then written twice over; a clean transcript was then delivered to Mr. Pope, and when he afterwards sent it to Mr. Dodsley to be printed, he found every line had been written twice over a second time'.[1] The surviving manuscripts are sufficiently close to the printed versions (and to one another when there is more than one) to make a further stage of rewriting unnecessary, but a copyist might conceivably have changed italics and capitals either on his own initiative or on Pope's instructions. More likely Pope himself made minor changes either in the transcript or in proof.

We can see the normal pattern of Pope's later works by looking at a specimen of the first ethic epistle, the *Epistle to Burlington*, 'Of false taste', printed at the end of 1731 (Figure 88). The first two pages of the manuscript, which survive in the Pierpont Morgan Library,[2] are very close in accidentals to the first edition even though the whole order of passages has been changed. I have rearranged photographs of the first edition so that comparisons can be made easily. The chief difference is that the manuscript has few underlinings for italic for proper names (see 'Topham', 'Fountain', 'Curio', 'Herne', 'Mead', and 'Sloan' in lines 5–10) and although in the second paragraph the manuscript italicizes '*how many*' (line 14), '*Knights*', and '*Taste*' (line 16), in the third paragraph it does not italicize 'sense', 'yourself', or 'Nature'. Although there are some changes to capitals and punctuation, they are comparatively minor. The second page (Figure 89) tells much the same story; the capitalization is very heavy in both versions. In fact the nouns in this passage which do not begin with a capital in the first edition, 'hands' (at 1) in

> The vast *Parterres* a thousand hands shall make

(note the italic for the French word) and 'strength' (at 3) in

> And strength of Shade contends with strength of Light

are so written in the manuscript. Notice the double underline for the honorific small capitals of 'Stow'. Again italic is added for emphasis, as in '*Poor*', '*Hungry*', '*Health*', '*Bread*' (at 2).

Figure 90 shows a passage as printed in the first edition, in the quarto *Works* of 1735, and in the octavo of the same year. This is not a proper demonstration of the typography, as I have reduced the type of folio and quarto to approximately the size of the octavo, and obliterated the page divisions, but it is still possible to see that the *Works* close up the white lines between paragraphs, and that there is less leading between the lines in each succeeding edition. The octavo adds six lines after line 50, which I have cut; it also has footnotes, of which only a token remains. I don't want to go into the detailed changes, but to establish the pattern of all Pope's later works: the separately published folios use italic freely in the tra-

---

[1] *Works of Alexander Pope*, ed. J. Warton (1822), iv. 294; quoted by Butt, 'Pope's poetical manuscripts', p. 30.
[2] Reproduced by Mack, *The last and greatest art*, pp. 156–64.

OF TASTE:

AN

EPISTLE

To The

EARL of BURLINGTON.

'TIS strange, the Miser still employs ^ what he ne'er enjoys
His cares to gain the wealth he can't enjoy:
Is it less strange, the Prodigal shou'd waste
His Wealth to purchase what he never can Taste?
Not for himself, but Fountain, Gem, he buys;
Pictures, to raise the noble Thoughts of Arise;
For Topham, Drawings & far-fetched Designs;
For Pembroke, Statues, brazen Gods, and Coins;
Rare Monkish Manuscripts for Herne alone;
And Books for Mead, & Rarities for Sloan.
Think we all these, are for Himself? No more
Than his fine Wife, or finer Whore.
    For what has Virro painted, planted?
Only to show how many Tastes he wanted.
What brought Sir Shylock's ill-got wealth to waste?
Some Dæmon whisper'd, "Knights shou'd have a Taste"
Heav'n visits with a Taste, the wealthy Fool,
And needs no Rod, but Mo's with a Rule.
See sporting Fates to punish aukward Pride,
Bids Babo build, and such a Guide;
A Sermon! at each Years Expence,
That never Coxcomb reach'd Magnificence.
    Oft' have You hinted to your Brother Peer,
A certain truth, which many buy too dear:
Something there is, that shou'd precede Expence,
Something to govern Taste itself — 'tis Sense;
Good Sense, which only is the Gift of Heav'n,
And tho' no Science, fairly is worth the Seven;
To build, to plant, whatever you intend,
To rear the Column, or the Arch to bend,
To swell the Terras, or to sink the Grot;
In all, let Nature never be forgot:
                              Con-

(a)

'TIS strange, the Miser should his Cares imploy
     To *gain* those Riches he can ne'er *enjoy*:
Is it less strange, the Prodigal should *waste*
His Wealth to purchase what he ne'er can *taste*?
Not for himself he sees, or hears, or eats ;
Artists must chuse his Pictures, Music, Meats :
He buys for *Topham* Drawings and Designs,
For *Fountain* Statues, and for *Curio* Coins,
Rare Monkish Manuscripts for *Hearne* alone,
And Books for *Mead*, and Rarities for *Sloan*.

Think we all these are for himself? no more
Than his fine Wife (my Lord) or finer Whore.

     For what has *Virro* painted, built, and planted?
Only to shew *how many* Tastes he wanted.
What brought Sir *Shylock*'s ill-got Wealth to waste ?
Some Dæmon whisper'd, " *Knights* shou'd have a *Taste*."
Heav'n visits with a *Taste* the wealthy Fool,
And needs no Rod, but S——d with a Rule.
See sportive Fate, to punish aukward Pride,
Bids *Babo* build, and sends him such a Guide :
A standing Sermon! at each Year's expence,
That never Coxcomb reach'd Magnificence.

     Oft have have you hinted to your Brother Peer,
A certain Truth, which many buy too dear:
Something there is, more needful than Expence,
And something previous ev'n to Taste — 'Tis *Sense* ;
Good Sense, which only is the Gift of Heav'n,
And tho' no Science, fairly worth the Seven.
A Light, which in *yourself* you must perceive ;
* *Jones* and † *Le Nôtre* have it not to give.

     To build, to plant, whatever you intend,
To rear the Column, or the Arch to bend,
To swell the Terras, or to sink the Grot ;
In all, let *Nature* never be forgot.

(b)

FIG. 88. Beginning of *Epistle to Burlington* in autograph manuscript (Pierpont Morgan MS MA 352; 302 × 184) and first edition (1731), pp. 5–7 (Bodl. G. Pamph. 1661(1); 359 × 219)

Consult the Genius of the Place in all;
~~That~~
That tells your Waters or to rise, or fall,
Here ~~shall it~~ point the future Mount, & here
Invites to scoop the circling Theatre.

Begin with ~~this~~ Sense, of ev'ry Art the Soul,
Parts answ'ring Parts, will slide into a whole,
Nature, shall join you; time shall make it ~~and y work shall gro~~
    A work!
40 Something to wonder at — perhaps a *Stow.
Without it, ~~Babylons proud Garden~~ thy Glory falls,
And Nero's Terraces desert their Walls.
The vast Parterres a thous'd hands shall make,
Lo! ~~Bridgman~~ comes, and floats y⁻ w⁻ a Lake
Or cut wide Views thro' mountains to y⁻ Plain.
You'l wish your Hill & sheltred Seat again.
~~thro' his young woods &c~~
[3.] Yet hence the Poor are cloath'd, y⁻ Hungry fed, Behold,
Health to himself, and to his Infants Bread
The Lab'rer bears: what Thy hard Heart denies
60 Thy charitable Vanity supplies.
~~What better can be~~ set the golden Ear
    another age shall see;
Imbrown thy Slope, & nod on thy Parterre,
Deep Harvests bury all thy Pride has plann'd
And laughing Ceres re-assume the Land.

~~Nor ...~~
first. (1) ~~Behold Villario's~~ ten years toil compleat!
His Arbours darken, and his thickets meet;
The Wood supports the Plain; the Parts unite,
70 And strength of Shade contends with strength of Light,
His blooming Beds a waving Glow display,
Blushing in bright Diversities of Day,
With Silver-qu iv'ring Rills maanderd o'er:
— Enjoy them You! the Owner can no more:
~~The goaty Owner on his Couch is thrown~~
And humbly begs you that yo⁻ walk alone.
    2. Thro'

× The Seat & Gardens of the
Lord Viscount Cobham in
Buckinghamshire.

Ey Thames's Stow why adds a creeping Rill,
This year dig it, ky next shall fill.
Ev'n in an Ornament, its Place remark,
Nor in an Hermitage set D⁻ Clark.

Confult the *Genius* of the *Place* in all,

That tells the Waters or to rife, or fall,

Or helps th' ambitious Hill the Heav'ns to fcale,

Or fcoops in circling Theatres the Vale,

    Begin with *Senfe*, of ev'ry Art the Soul,

Parts anfw'ring Parts, fhall flide into a Whole,

Nature fhall join you; *Time* fhall make it grow

A Work to wonder at — perhaps a * S T O W.

    Without it, proud *Verfailles!* thy Glory falls,

And *Nero*'s Terraffes defert their Walls:

1   The vaft *Parterres* a thoufand hands fhall make,

Lo! *Bridgman* comes, and floats them with a *Lake*:

Or cut wide *Views* thro' Mountains to the Plain,

You'll wifh your Hill, and fhelter'd Seat, again.

2   Yet hence the *Poor* are cloth'd, the *Hungry* fed;

Health to himfelf, and to his Infants *Bread*

The Lab'rer bears; What thy hard Heart denies,

Thy charitable Vanity supplies.

Another Age fhall fee the golden Ear

Imbrown thy Slope, and nod on thy Parterre,

Deep Harvefts bury all thy Pride has plann'd,

And laughing *Ceres* re-affume the Land.

    Behold *Villario*'s ten-years Toil compleat,

His *Quincunx* darkens, his Efpaliers meet,

The Wood fupports the Plain; the Parts unite,

3   And ftrength of Shade contends with ftrength of Light;

His bloomy Beds a waving Glow difplay,

Blufhing in bright Diverfities of Day,

With filver-quiv'ring Rills mæander'd o'er —

— Enjoy them, you! *Villario* can no more;

Tir'd of the Scene Parterres and Fountains yield,

He finds at laft he better likes a Field.

FIG. 89. Continuation of *Epistle to Burlington* in manuscript (Pierpont Morgan MS MA 352; 302 × 184) and first edition (1731), pp. 7–9 (Bodl. G. Pamph. 1661(1); 359 × 219)

FIG. 90. Three editions of *Epistle to Burlington*: (*a*) first edition (1731), pp. 6–8 (Bodl. G. Pamph. 1661(1); 359 × 219); (*b*) quarto *Works* II (1735), pp. ²41–2 (Bodl. Vet. A4 d. 141; 290 × 224); (*c*) octavo *Works* II (1735), pp. 48–9 (Bodl. 280 n. 415(2); 159 × 94)

Oft have have you hinted to your Brother Peer,
A certain Truth, which many buy too dear:
Something there is, more needful than Expence,
And something previous ev'n to Taſte — 'Tis *Senſe*;
Good Senſe, which only is the Gift of Heav'n,
And tho' no Science, fairly worth the Seven.
A Light, which in *yourſelf* you muſt perceive;
* *Jones* and † *Le Nôtre* have it not to give.

To build, to plant, whatever you intend,
To rear the Column, or the Arch to bend,
To ſwell the Terras, or to ſink the Grot;
In all, let *Nature* never be forgot.
Conſult the *Genius* of the *Place* in all,
That tells the Waters or to riſe, or fall,
Or helps th' ambitious Hill the Heav'ns to ſcale,
Or ſcoops in circling Theatres the Vale,
Calls in the Country, catches opening Glades,
Joins willing Woods, and varies Shades from Shades,
Now breaks, or now directs, th' intending Lines;
Paints as you plant, and as you work, *Deſigns*.

Begin with *Senſe*, of ev'ry Art the Soul,
Parts anſw'ring Parts, ſhall ſlide into a Whole,
Spontaneous Beauties all around advance,
Start, ev'n from *Difficulty*, ſtrike, from *Chance*;
Nature ſhall join you; *Time* ſhall make it grow
A Work to wonder at — perhaps a * S T O W.

Without it, proud *Verſailles!* thy Glory falls,
And *Nero's* Terraſſes deſert their Walls:

The vaſt *Parterres* a thouſand hands ſhall make,
Lo! *Bridgman* comes, and floats them with a *Lake*:

(*a*)

Oft have you hinted to your Brother Peer,
A certain Truth, which many buy too dear: 40
Something there is, more needful than Expence,
And something previous ev'n to Taste — 'Tis *Sense*:
Good Sense, which only is the Gift of Heav'n,
And tho' no Science, fairly worth the seven:
A Light, which in yourself you must perceive; 45
Jones and Le Nôtre have it not to give.

To build, to plant, whatever you intend,
To rear the Column, or the Arch to bend,
To swell the Terras, or to sink the Grot;
In all, let *Nature* never be forgot. 50
Consult the Genius of the Place in all;
That tells the Waters or to rise, or fall,
Or helps th' ambitious Hill the heav'ns to scale,
Or scoops in circling Theatres the Vale;

Calls in the Country, catches opening Glades, 55
Joins willing Woods, and varies Shades from Shades,
Now breaks, or now directs, th' intending Lines;
Paints as you plant, and as you work, designs.

Begin with *Sense*, of ev'ry Art the Soul,
Parts answ'ring Parts shall slide into a Whole, 60
Spontaneous Beauties all around advance,
Start ev'n from Difficulty, strike, from Chance;
Nature shall join you; Time shall make it grow
A Work to wonder at — perhaps a STOW.

· Without it, proud Versailles! thy Glory falls, 65
And Nero's Terraces desert their Walls:
The vast Parterres a thousand hands shall make,
Lo! COBHAM comes, and floats them with a Lake:

(b)

Oft have you hinted to your brother Peer,
A certain truth, which many buy too dear: 40
Something there is, more needful than Expence,
And something previous ev'n to Taste——'Tis *Sense*:
Good Sense, which only is the gift of heav'n,
And tho' no science, fairly worth the seven:
A Light, which in yourself you must perceive; 45
Jones and *Le Nôtre* have it not to give.

To build, to plant, whatever you intend,
To rear the Column, or the Arch to bend,
To swell the Terras, or to sink the Grot;
In all, let *Nature* never be forgot. 50

Consult the *Genius* of the *Place* in all;
That tells the Waters or to rise, or fall,
Or helps th' ambitious Hill the heav'ns to scale,
Or scoops in circling Theatres the Vale, 60
Calls in the Country, catches opening glades,
Joins willing woods, and varies shades from shades,
Now breaks, or now directs, th' intending Lines;
Paints as you plant, and as you work, designs.

Begin with *Sense*, of ev'ry Art the Soul, 65
Parts answ'ring parts shall slide into a Whole,
Spontaneous beauties all around advance,
Start ev'n from *Difficulty*, strike, from *Chance*;
Nature shall join you, *Time* shall make it grow
A Work to wonder at——perhaps a STOW. 70

Without it, proud *Versailles!* thy glory falls,
And *Nero*'s Terraces desert their walls:
The vast Parterres a thousand hands shall make,
Lo! COBHAM comes, and floats them with a Lake:

V. 57, &c. The first Rule, to adapt all to the *Nature*
and *Use* of the *Place*, and the Beauties not forced into it,
but resulting from it.

V. 70. The Seat and Gardens of the Lord Viscount
Cobham in Buckinghamshire.

V. 71, &c. For want of this *Sense*, and thro' neglect
of this *Rule*, men are disappointed in the most expensive
Undertakings. Nothing without this will ever please *long*,
if it pleases *at all*.

D   O†

(c)

ditional way; the collected poems in the large formats which sold to a limited public, mainly in Pope's own circles, abandon italic for proper names and for stress, keeping it only for key words (here '*Nature*' and '*Sense*'); and the popular octavo editions of collected poems go back to the conventional use of italic for proper names, and more freely for '*Genius* of the *Place*' in line 51, '*Difficulty*', '*Chance*', '*Nature*', and '*Time*' in lines 68–9. But the octavo *Works* does not use italic purely for stress in 'yourself' (line 45) or to point the antithesis 'Paints' : 'designs' (line 64) as the folio first edition does. What I do not show here is the abandonment of italic again in the death-bed editions, prepared in association with Warburton. I want to stress again the alternation between the use of italic in trade editions, and its absence in the 'definitive' editions.

As for capitals, many in this passage like 'Column', 'Arch', 'Terras', 'Grot' in lines 48–9 are generic and undergo no change, but the first eight lines give some idea of the sort of change that is typical of these stages of Pope's later poems. The collected *Works* merely tidy up, here removing the capital from 'Seven' in line 44; by contrast the octavo is much more systematic in considering which nouns really merit capitals in this context. As a result we have lower-case 'brother', 'truth', 'gift of heav'n', and 'science' in the first paragraph. I must stress the point that these changes are highly selective and well-considered; if they are not (as I believe) the work of Pope himself, they are the work of someone working under his instructions and supervision. They are not the work of his printer John Wright, who printed all these editions (and more) and has shown no desire to rationalize in the earlier ones.

The next epistle, *To Bathurst*, 'Of the use of riches', printed a year later in the winter of 1732–3,[3] has a more complex history. I reproduce a page of the second and most legible of three manuscripts (the other two have a very similar texture of accidentals) for comparison with the first edition (Figure 91).[4] The first edition was printed by Wright at about the same time as the *Dunciad* for the *Works* of 1735, and Pope and Wright were removing all italic from that text, which was already very lightly capitalized. It is not therefore surprising to find that both capitals and italics are reduced in the first edition. The reduction of capitals can be seen in the first four lines of the printed text—'heav'n', 'babe', 'market-place', 'poor', 'bread' (the last was taken down in the third manuscript); the punctuation is also changed, sometimes following the third manuscript but sometimes ignoring it or making independent changes. The italic is more interesting: not so much that 'Man of Ross' loses italics (as does 'all' in line 14 of the printed page, follow-

---

[3] Pope's letter to Jonathan Richardson the elder of 2 Nov. 1732 probably refers to this work, but when he says, 'You will see another poem next week', he could well refer to a manuscript rather than the printed copy Sherburn suggests. It may be that the Richardsons were at this time acting as Pope's scribes, hence Pope's 'I thank you for all. Your prudence I never doubt, nor your son's' (*Correspondence*, iii, 327).

[4] Huntington MS 6007. The 3rd manuscript (HM 6008) has been folded and sealed; if my speculation in the preceding note is correct, it could be the copy sent to Richardson. The manuscripts are reproduced by E. R. Wasserman, *Pope's 'Epistle to Bathurst'* (Baltimore, 1960).

ing the third manuscript) but that '*Stars*' in the last line is exceptional in keeping them. It must have a special significance, which is clearly as part of the insignia of knighthood, as in star and garter: Pope is contrasting the good done by the modest Man of Ross with the great who frequent the 'proud Courts'.[5]

If we compare the first edition with the two others in Figure 92 (both calling themselves 'second edition', but the second of these at least a couple of months later than the first), we can see the evidence of Pope's preference for italics in his trade editions. In the two later editions italic is added to all proper nouns, and it seems probable that instructions were given to Wright in those terms, so that he only italicized '*Ross*' in 'Man of *Ross*' in the second. This is not what Pope intended, as we can see from his manuscript, and so he intervenes in the third edition to achieve '*Man of Ross*' in italic. At the same time capitals are added to 'Medicine' and 'Courts' (at 1 and 2), and we must, I think, attribute these to Pope. In the quarto *Works* of 1735, where italics are as usual removed, 'MAN of ROSS' achieves the dignity of caps. and small caps.,[6] while both words '*little Stars*' are printed in italic. Since the first edition was lightly capitalized there is not so much reduction of capitals in the later editions as there was in the *Epistle to Burlington*, but some adjustments are made, both up and down. In the octavos, '*Quacks*' and '*Attornies*' appear in italics, presumably to take up the 'Medicine' and 'Courts' capitalized before, though the capital for 'Medicine' drops out here, probably because it is specific and not generic as 'Courts' is: the Man of Ross is described preparing bottles of medicine rather than practising the profession. The death-bed edition naturally removes the italics from '*Quacks*' and '*Attornies*'; the capital for 'medicine' is still absent.

AN ESSAY ON MAN

The first epistle of the *Essay on man* again presents special circumstances; for this major work Pope seems to have worked more closely than usual on representing his final typographic intentions in his fair copies; at the same time he intended the type-setting of the first edition to be used for printing the three formats for the *Works* of 1735, and so italic is not used for proper names.

The draft and fair copy of the opening of the second book show us in exaggerated form Pope's weakness for italic and his deliberate abandonment of it.[7] The earlier draft (Figure 93) is written as recommended in John Smith, *The printer's grammar* (1735), p. 167: 'But some authors who are still better Methodists in writing for the Press, divide each side of the paper into two Columns, filling one

---

[5] Cf. *Windsor Forest*, line 290, 'And add new Lustre to her Silver *Star*', italicized in both manuscript and 1st edn., with the same meaning.

[6] The 1st manuscript reads '*Man* of Ross', so this is something of a return to the earliest form.

[7] The manuscripts of the *Essay on man* are reproduced by Mack in *The last and greatest art*, pp. 190–418.

(a)

with Text matter, and leaving the other Column for Insertions, Alterations, Notes, &c.' The opening paragraph in the left-hand column shows how heavily Pope could italicize later in his life—it is as heavy as the *Essay on criticism*. The same passage at the top of the second page of fair copy (Figure 94) is entirely free from italic; only the first line on the preceding page has the italic 'But know, the Study of Mankind is *Man*'.

When we turn from the second book to the first, we find that the two surviving manuscripts are successive fair copies,[8] so the earlier is already light in its under-

[8] The Morgan MS of this book is a fair copy, unlike the rest of the Morgan MS, but it was superseded by the Houghton MS. As noted in Ch. 4, n. 12 above, it was marked off in 14- and 16-line sections.

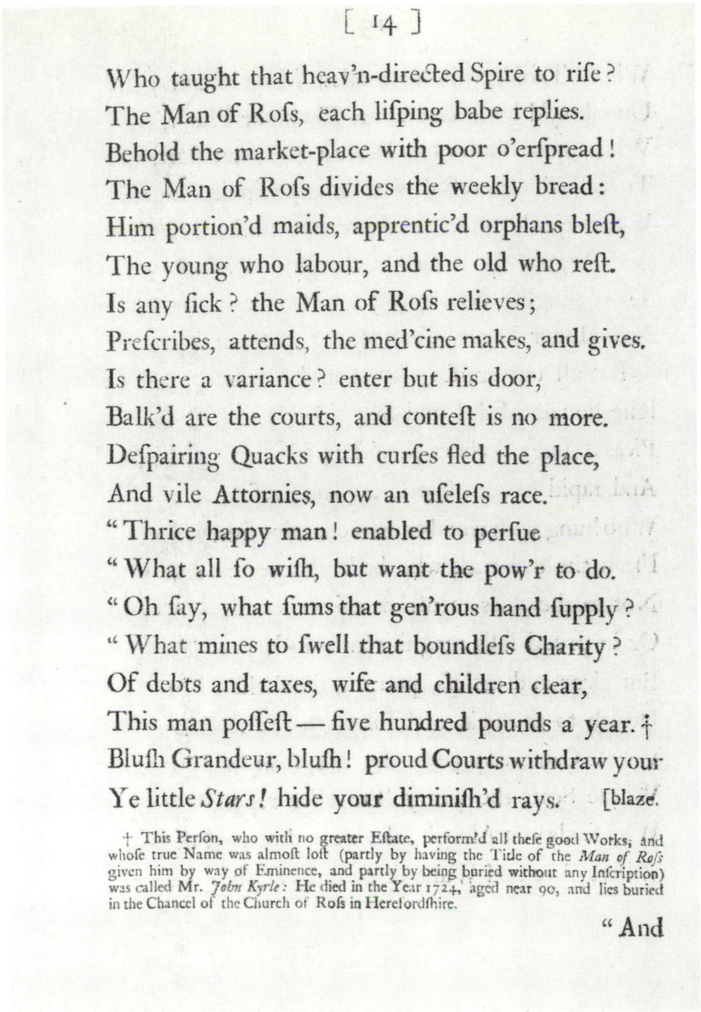

[14]

Who taught that heav'n-directed Spire to rise?
The Man of Ross, each lisping babe replies.
Behold the market-place with poor o'erspread!
The Man of Ross divides the weekly bread:
Him portion'd maids, apprentic'd orphans blest,
The young who labour, and the old who rest.
Is any sick? the Man of Ross relieves;
Prescribes, attends, the med'cine makes, and gives.
Is there a variance? enter but his door,
Balk'd are the courts, and contest is no more.
Despairing Quacks with curses fled the place,
And vile Attornies, now an useless race.
"Thrice happy man! enabled to persue
"What all so wish, but want the pow'r to do.
"Oh say, what sums that gen'rous hand supply?
"What mines to swell that boundless Charity?
Of debts and taxes, wife and children clear,
This man possest — five hundred pounds a year. †
Blush Grandeur, blush! proud Courts withdraw your
Ye little *Stars!* hide your diminish'd rays.   [blaze.

14

† This Person, who with no greater Estate, perform'd all these good Works, and whose true Name was almost lost (partly by having the Title of the *Man of Ross* given him by way of Eminence, and partly by being buried without any Inscription) was called Mr. *John Kyrle*: He died in the Year 1724, aged near 90, and lies buried in the Chancel of the Church of Ross in Herefordshire.

"And

(b)

FIG. 91. Autograph manuscript of *Epistle to Bathurst*, fo. 349ʳ (Huntington HM 6007; 238 × 187) and first edition (1732), p. 14 (Bodl. M 3.19(3) Art; 365 × 219)

lining. Figure 95 shows two pages which overlap for the passage beginning 'Presumptuous Man!' It is, however, the following paragraph beginning 'Respecting Man' which is interesting, since the second draft drops all the italics found in the first. But the passage inserted in the margin, 'Of Systems possible', reverts to Pope's natural use of italic in '*full*', '*coherent*', '*degree*', '*plac'd*'. There is also some change in capitalization of the 'Respecting Man' paragraph: the second draft capitalizes 'Wrong' in the first line and 'Movements' in the fourth; it also removes the capital from 'use' two lines later. Notice the running title 'EPISTLES' 'Book I.' (actually 'misprinted') added by Pope at the head of the second manuscript when he inserted four lines above the first page number 3: this was intended for the collected printing of the whole *Essay*, as we shall see.

[14]

Who taught that heav'n-directed Spire to rife?
The Man of Rofs, each lifping babe replies.
Behold the market-place with poor o'erfpread!
The Man of Rofs divides the weekly bread:
Him portion'd maids, apprentic'd orphans bleft,
The young who labour, and the old who reft.
Is any fick? the Man of Rofs relieves;
Prefcribes, attends, the med'cine makes, and gives.
Is there a variance? enter but his door,
Balk'd are the courts, and conteft is no more.
Defpairing Quacks with curfes fled the place,
And vile Attornies, now an ufelefs race.
"Thrice happy man! enabled to perfue
"What all fo wifh, but want the pow'r to do.
"Oh fay, what fums that gen'rous hand fupply?
"What mines to fwell that boundlefs Charity?
Of debts and taxes, wife and children clear,
This man poffeft —— five hundred pounds a year. †
Blufh Grandeur, blufh! proud Courts withdraw your
Ye little *Stars!* hide your diminifh'd rays.　[blaze.

† This Perfon, who with no greater Eftate, perform'd all thefe good Works, and whofe true Name was almoft loft (partly by having the Title of the *Man of Rofs* given him by way of Eminence, and partly by being buried without any Infcription) was called Mr. *John Kyrle*: He died in the Year 1724, aged near 90, and lies buried in the Chancel of the Church of Rofs in Herefordfhire.

"And

[16]

Who taught that heav'n-directed Spire to rife?
The Man of *Rofs*, each lifping babe replies.
Behold the market-place with poor o'erfpread!
The Man of *Rofs* divides the weekly bread:
Him portion'd-maids, apprentic'd-orphans bleft,
The young who labour, and the old who reft.
Is any fick? the Man of *Rofs* relieves;
Prefcribes, attends, the med'cine makes, and gives.
Is there a variance? enter but his door,
Balk'd are the courts, and conteft is no more.
Defpairing Quacks with curfes fled the place,
And vile Attornies, now an ufelefs race.
"Thrice happy man! enabled to perfue
"What all fo wifh, but want the pow'r to do.
"Oh fay, what fums that gen'rous hand fupply?
"What mines to fwell that boundlefs Charity?
Of debts and taxes, wife and children clear,
This man poffeft —— five hundred pounds a year. †
Blufh Grandeur, blufh! proud Courts withdraw your
Ye little *Stars!* hide your diminifh'd rays.　[blaze.

† This Perfon, who with no greater Eftate, perform'd all thefe good Works, and whofe true Name was almoft loft (partly by having the Title of the *Man of Rofs* given him by way of Eminence, and partly by being buried without any Infcription) was called Mr. *John Kyrle*: He died in the Year 1724, aged near 90, and lies buried in the Chancel of the Church of Rofs in *Herefordfhire.*

"And

When we compare this manuscript with the first edition (Figure 96) we see that some of the italics in the marginal paragraph, 'Of Systems possible', have been taken over: '*full*' and '*coherent*', as well as '*plac'd*' extended to '*plac'd him wrong*'. But 'degree' is not in italic, while '*some where*' is. It looks as though when Pope came across this italicized passage, whether in a transcript or in proof, he revised the italicization rather than abandoning it as in consistency with the rest of the text he should. What is more, in the following paragraph, 'Respecting Man', he carries on using italics, reverting to the '*Man*' and '*All*' found in the earlier manuscript and adding some different ones. Capitalization is also changed— 'Reason' capitalized in line 43 and 'fields' reduced in line 49, which seems a sensible change. In lines 59 and 62, 'wrong' and 'movements' return to the lower case of the earlier draft, while '*Use*' (line 64) regains the capital it had there; but the capital to '*End*' (line 63) is new and other capitals of the earlier draft are

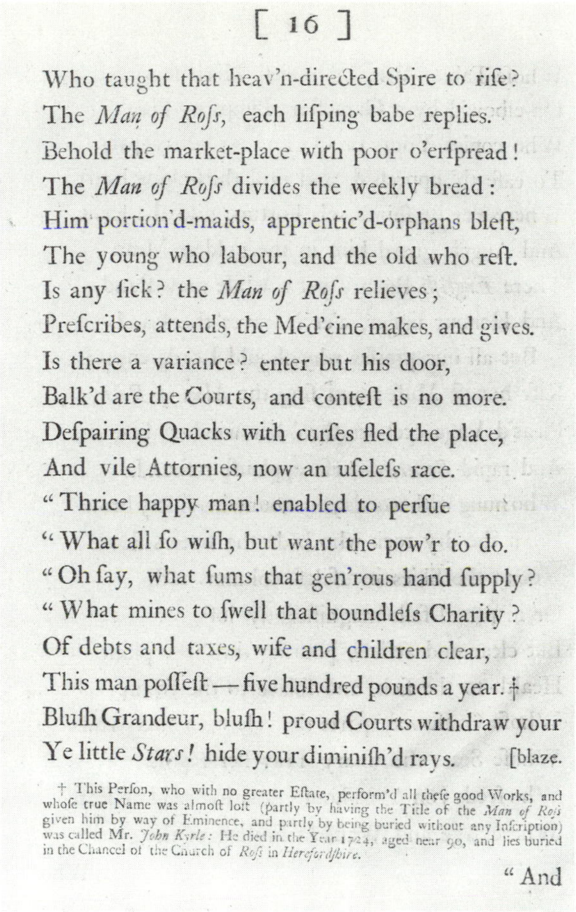

[16]

Who taught that heav'n-directed Spire to rife?
The *Man of Rofs*, each lifping babe replies.
Behold the market-place with poor o'erfpread!
The *Man of Rofs* divides the weekly bread:
Him portion'd-maids, apprentic'd-orphans bleft,
The young who labour, and the old who reft.
Is any fick? the *Man of Rofs* relieves;

1

Prefcribes, attends, the Med'cine makes, and gives.
Is there a variance? enter but his door,

2

Balk'd are the Courts, and conteft is no more.
Defpairing Quacks with curfes fled the place,
And vile Attornies, now an ufelefs race.
"Thrice happy man! enabled to perfue
"What all fo wifh, but want the pow'r to do.
"Oh fay, what fums that gen'rous hand fupply?
"What mines to fwell that boundlefs Charity?
Of debts and taxes, wife and children clear,
This man poffeft — five hundred pounds a year. †
Blufh Grandeur, blufh! proud Courts withdraw your
Ye little *Stars*! hide your diminifh'd rays.   [blaze.

† This Perfon, who with no greater Eftate, perform'd all thefe good Works, and
whofe true Name was almoft loft (partly by having the Title of the *Man of Rofs*
given him by way of Eminence, and partly by being buried without any Infcription)
was called Mr. *John Kyrle*: He died in the Year 1724, aged near 90, and lies buried
in the Chancel of the Church of *Rofs* in *Herefordfhire*.

"And

FIG. 92. First three editions of *Epistle to Bathurst*: (*a*) first edition (1732), p. 14 (Bodl. M 3.19(3) Art; 365 × 219); (*b*) second edition (1733), p. 16 (Bodl. Don. c. 43; 352 × 220); (*c*) third [called second] edition (1733), p. 16 (Bodl. Vet. A4 c. 291; 326 × 214)

ignored. The revision is a selective one. Note also that 'JOVE' (line 50) is honoured by caps. and small caps.

This passage is of particular interest because Pope made revisions in standing type between the first two folio printings (Figure 97). Because the same setting of type is used, we can be sure that the changes made are the result of a deliberate intervention rather than due to a whim of a compositor in resetting the passage. The third folio printing, 'corrected by the author', is from a new setting of type, but as the text is based on the first, Pope's revisions to accidentals in the second are lost.[9]

I have argued in Chapter 3 that the second version was intended to form part of the second volume of Pope's *Works*, and this is confirmed by the insertion of the

[9] The printer of all three was John Huggonson.

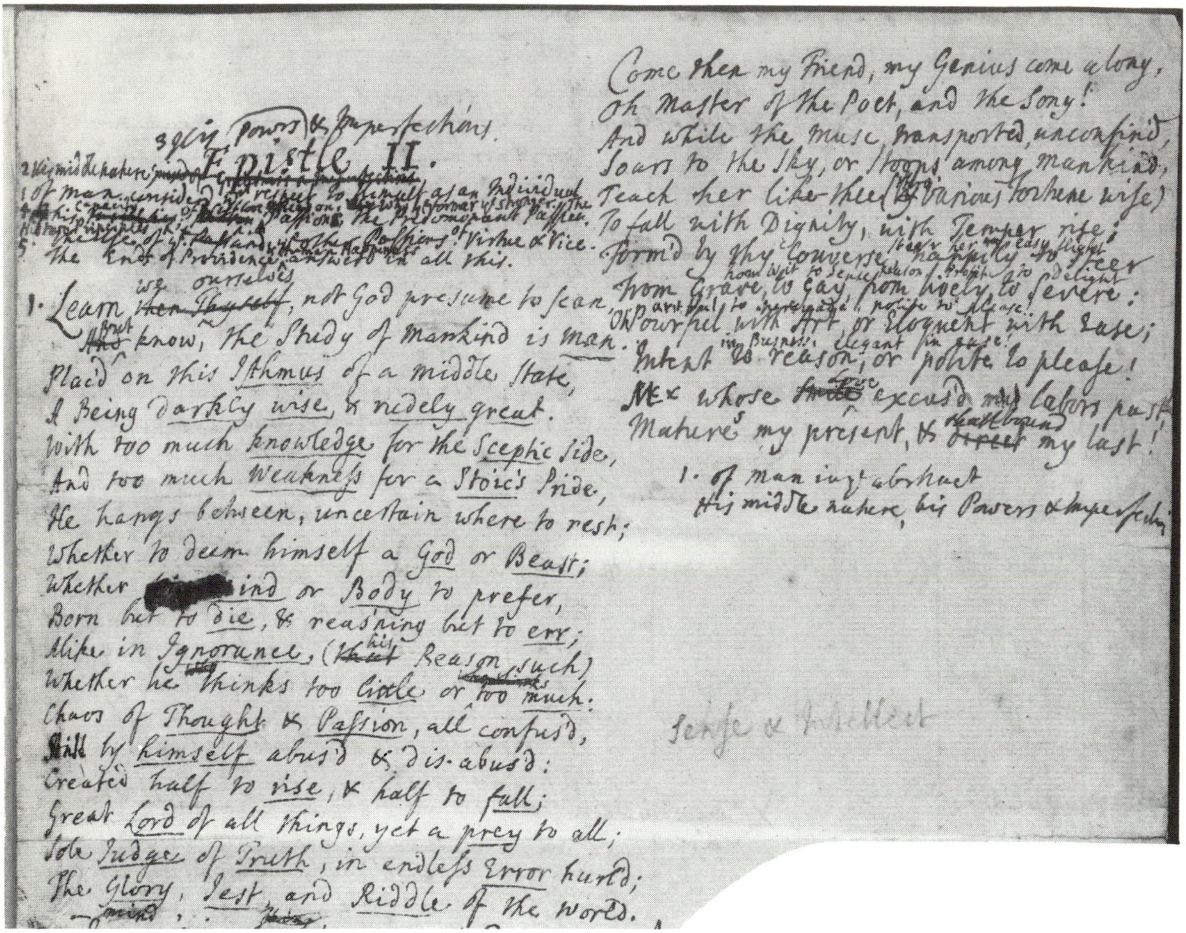

FIG. 93. Early draft of opening of *Essay on man*, book II (Pierpont Morgan MS MA 348; 318 × 203)

running-title 'Epistles' which Pope had already put in his second manuscript; it is removed from the third printing. The most obvious change in the first two is that Pope has closed up the white lines between his verse paragraphs to bring this text into line with the typography of the planned *Works* II, and has made a separate paragraph of the couplet in lines 59–60. In the paragraph we saw inserted in the manuscript (lines 51–8), he has removed italics from 'full', 'coherent', 'Man', and 'him wrong', in the last case reverting to the manuscript reading. As for capitals, they are removed from 'weeds' (line 48) and 'scale' (line 55). 'Jove' is printed in roman, not small caps. (line 50), and the spelling of 'e'er' corrected in line 57.

In almost all respects the third version (which transposes the verse paragraph preceding our extract to follow it) reverts to the first; the exceptions are that in line 41 (first edition, line 49) 'Fields' has a capital and in line 46 (first edition, line

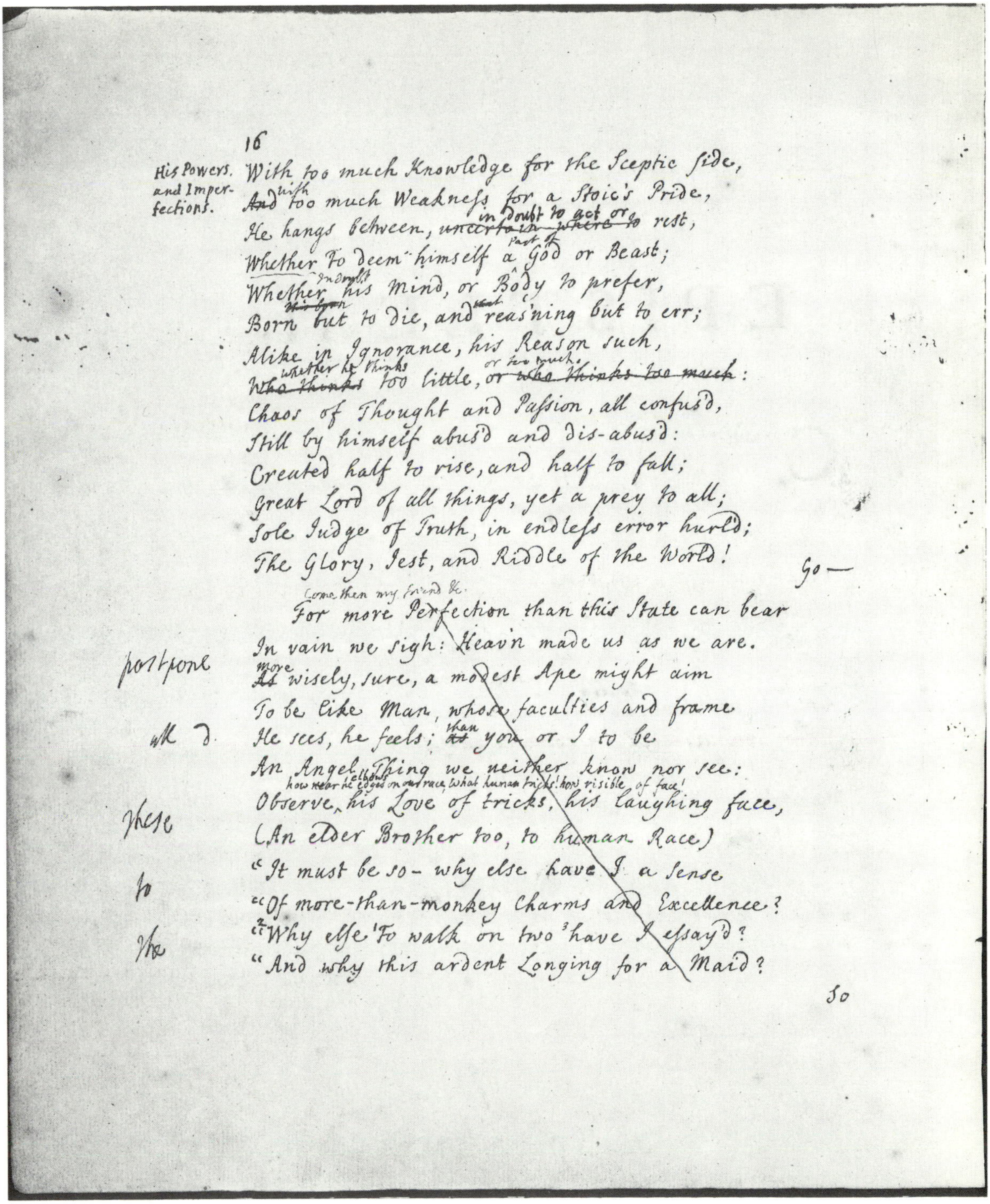

16

His Powers.
and Imper-
fections.

With too much Knowledge for the Sceptic side,
And too much Weakness for a Stoic's Pride,
He hangs between, uncertain where to rest,
Whether to deem himself a God or Beast;
Whether his Mind, or Body to prefer,
Born but to die, and reas'ning but to err;
Alike in Ignorance, his Reason such,
Who thinks too little, or who thinks too much:
Chaos of Thought and Passion, all confus'd,
Still by himself abus'd and dis-abus'd:
Created half to rise, and half to fall;
Great Lord of all things, yet a prey to all;
Sole Judge of Truth, in endless error hurl'd;
The Glory, Jest, and Riddle of the World!   Go —

For more Perfection than this State can bear
In vain we sigh: Heav'n made us as we are.
As wisely, sure, a modest Ape might aim
To be like Man, whose faculties and frame
He sees, he feels; as you or I to be
An Angel Thing we neither know nor see:
Observe, his Love of tricks, his laughing face,
(An elder Brother too, to human Race)
"It must be so — why else have I a Sense
"Of more-than-monkey Charms and Excellence?
"Why else 'To walk on two' have I essay'd?
"And why this ardent Longing for a Maid?

postpone

w.th ?

these

to

the

50

FIG. 94. Fair copy of part of opening of *Essay on man*, book II (Houghton MS Eng. 233.1;
297 × 241)

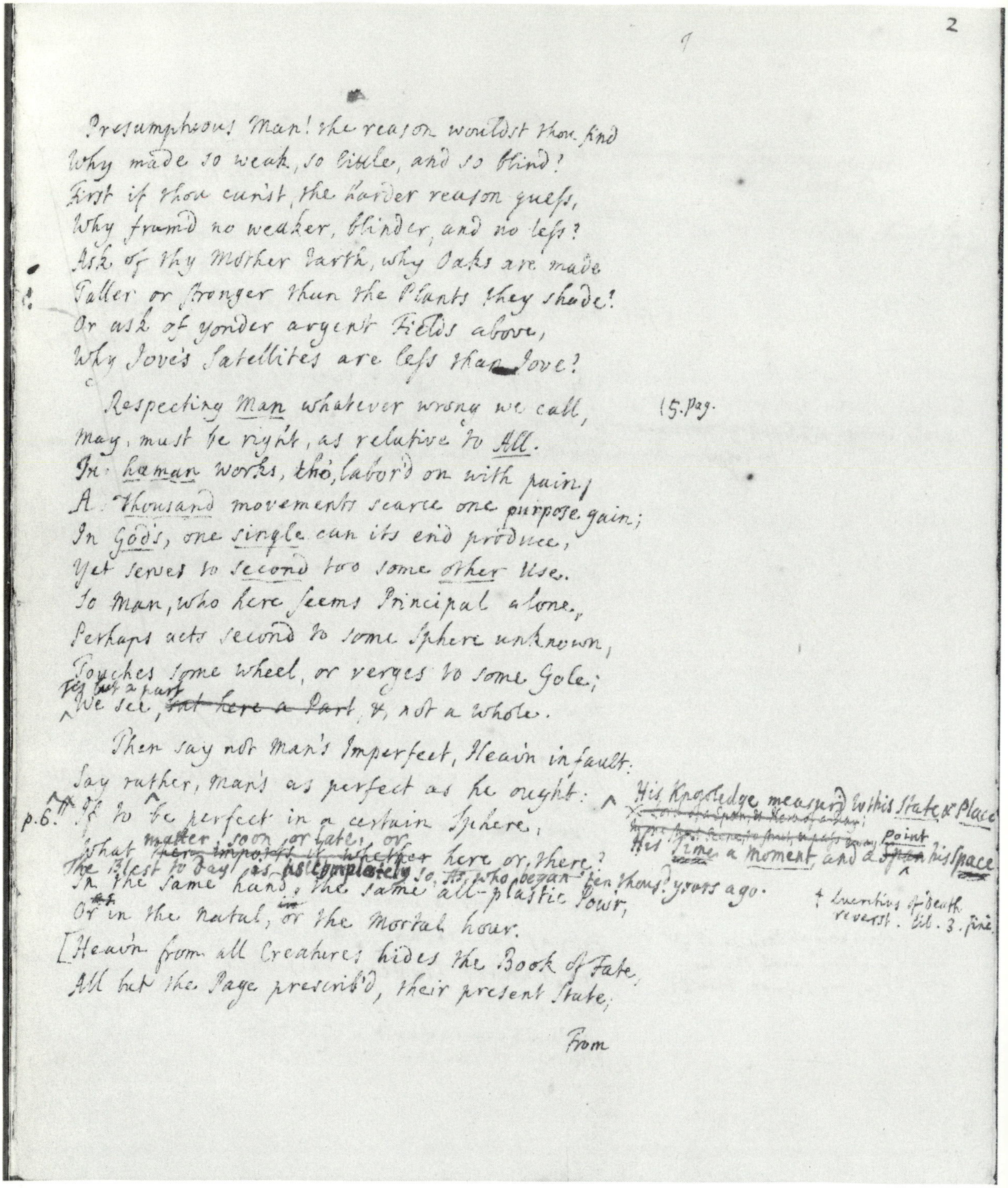

Presumptuous Man! the reason wouldst thou find
Why made so weak, so little, and so blind?
First if thou canst, the harder reason guess,
Why framd no weaker, blinder, and no less?
Ask of thy Mother Earth, why Oaks are made
Taller or stronger than the Plants they shade!
Or ask of yonder argent Fields above,
Why Jove's Satellites are less than Jove?

Respecting Man whatever wrong we call,
May, must be right, as relative to All.
In human works, tho, labor'd on with pain,
A thousand movements scarce one purpose gain;
In God's, one single can its end produce,
Yet serves to second too some other Use.
So man, who here seems Principal alone,
Perhaps acts second to some Sphere unknown,
Touches some wheel, or verges to some Gole;
We see, but here a Part, & not a Whole.

Then say not Man's Imperfect, Heav'n in fault.
Say rather, Man's as perfect as he ought:
If so be perfect in a certain Sphere,
What matter soon or later, or whether here or there?
The Blest to day is as compleatly so,
In the same hand, the same all-plastic Powr,
Or in the natal, or the Mortal hour.
[Heav'n from all Creatures hides the Book of Fate,
All but the Page prescrib'd, their present State;

From

His Knowledge measurd to this State & Place
His Time a moment, and a point his Space

FIG. 95. Two fair copies of *Essay on man*, book I: (*a*) Pierpont Morgan MS MΛ 348 (225 × 183), (*b*) Houghton MS Eng. 233.1 (297 × 241)

He who thro' vast Im
See worlds on worlds
Observe how System

What other
What vary'd Being peoples ev'ry Star;
May tell, why Heav'n has made us as we are.

  When the proud ~~Steed~~ shall know; why Man ~~now rein~~ restrains
His ~~stubborn neck~~ fiery course, ~~now~~ or urges drives him o'er the Plains;
When the dull Oxe, why now he breaks the Clod,
Now wears a Garland, an Ægyptian God;
Then shall Man's Pride and Dulness comprehend
His Action's, Passion's, Being's, Use and End;
Why doing, suffering, check'd, impell'd; and why
This hour a Slave, the next a Deity.

  Presumptuous Man! the reason wouldst thou find
Why made so weak, so little, and so blind?
First, if thou can'st, the harder reason guess,
Why form'd no weaker, blinder, and no less?
Ask of thy Mother Earth, why Oaks are made
Taller or stronger than the Plants they shade?
Or ask, of yonder argent Fields above,
Why Jove's Satellites are less than Jove?

  Respecting Man, whatever wrong we call
May, must be, right, as relative to All.
In human works, tho' labour'd on with pain
A thousand Movements scarce one purpose gain;
In God's, one single can its end produce,
Yet serves to second too some other use:

Of Systems possible, if 'tis confest
That Wisdom infinite must form the best,
Where all must full, or not coherent be,
And all that rises, rise in due degree;
Then, in the scale of Life & Sense, 'tis plain
There must be some where, such a Rank as Man:
And all the question (argue we so long)
× Is only this, If God has plac'd him wrong?
× Is but if God has plac'd his
   creature wrong?

So

When the proud Steed shall know; why Man restrains
His fiery course, or urges, now drives him o'er the Plains;
When the dull Oxe, why now he breaks the clod,
Now wears a Garland, an Ægyptian God;
Then shall Man's Pride and Dulness comprehend
His Action's, Passion's, Being's, Use and End;
Why doing, suffering, check'd, impell'd; and why
This hour a Slave, the next a Deity.

Presumptuous Man! the reason wouldst thou find
Why made so weak, so little, and so blind?
(45) First, if thou can'st, the harder reason guess,
Why form'd no weaker, blinder, and no less?
Ask of thy Mother Earth, why Oaks are made
Taller or stronger than the Plants they shade?
Or ask, of yonder argent Fields above,
(50) Why Jove's Satellites are less than Jove?

Respecting Man, whatever Wrong we call,
(60) May, must be, right, as relative to All.
In human works, tho' labour'd on with pain
A thousand Movements scarce one purpose gain;
In God's, one single can its end produce,
Yet serves to second too some other use:

Of Systems possible, if 'tis confest
That wisdom infinite must form ye best
Where all must full, or not coherent be,
And all that rises, rise in due degree;
Then, in the scale of Life & Sense, 'tis plain
There must be somewhere such a Race
And all the question (argue 'e'er so long)
×Is only this, If God has plac'd him wrong?
×Is but if God has plac'd his
     creature wrong?

So

54) ‘*degree*’ is italicized, both as they were in the manuscript. One is led to wonder whether Pope kept the manuscript at hand when revising accidentals or whether the transcript used for the first edition was used again here, with slightly different changes made in proof. There are two related substantive revisions, ‘form'd’ for ‘made’ in lines 36 and 38: the first manuscript reads ‘made’ and ‘fram'd’, the second ‘made’ and ‘form'd’. Clearly Pope decided that elegant variation was inappropriate, but changed his decision on which word to repeat. (The misspelling ‘‘ere’ reappears in line 49, and is copied by the quarto editions.)

The next stage in the revision of the text was the collected edition of the four epistles, which was incorporated in the 1735 *Works* (with the exception of the

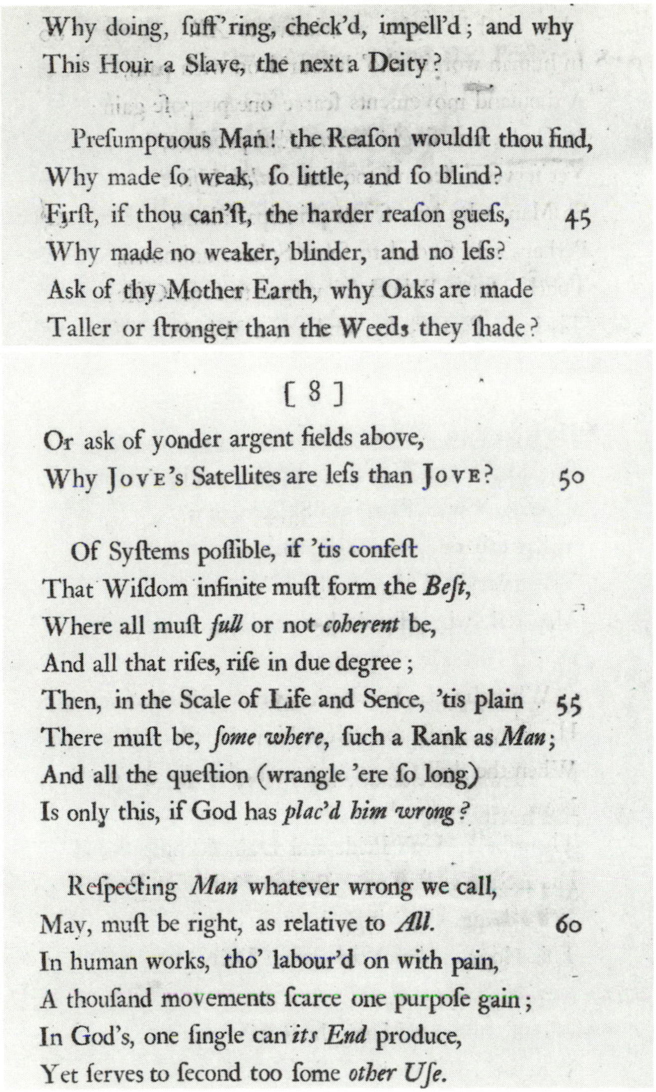

Why doing, fuff'ring, check'd, impell'd; and why
This Hour a Slave, the next a Deity?

Prefumptuous Man! the Reafon wouldft thou find,
Why made fo weak, fo little, and fo blind?
Firft, if thou can'ft, the harder reafon guefs,   45
Why made no weaker, blinder, and no lefs?
Ask of thy Mother Earth, why Oaks are made
Taller or ftronger than the Weeds they fhade?

[8]

Or ask of yonder argent fields above,
Why J o v e's Satellites are lefs than J o v e?   50

Of Syftems poffible, if 'tis confeft
That Wifdom infinite muft form the *Beft*,
Where all muft *full* or not *coherent* be,
And all that rifes, rife in due degree;
Then, in the Scale of Life and Sence, 'tis plain   55
There muft be, *fome where*, fuch a Rank as *Man*;
And all the queftion (wrangle 'ere fo long)
Is only this, if God has *plac'd him wrong?*

Refpecting *Man* whatever wrong we call,
May, muft be right, as relative to *All*.   60
In human works, tho' labour'd on with pain,
A thoufand movements fcarce one purpofe gain;
In God's, one fingle can *its End* produce,
Yet ferves to fecond too fome *other Ufe*.

FIG. 96. Fair copy of *Essay on man*, book I (Houghton MS Eng. 233.1; 297 × 241) and first edition (1733), pp. 7–8 (Bodl. M 3.19(8) Art; 364 × 218)

ordinary-paper quartos for which a new quarto edition was printed; it shows no revisions in this passage). The collected edition (Figure 98) carries on the revision found in the second folio to its logical conclusion, closing up the verse paragraphs, removing all italics, and reducing 'Jove' to lower case again. It also goes further in removing initial capitals: in lines 39–41 we have 'mother Earth', 'oaks', 'weeds', and 'fields'; in the next paragraph 'best', 'scale of life and sense' (with the spelling normalized), and 'rank'; and in lines 59–60 'wheel' 'gole', 'part', and 'whole'. I would regard this as the triumph of Pope's desire to classicize over the tendency to capitals and italics found in the intermediate editions.

Prefumptuous Man! the Reafon wouldft thou find,
Why made fo weak, fo little, and fo blind?
Firft, if thou can'ft, the harder reafon guefs,   45
Why made no weaker, blinder, and no lefs?
Ask of thy Mother Earth, why Oaks are made
Taller or ftronger than the Weeds they fhade?

[8]

Or ask of yonder argent fields above,
Why JOVE's Satellites are lefs than JOVE?   50

Of Syftems poffible, if 'tis confeft
That Wifdom infinite muft form the *Beft,*
Where all muft *full* or not *coherent* be,
And all that rifes, rife in due degree;
Then, in the Scale of Life and Sence, 'tis plain   55
There muft be, *fome where,* fuch a Rank as *Man;*
And all the queftion (wrangle 'ere fo long)
Is only this, if God has *plac'd him wrong?*

Refpecting *Man* whatever wrong we call,
May, muft be right, as relative to *All.*   60
In human works, tho' labour'd on with pain,
A thoufand movements fcarce one purpofe gain;
In God's, one fingle can *its End* produce,
Yet ferves to fecond too fome *other Ufe.*
So Man, who here feems principal alone,   65
Perhaps acts fecond to fome Sphere unknown,
Touches fome Wheel, or verges to fome Gole;
'Tis but a Part we fee, and not a Whole.

Prefumptuous Man! the Reafon wouldft thou find,
Why made fo weak, fo little, and fo blind?
Firft, if thou can'ft, the harder reafon guefs,   45
Why made no weaker, blinder, and no lefs?
Ask of thy Mother Earth, why Oaks are made
Taller or ftronger than the weeds they fhade?
Or ask of yonder argent fields above,
Why Jove's Satellites are lefs than Jove?   50

10   *EPISTLES.*

Of Syftems poffible, if 'tis confeft
That Wifdom infinite muft form the *Beft,*
Where all muft full or not coherent be,
And all that rifes, rife in due degree;
Then, in the fcale of Life and Sence, 'tis plain   55
There muft be, *fome where,* fuch a Rank as Man;
And all the queftion (wrangle e'er fo long)
Is only this, if God has *plac'd* him wrong?
Refpecting *Man* whatever wrong we call,
May, muft be right, as relative to *All.*   60
In human works, tho' labour'd on with pain,
A thoufand movements fcarce one purpofe gain;
In God's, one fingle can *its End* produce,
Yet ferves to fecond too fome *other Ufe.*
So Man, who here feems principal alone,   65
Perhaps acts fecond to fome Sphere unknown,
Touches fome Wheel, or verges to fome Gole;
'Tis but a Part we fee, and not a Whole.

We can also see very clearly how Pope reverts to the use of italic in trade editions by comparing this with the first edition of the octavo *Works* of 1735 and then with the death-bed edition (Figure 98). In lines 35 to 60 of the octavo the italics revert to those of the third folio with the exception of roman 'Man' in lines 48 and 51 ('man'). 'JOVE' reverts to the caps. and small caps. of the first and third folios. Typically the octavo continues to reduce capitals, removing nine from this passage; the capital for 'Satellites' in line 42 is kept because Pope is using the Latin form for the technical astronomical term. It is worth noting that Pope later

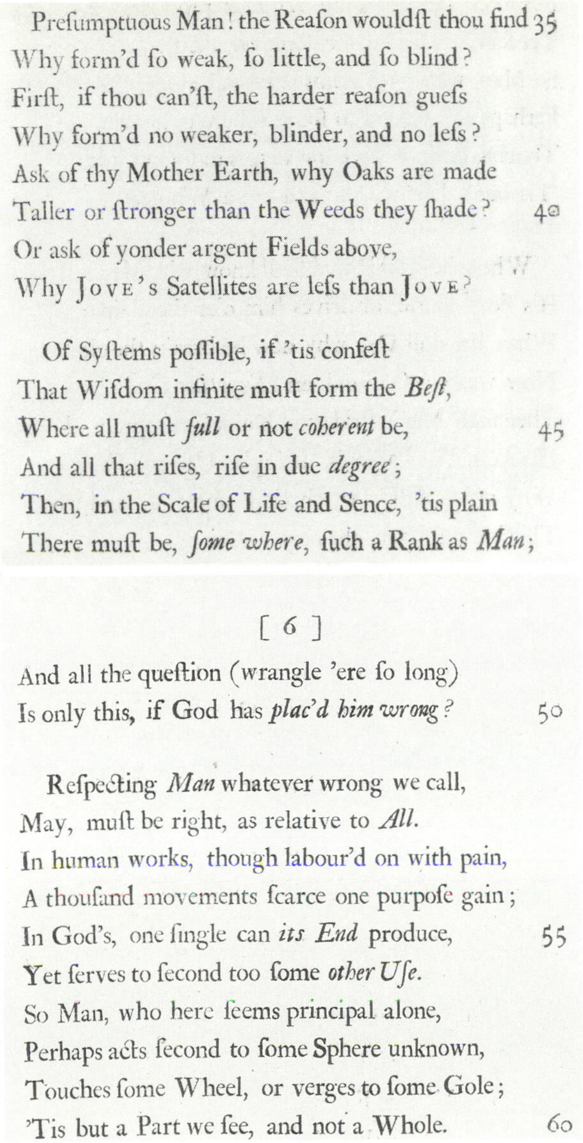

Prefumptuous Man! the Reafon wouldft thou find 35
Why form'd fo weak, fo little, and fo blind?
Firft, if thou can'ft, the harder reafon guefs
Why form'd no weaker, blinder, and no lefs?
Ask of thy Mother Earth, why Oaks are made
Taller or ftronger than the Weeds they fhade? 40
Or ask of yonder argent Fields above,
Why JOVE's Satellites are lefs than JOVE?

Of Syftems poffible, if 'tis confeft
That Wifdom infinite muft form the *Beft*,
Where all muft *full* or not *coherent* be, 45
And all that rifes, rife in due *degree*;
Then, in the Scale of Life and Sence, 'tis plain
There muft be, *fome where*, fuch a Rank as *Man*;

[6]

And all the queftion (wrangle 'ere fo long)
Is only this, if God has *plac'd him wrong?* 50

Refpecting *Man* whatever wrong we call,
May, muft be right, as relative to *All*.
In human works, though labour'd on with pain,
A thoufand movements fcarce one purpofe gain;
In God's, one fingle can *its End* produce, 55
Yet ferves to fecond too fome *other Ufe*.
So Man, who here feems principal alone,
Perhaps acts fecond to fome Sphere unknown,
Touches fome Wheel, or verges to fome Gole;
'Tis but a Part we fee, and not a Whole. 60

FIG. 97. First three editions of *Essay on man*, book I: (*a*) first edition (1733), pp. 7–8 (Bodl. M 3.19(8) Art; 364 × 218); (*b*) second [revised standing type], pp. 9–10 (Bodl. S 3.5(50) Jur; 295 × 177); (*c*) third, pp. 5–6 (Bodl. M 3.19(9) Art; 364 × 218)

seems to have felt that in places he removed too many capitals, and restored some in a marked octavo of 1736 I discuss later, though not in this passage.

With the death-bed edition on the right of Figure 98 we are faced—as we were with the editions of the *Essay on criticism* and the *Dunciad* in the last chapter— with the likelihood of Warburton at work, and a consequent doubt as to which changes are due to Pope. What complicates matters further is that Maynard Mack has shown that in a passage following shortly on the one illustrated the substitution of 'Knowledge' for 'Being' (line 71), which might well have been

Is the great Chain that draws all to agree,
And drawn fupports, upheld by God, or thee?
  Prefumptuous Man! the Reafon would'ft thou find
Why form'd fo weak, fo little, and fo blind? 36
Firft, if thou can'ft, the harder reafon guefs
Why form'd no weaker, blinder, and no lefs?
Ask of thy mother Earth, why oaks are made
Taller or ftronger than the weeds they fhade? 40
Or ask of yonder argent fields above,
Why Jove's Satellites are lefs than Jove?
  Of Syftems poffible, if 'tis confeft
That Wifdom infinite muft form the beft,
Where all muft full, or not coherent be, 45
And all that rifes, rife in due degree;
Then, in the fcale of life and fenfe, 'tis plain
There muft be, fome where, fuch a rank as Man;

And all the queftion (wrangle 'ere fo long)
Is only this, if God has plac'd him wrong? 50
  Refpecting Man whatever wrong we call,
May, muft be right, as relative to All.
In human works, though labour'd on with pain,
A thoufand movements fcarce one purpofe gain;
In God's, one fingle can its End produce, 55
Yet ferves to fecond too fome other Ufe.
So Man, who here feems principal alone,
Perhaps acts fecond to a Sphere unknown,
Touches fome wheel, or verges to fome gole;
'Tis but a part we fee, and not a whole. 60

Is the great Chain that draws all to agree,
And drawn fupports, upheld by God, or thee?
  Prefumptuous man! the reafon wouldft thou find 35
Why form'd fo weak, fo little, and fo blind?
Firft, if thou canft, the harder reafon guefs
Why form'd no weaker, blinder, and no lefs?
Ask of thy mother earth, why oaks are made
Taller or ftronger than the weeds they fhade? 40
Or ask of yonder argent fields above,
Why Jove's Satellites are lefs than Jove?
  Of Syftems poffible, if 'tis confeft
That Wifdom infinite muft form the *beft*,
Where all muft *full* or not *coherent* be, 45
And all that rifes, rife in due *degree*;
Then, in the fcale of life and fenfe, 'tis plain
There muft be, *fome where*, fuch a rank as Man;
And all the queftion (wrangle e're fo long)
Is only this, if God has *plac'd him wrong*? 50
  Refpecting man whatever wrong we call,
May, muft be right, as relative to *all*.
In human works, though labour'd on with pain,
A thoufand movements fcarce one purpofe gain;
In God's, one fingle can *its end* produce, 55
Yet ferves to fecond too fome *other ufe*.
So man, who here feems principal alone,
Perhaps acts fecond to fome fphere unknown,
Touches fome wheel, or verges to fome gole;
'Tis but a part we fee, and not a whole. 60

suggested by Warburton, is actually a return to Pope's original reading found in the manuscript—so Pope's original manuscript may also be a source here.[10]

I am sure we can attribute to Pope the general decision to remove the italics once more—this is another 'classic' edition. 'Man' is capitalized again in lines 35, 51, and 57, and this could be a return to the manuscript. But when we come to the

---

[10] Mack, *The last and greatest art*, p. 198. Mack also points out that an additional couplet in epistle III (lines 307–8) is to be found in the Morgan MS. In *Collected in himself*, pp. 337–8, he says the manuscript was also consulted in revision of the passage on man as child in epistle II. For an interesting discussion of the death-bed changes (but without reference to the manuscripts), see R. W. Rogers, 'Notes on Pope's collaboration with Warburton', *PQ* xxvi (1947), 358–66.

Is the great chain, that draws all to agree,
And drawn fupports, upheld by God, or thee?
35   II. Prefumptuous Man! the reafon wouldft thou find,
Why form'd fo weak, fo little, and fo blind!
Firft, if thou canft, the .harder reafon guefs,
Why form'd no weaker, blinder, and no lefs!
Afk of thy mother earth, why oaks are made
40   Taller or ftronger than the weeds they fhade?
Or afk of yonder argent fields above,
Why JOVE's Satellites are lefs than JOVE?
Of Syftems poffible, if 'tis confeft
That Wifdom infinite muft form the beft,

45   Where all muft full or not coherent be,
And all that rifes, rife in due degree;
Then, in the fcale of reas'ning life, 'tis plain
There muft be, fomewhere, fuch a rank as Man;
And all the queftion (wrangle e'er fo long)
50   Is only this, if God has plac'd him wrong?
Refpecting Man, whatever wrong we call,
May, muft be right, as relative to all.

In human works, tho' labour'd on with pain,
A thoufand movements fcarce one purpofe gain;
55   In God's, one fingle can its end produce;
Yet ferves to fecond too fome other ufe.
So Man, who here feems principal alone,
Perhaps acts fecond to fome fphere unknown,
Touches fome wheel, or verges to fome goal;
60   'Tis but a part we fee, and not a whole.

FIG. 98. Three later editions of *Essay on man*: (*a*) collected quarto (1734), pp. 9–10 (Bodl. G. Pamph. 1487(6); 262 × 204); (*b*) octavo *Works* II (1735), pp. 8–9 (Bodl. Radcliffe. f. 244; 170 × 104); (*c*) death-bed quarto (1743), pp. 4–6 (Bodl. CC 51(2) Art; 255 × 191)

substantive change in line 47 from 'the scale of life and sense' to 'the scale of reas'ning life' there is no manuscript precedent, and we can suspect Warburton's hand, as in such formal matters as the substitution of exclamation for question marks in lines 36 and 38 and the elimination of the archaic spelling 'gole'—to rhyme with 'whole' (line 59)—which had been preserved from Pope's manuscript through all editions. The trouble is that we have no way of being sure who was responsible.

When we consider the changes to capitals and italics I have shown, it is worth stressing that a process of piecemeal revision of this kind, especially when directed by an author whose preferences vary according to the nature of the edition, takes a long time even to approach consistency. If we look at the use of italics in the *Works* of 1735, we see that the oldest poem, the *Dunciad*, is completely free of italic, while in the *Essay on man* it is the first and most reprinted epistle

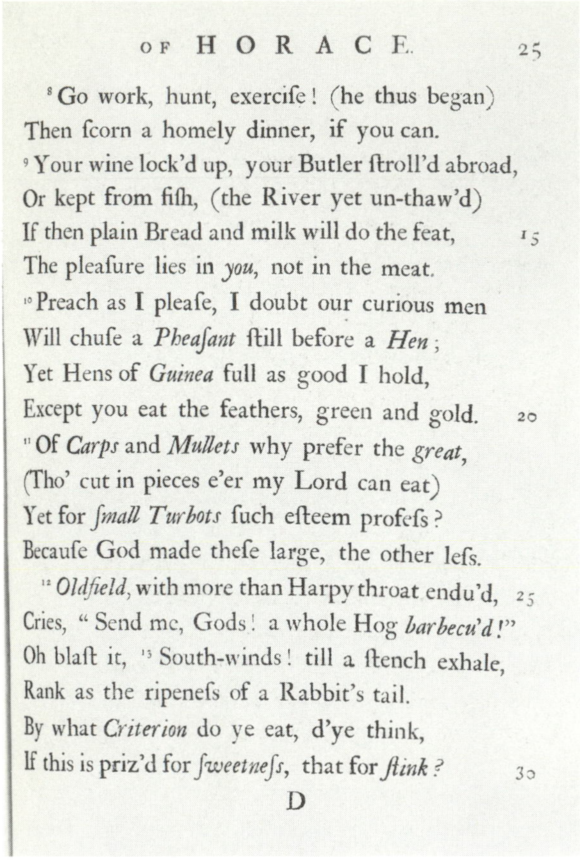

OF HORACE. 25

⁸Go work, hunt, exercife! (he thus began)
Then fcorn a homely dinner, if you can.
⁹Your wine lock'd up, your Butler ftroll'd abroad,
Or kept from fifh, (the River yet un-thaw'd)
If then plain Bread and milk will do the feat, 15
The pleafure lies in *you*, not in the meat.
¹⁰Preach as I pleafe, I doubt our curious men
Will chufe a *Pheafant* ftill before a *Hen*;
Yet Hens of *Guinea* full as good I hold,
Except you eat the feathers, green and gold. 20
¹¹Of *Carps* and *Mullets* why prefer the *great*,
(Tho' cut in pieces e'er my Lord can eat)
Yet for *fmall Turbots* fuch efteem profefs?
Becaufe God made thefe large, the other lefs.
¹²*Oldfield*, with more than Harpy throat endu'd, 25
Cries, "Send me, Gods! a whole Hog *barbecu'd!*"
Oh blaft it, ¹³South-winds! till a ftench exhale,
Rank as the ripenefs of a Rabbit's tail.
By what *Criterion* do ye eat, d'ye think,
If this is priz'd for *fweetnefs*, that for *ftink*? 30

D

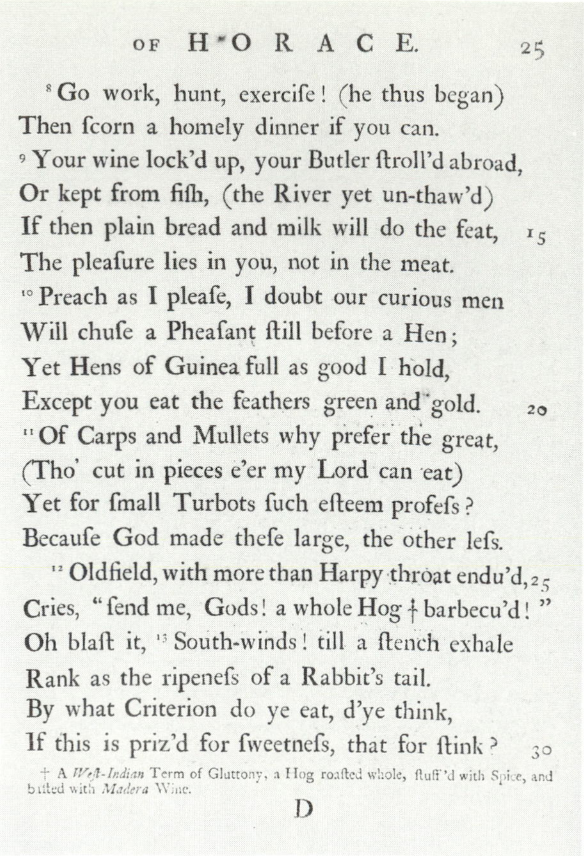

OF HORACE. 25

⁸Go work, hunt, exercife! (he thus began)
Then fcorn a homely dinner if you can.
⁹Your wine lock'd up, your Butler ftroll'd abroad,
Or kept from fifh, (the River yet un-thaw'd)
If then plain bread and milk will do the feat, 15
The pleafure lies in you, not in the meat.
¹⁰Preach as I pleafe, I doubt our curious men
Will chufe a Pheafant ftill before a Hen;
Yet Hens of Guinea full as good I hold,
Except you eat the feathers green and gold. 20
¹¹Of Carps and Mullets why prefer the great,
(Tho' cut in pieces e'er my Lord can eat)
Yet for fmall Turbots fuch efteem profefs?
Becaufe God made thefe large, the other lefs.
¹²Oldfield, with more than Harpy throat endu'd, 25
Cries, "fend me, Gods! a whole Hog † barbecu'd!"
Oh blaft it, ¹³South-winds! till a ftench exhale
Rank as the ripenefs of a Rabbit's tail.
By what Criterion do ye eat, d'ye think,
If this is priz'd for fweetnefs, that for ftink? 30

† A *Weſt-Indian* Term of Gluttony, a Hog roafted whole, ftuff'd with Spice, and baſted with *Madera* Wine.

D

FIG. 99. Removal of italic from *Second satire of the second book of Horace* between first edition (1734), p. 25 (Bodl. S 3.5(46) Jur; 291 × 197) and quarto *Works* II (1735), p. 25 (Bodl. Vet. A4 d. 141; 290 × 224)

from which most italic has been removed. The possibility of simple oversight is shown in the section containing the satires of Horace and Donne. *The first satire of the second book of Horace* had been published separately with heavy use of italic, and in the summer of 1734 Pope had it reprinted with the new second satire ready for inclusion in the collected *Works*. Italics were systematically removed from the first satire, but the early pages of the second satire again show heavy italic which must derive from Pope's manuscript. Pope had to reprint the second satire for the *Works*, but though he successfully removed the italics from one page (Figure 99), the second still escaped him (Figure 100) even though the second line was revised and two superfluous capitals in 'Friend' (line 44) and 'Vice' (line 46) were removed. It may be that the other revisions led him to neglect the italics—it is a well-known phenomenon of proof-reading that correcting

OF HORACE. 27

When the tir'd Glutton labours thro' a Treat,
The fweeteft thing will ftink that he can eat;
He calls for fomething bitter, fomething four,
And the rich feaft concludes extremely poor:
¹⁵Cheap eggs, and herbs, and olives ftill we fee,   35
Thus much is left of old Simplicity!
  ¹⁶The *Robin-red-breaft* till of late had reft,
And children facred held a *Martin*'s neft,
Till *Becca-ficos* fold fo dev'lifh dear
To one that was, or would have been a Peer.   40
  ¹⁷Let me extoll a *Cat* on Oyfters fed,
I'll have a Party at the *Bedford Head*,
Or ev'n to crack live *Crawfifh* recommend,
I'd never doubt at Court to make a Friend.
  ¹⁸'Tis yet in vain, I own, to keep a pother   45
About one Vice, and fall into the other:
Between Excefs and Famine lies a mean,
Plain, but not fordid, tho' not fplendid, clean.
¹⁹*Avidien* or his Wife (no matter which,
For him you'll call a ²⁰dog, and her a bitch)   50

D 2

OF HORACE. 27

When the tir'd Glutton labours thro' a Treat,
He'll find no relifh in the fweeteft Meat,
He calls for fomething bitter, fomething four,
And the rich feaft concludes extremely poor:
¹⁵Cheap eggs, and herbs, and olives ftill we fee,   35
Thus much is left of old Simplicity!
  ¹⁶The *Robin-red-breaft* till of late had reft,
And children facred held a *Martin*'s neft,
Till *Becca-ficos* fold fo dev'lifh dear
To one that was, or would have been a Peer.   40
¹⁷Let me extoll a *Cat* on Oyfters fed,
I'll have a Party at the *Bedford Head*,
Or ev'n to crack live *Crawfifh* recommend,
I'd never doubt at Court to make a friend.
  ¹⁸'Tis yet in vain, I own, to keep a pother   45
About one vice, and fall into the other:
Between Excefs and Famine lies a mean,
Plain, but not fordid, tho' not fplendid, clean.
¹⁹*Avidien* or his Wife (no matter which,
For him you'll call a ²⁰dog, and her a bitch)   50

D 2

FIG. 100. Failure to remove italic from *Second satire of the second book of Horace* between first edition (1734), p. 27 (Bodl. S 3.5(46) Jur; 291 × 197) and quarto *Works* II (1735), p. 27 (Bodl. Vet. A4 d. 141; 290 × 224)

one type of error leads one to ignore another. One wonders whether Pope had this in mind when he wrote the same summer to Mallet about Harte's *Essay on reason*: 'I undertook to correct the press, but find myself so bad a Reviser by what I see has escaped me in my last thing, that I believe he had best have it sent him to Oxford' (*Correspondence*, iii. 408). But I am not sure Oxford would have done better than Pope.

THE OCTAVO *WORKS*

In distinction to the *Works* of 1735 in quarto and folio, I have spoken generically of the octavo *Works*, but that hardly does justice to what became six volumes of verse which went through from four to six editions each. Apart from restoring

the traditional use of italic in them, Pope used successive editions to make significant revisions in the accidentals as well as the substantives of his text, and we
know that he read proof for the volumes produced for Lintot as well as those of
his own printer and publishers. I have not collated the accidentals of all these editions; instead I propose to discuss a volume marked with Pope's revisions which
gives some idea of his attention to detail. It is a copy of the 1736 third edition of
the second volume of the octavo works (Griffith 430), containing the *Essay on
man*, 'Ethic epistles', and the other poems added to the *Dunciad* to make up the
Works of 1735. The revisions were never incorporated in later editions, and it is
worth asking why, even if the answer must be tentative. This is the last of the
three editions of this volume published by Gilliver; the fourth edition was published by Dodsley and Cooper in January 1739 after Pope had severed his relationship with Gilliver. The most plausible explanation for the fate of this
annotated volume then is that Pope prepared it for Gilliver's use in later editions,
and that after the break it remained in Gilliver's hands.[11]

Pope may have been brought to work on the volume because it contained a
number of errors. Many, though not all, of these were corrected by a list of errata,
but the annotated copy was prepared independently and presumably later. The
markings are heavier in the earlier part of the volume, perhaps because the *Essay
on man* and 'Ethic epistles' were more important to Pope. The pages in Figure
101 are typical of those displaying the heavier markings, with most changes to
punctuation, followed by changes to capitalization. In the text the movement is
generally to increase capitals ('Reason', line 224, 'Infinite', line 232, 'Whole', line
242) though this is usually accompanied by some reduction as here in line 235
with 'void'. In the footnotes the movement is always to reduce capitals, since the
notes are a later addition and have not yet been fully purged of Pope's manuscript
capitalization. The note on p. 16 is older than most, having originated in the collected *Essay on man* of 1734; eight capitals have already been removed in the
earlier octavos, and the other three lower-case letters marked here bring the
whole note into consistency. Other pages show Pope changing the use of italic,
and in Figure 102 marking Bethel's name with the double underline for honorific
small caps., to bring him up to the level of those at the head of the page.

I hope these examples will give a fair idea of how Pope worked on his accidentals, though I certainly do not claim that his work was punctiliously systematic or consistent. The substantive changes he made demand a closer scrutiny; let
us look first at the prose 'design' to the *Essay on man* (Figure 103) where the
changes are so clearly for the better that it is hard to imagine how Pope could
have let the existing text stand so long. The first removes the preposition 'in'

---

[11] The earliest mark of provenance is 'Wm. Awmack/His Book/1770' on the lower paste-down; he repeated this without date on the first surviving and last leaves of text. The volume lacks the first four
leaves of preliminaries. The upper paste-down has a cutting from an unidentified bookseller's catalogue
offering the volume for £20. It was subsequently in the possession of John Murray and was acquired by
the BL in 1971 (C. 122 e. 31). Mack lists the changes in substantives in *Collected in himself*, pp. 344–5.

from its hanging position at the end of the clause; the death-bed edition retains the old reading. The second change is less obviously necessary, but it is still good to be rid of 'much such'; the death-bed edition retains 'such' but revises the following words to read 'such finer nerves and vessels, the conformation and uses of which will for ever escape our observation'. The result is that 'such' is deprived of its companion 'as', and we need to change it to 'those'. Over the page (Figure 104) the change from 'truer' to 'more certain' is most necessary of all; it restores the reading of earlier editions and is also made in the death-bed edition. I think consideration of these changes will show that this last change was made independently, or on the basis of an earlier edition, for the death-bed edition; if Pope had had this marked copy beside him, all these changes would have been adopted.

Changes to the text of the poems are more difficult to evaluate. We can say that the change from 'that' to 'which' (Figure 105, line 89) is typical of Pope's revision, though he failed to make it subsequently when he had the chance. On the other hand the transposition to 'Fix, or describe, one movement of the mind?' (Figure 106, line 36) can only be judged subjectively; I feel it is an improvement, and could rationalize my preference without convincing anyone who felt the converse. And what are we to make of the change to the character of Sporus in the *Epistle to Arbuthnot* from 'bug' to 'Fly' (Figure 107, line 309)? Certainly (to argue as Bentley might have done), although the bedbug stinks and stings, it is not notable for gilded wings, nor does it buzz; it may be that Pope had in mind both the general sense of bug as an insect as well as the particular use for bedbug; I am tempted to suspect that he may also have had the word 'bugger' at the back of his mind. 'Fly', though less offensive, fits the passage better and alliterates well; whether the link with 'butterfly' in the preceding line is good or bad is open to argument. One cannot be sure that Pope would have made this change in later editions if he had had this annotation before him (or, indeed, whether having made it he would have changed it back again subsequently), but I think there is a strong case for accepting it. The overall revision of the poem is clearly purposive and it is the most thorough revision it was to undergo; if Pope had subsequently made another independent revision of the text, one might wish to be more cautious.[12] One problem does remain—whether Pope wished 'Fly' to be capitalized, or whether he wrote it with a capital according to his habit. From the texture of the passage I would opt for lower case, since the capitals of the early editions have all been removed with the exception of 'Ass's milk', in line 306; that remains to point the joke about the foolishness of asses.

This is not the only such volume Pope prepared, for in the preface to his edition of 1751 Warburton writes:

---

[12] According to the *Twickenham*'s collation (iv. 120), the octavo of 1739 changed 'Nor Lucre' to 'Not Lucre' in line 335. This could as easily be a compositor's error as a revision by Pope, though a similar change was made to the following line for the *Works* in 1735.

FIG. 101. Pope's changes marked in a copy of octavo *Works* II (1736), pp. 16–17 (BL C. 122. e. 31; 182 × 110)

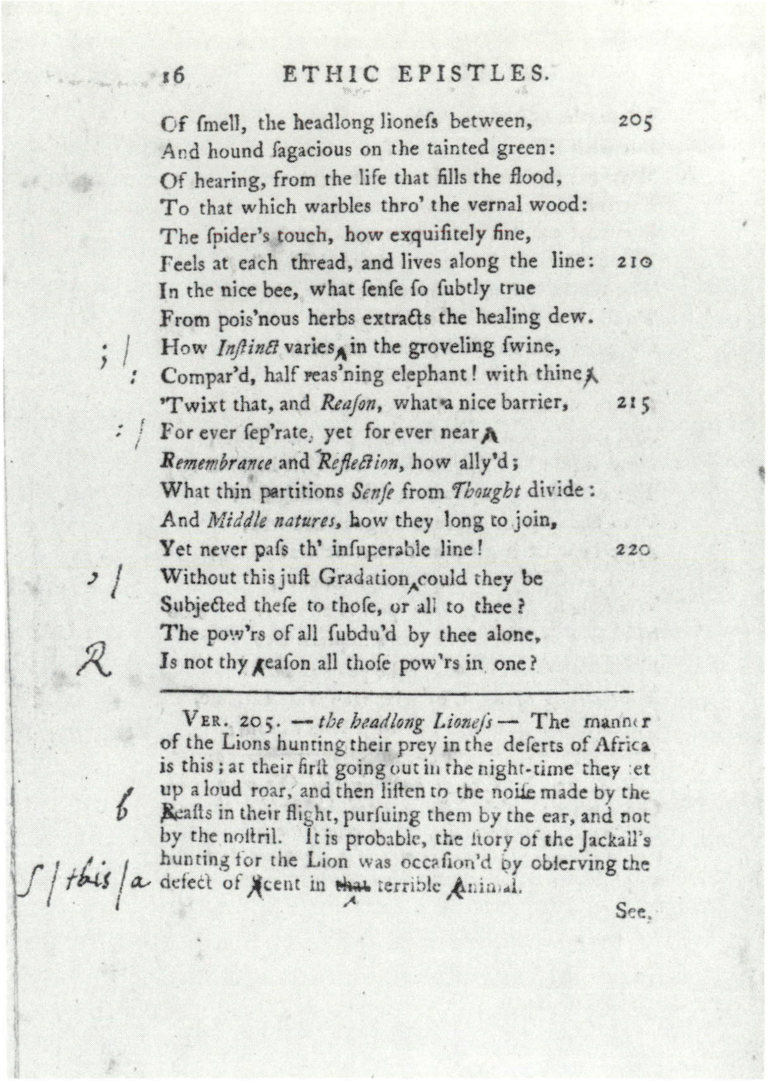

16 ETHIC EPISTLES.

Of ſmell, the headlong lioneſs between,
And hound ſagacious on the tainted green:
Of hearing, from the life that fills the flood,
To that which warbles thro' the vernal wood:
The ſpider's touch, how exquiſitely fine,
Feels at each thread, and lives along the line:
In the nice bee, what ſenſe ſo ſubtly true
From pois'nous herbs extracts the healing dew.
How *Inſtinct* varies in the groveling ſwine,
Compar'd, half reas'ning elephant! with thine
'Twixt that, and *Reaſon*, what a nice barrier,
For ever ſep'rate, yet for ever near
Remembrance and *Reflection*, how ally'd;
What thin partitions *Senſe* from *Thought* divide:
And *Middle natures*, how they long to join,
Yet never paſs th' inſuperable line!
Without this juſt Gradation could they be
Subjected theſe to thoſe, or all to thee?
The pow'rs of all ſubdu'd by thee alone,
Is not thy reaſon all thoſe pow'rs in one?

VER. 205. —*the headlong Lioneſs*— The manner of the Lions hunting their prey in the deſerts of Africa is this; at their firſt going out in the night-time they ſet up a loud roar, and then liſten to the noiſe made by the Beaſts in their flight, purſuing them by the ear, and not by the noſtril. It is probable, the ſtory of the Jackall's hunting for the Lion was occaſion'd by obſerving the defect of Scent in that terrible Animal.

See,

The FIRST Volume, and the original poems in the SECOND, are here printed from a copy corrected throughout by the Author himself, even to the very preface; Which, with several additional notes in his own hand, he delivered to the Editor a little before his death.

It seems clear from a cursory collation that this corrected volume was one of the octavo editions and not the original quarto *Works* of 1717—as would be expected, since the octavos had already made a good many revisions.

ETHIC EPISTLES.   17

See, thro' this air, this ocean, and this earth,   225
All matter quick, and burſting into birth.
Above, how high progreſſive life may go!
Around how wide! how deep extend below!
Vaſt chain of Being! which from God began,
Natures æthereal, human, angel, man,   230
Beaſt, bird, fiſh, inſect! what no eye can ſee,
No glaſs can reach! from Infinite to thee,
From thee to Nothing! On ſuperior pow'rs
Were we to preſs, inferior might on ours:
Or in the full creation leave a Void,   235
Where one ſtep broken, the great ſcale's deſtroy'd;
From nature's chain whatever link you ſtrike,
Tenth or ten thouſandth, breaks the chain alike.
   And if each Syſtem in gradation 'roll,
Alike eſſential to th' amazing whole;   240
The leaſt confuſion but in one, not all
That Syſtem only, but the whole muſt fall.
Let Earth unbalanc'd from her orbit fly,
Planets and ſuns ruſh lawleſs thro' the sky,
Let ruling Angels from their ſpheres be hurl'd,   245
Being on being wreck'd, and world on world,
Heav'n's whole foundations to their centre nod,
And Nature tremble, to the throne of God!

VER. 225.] *How much farther this* Gradation *and*
Subordination *may extend? were any part of which*
broken, the whole connected Creation *muſt be deſtroy'd.*
    B   All

CONCLUSION

It is time to try to sum up the situation which faces an editor of Pope. In the light
of Pope's continual revision of accidentals, and the changes in typography that he
pioneered, any idea of following the accidentals of a first edition copy-text (as
some of the *Twickenham* editors did) is clearly wrong; in some cases even the
punctuation is grossly unsatisfactory. Commonly accepted textual principles do
not work in this situation: Greg's assumption that compositors 'will normally
follow their own habits or inclination',[13] may be true of the Elizabethan printing

[13] Greg, 'Rationale of copy-text', *Collected papers*, p. 377.

FIG. 102. Pope's marking for caps. and small caps. in a copy of octavo *Works* II (1736), p. 48 (BL C. 122. e. 31; 182 × 110)

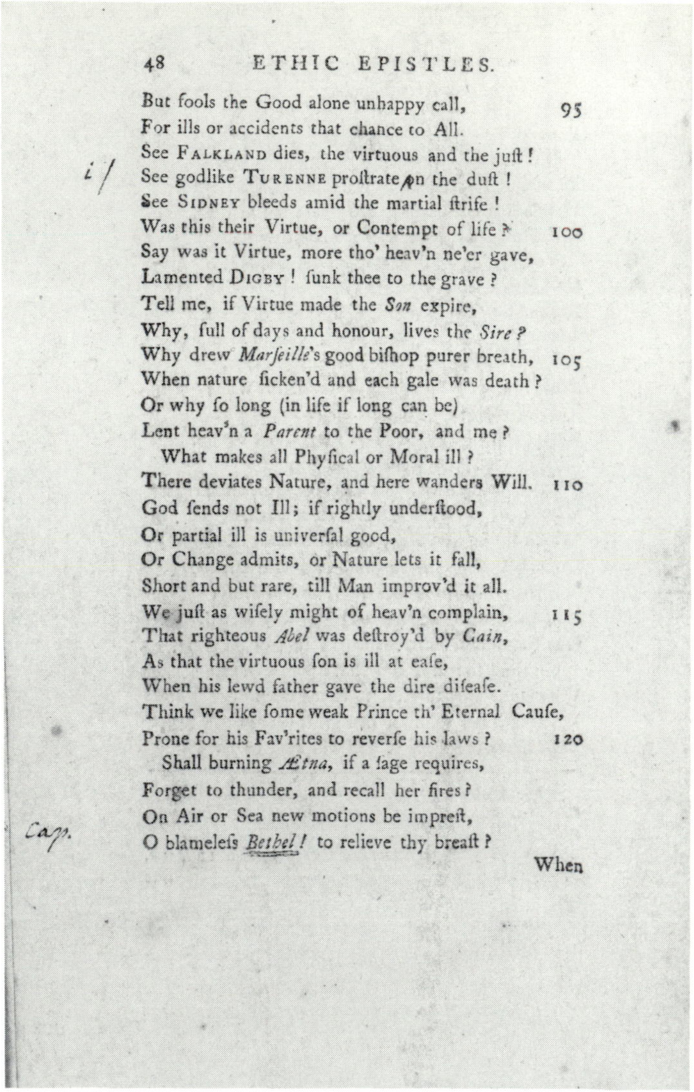

48   ETHIC EPISTLES.

But fools the Good alone unhappy call,   95
For ills or accidents that chance to All.
See FALKLAND dies, the virtuous and the juſt !
See godlike TURENNE proſtrate on the duſt !
See SIDNEY bleeds amid the martial ſtrife !
Was this their Virtue, or Contempt of life ?   100
Say was it Virtue, more tho' heav'n ne'er gave,
Lamented DIGBY ! ſunk thee to the grave ?
Tell me, if Virtue made the *Son* expire,
Why, full of days and honour, lives the *Sire ?*
Why drew *Marſeille's* good biſhop purer breath,   105
When nature ſicken'd and each gale was death ?
Or why ſo long (in life if long can be)
Lent heav'n a *Parent* to the Poor, and me ?
  What makes all Phyſical or Moral ill ?
There deviates Nature, and here wanders Will.   110
God ſends not Ill; if rightly underſtood,
Or partial ill is univerſal good,
Or Change admits, or Nature lets it fall,
Short and but rare, till Man improv'd it all.
We juſt as wiſely might of heav'n complain,   115
That righteous *Abel* was deſtroy'd by *Cain,*
As that the virtuous ſon is ill at eaſe,
When his lewd father gave the dire diſeaſe.
Think we like ſome weak Prince th' Eternal Cauſe,
Prone for his Fav'rites to reverſe his laws ?   120
  Shall burning *Ætna,* if a ſage requires,
Forget to thunder, and recall her fires ?
On Air or Sea new motions be impreſt,
O blameleſs *Bethel !* to relieve thy breaſt ?

                                                    When

house, but not of Pope's. Moreover, in the case of Pope the accidentals of his manuscripts, such as textual critics have thought it their duty to try and recover, are quite clearly not those which he wished to be printed, as we see by the changes he made in proof and in later editions.

If we return to the old textual principle of following the last edition revised by the author, we can see that here it has considerable merit—though, of course, we would not follow it blindly into error. But in this case we are faced with an additional problem in Pope's contrasting typographical styles—with italic for popular editions, without italic for what one might call the definitive editions. We would

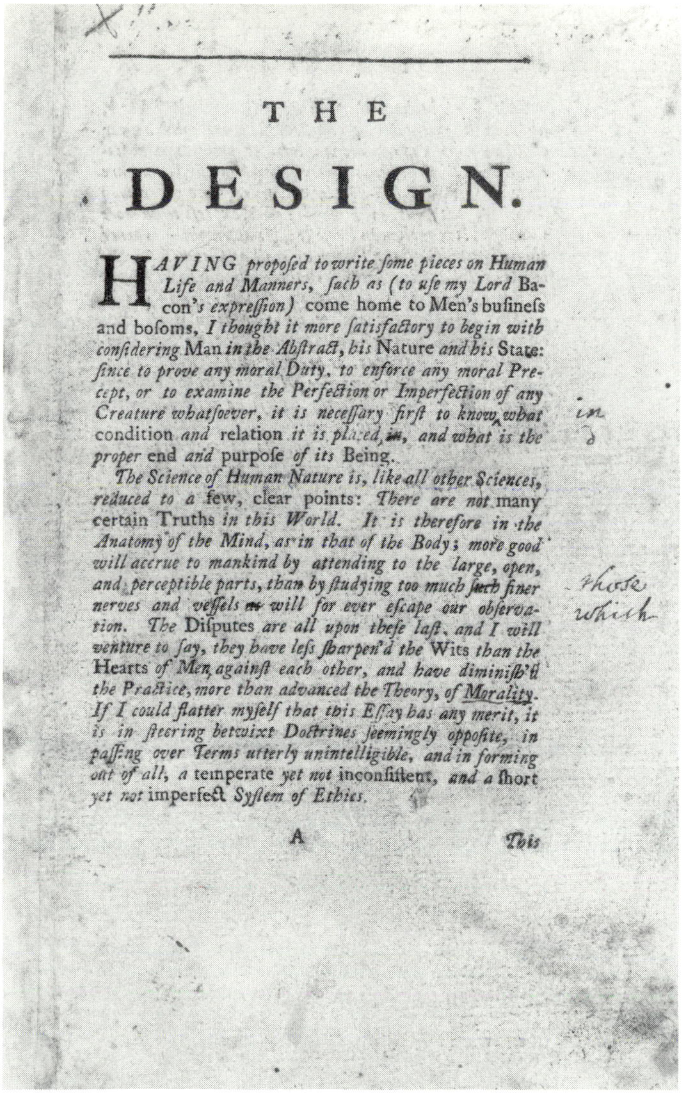

THE

DESIGN.

HAVING propos'd to write some pieces on Human Life and Manners, such as (to use my Lord Bacon's expression) come home to Men's business and bosoms, I thought it more satisfactory to begin with considering Man in the Abstract, his Nature and his State: since to prove any moral Duty, to enforce any moral Precept, or to examine the Perfection or Imperfection of any Creature whatsoever, it is necessary first to know what condition and relation it is placed in, and what is the proper end and purpose of its Being.

The Science of Human Nature is, like all other Sciences, reduced to a few, clear points: There are not many certain Truths in this World. It is therefore in the Anatomy of the Mind, as in that of the Body; more good will accrue to mankind by attending to the large, open, and perceptible parts, than by studying too much such finer nerves and vessels as will for ever escape our observation. The Disputes are all upon these last, and I will venture to say, they have less sharpen'd the Wits than the Hearts of Men against each other, and have diminish'd the Practice, more than advanced the Theory, of Morality. If I could flatter myself that this Essay has any merit, it is in steering betwixt Doctrines seemingly opposite, in passing over Terms utterly unintelligible, and in forming out of all, a temperate yet not inconsistent, and a short yet not imperfect System of Ethics.

A This

FIG. 103. Pope's substantive changes to 'The design' in a copy of octavo *Works* II (1736) (BL C. 122. e. 31; 182 × 110)

not be following Pope's intentions if we reprinted the four major works (the *Essay on criticism*, *Essay on man*, *Ethic epistles*, and *Dunciad*) in the classic style they took in the death-bed editions, but left those which he did not revise before his death in their popular style: Pope would certainly have wished for a consistent typography. The problem is made much more difficult by the fact that the death-bed editions, out of all Pope's works, are the very ones where it is almost certain that accidentals have been changed by Warburton and perhaps by their printer Bowyer.

The evidence that Bowyer had a role in the accidentals of the death-bed edi-

FIG. 104. Pope's substantive change to 'The design' in a copy of octavo *Works* II (1736) (BL C. 122. e. 31; 182 × 110)

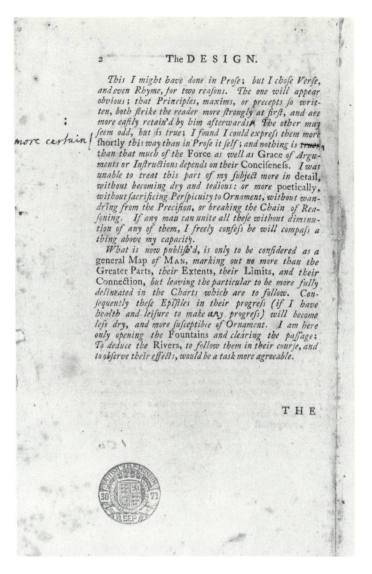

tions is found in Pope's letter to him of 3 March 1744 about the *Ethic epistles*: 'On Second thoughts, let the Proof of the Epistle to Lord Cobham, I, be done in the *Quarto*, not the *Octavo*, size: contrive the Capitals & evry thing exactly to correspond with that Edition.' (*Correspondence*, iv. 504.) Bowyer is being given general instructions to make the typography of the *Ethic epistles* correspond to the earlier quartos but this would still leave him a good deal of freedom.[14] Warburton's

[14] In unpub. research Richard Noble has shown that the *Essay on man* and *Essay on criticism* were printed by John Wright in octavo before Bowyer printed the death-bed quartos, though the octavos were not issued until 1748 and 1749. On a superficial check the quartos are very similar in accidentals to the

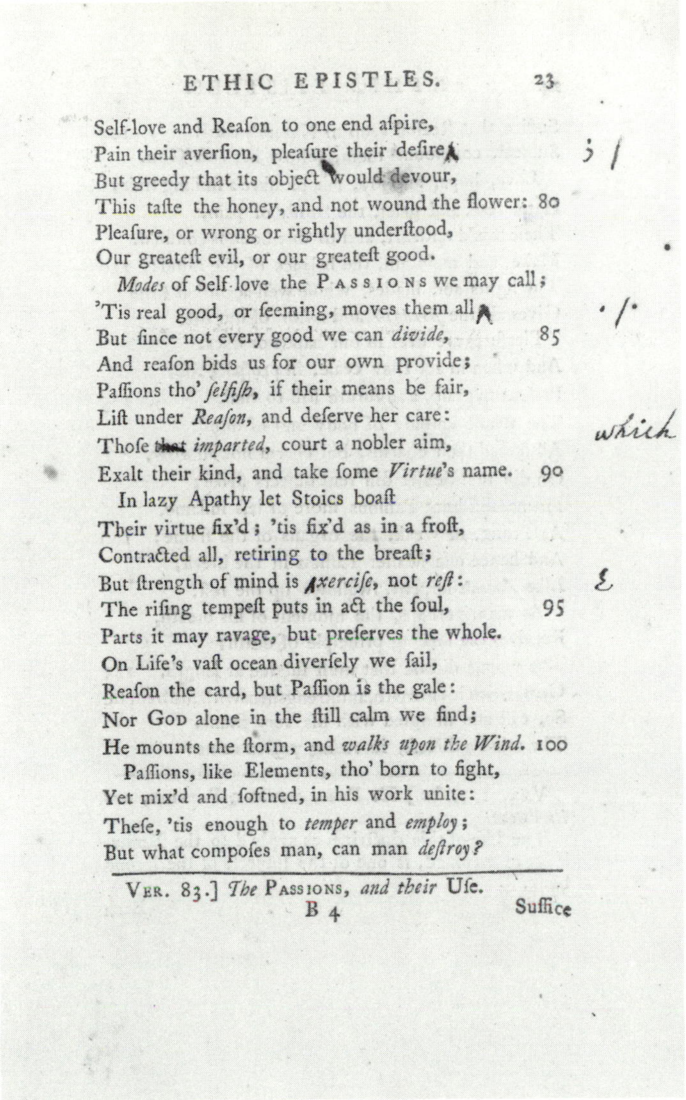

ETHIC EPISTLES. 23

Self-love and Reason to one end aspire,
Pain their aversion, pleasure their desire,
But greedy that its object would devour,
This taste the honey, and not wound the flower: 80
Pleasure, or wrong or rightly understood,
Our greatest evil, or our greatest good.
  Modes of Self-love the P A S S I O N S we may call;
'Tis real good, or seeming, moves them all,
But since not every good we can *divide*, 85
And reason bids us for our own provide;
Passions tho' *selfish*, if their means be fair,
List under *Reason*, and deserve her care:
Those that *imparted*, court a nobler aim,
Exalt their kind, and take some *Virtue*'s name. 90
  In lazy Apathy let Stoics boast
Their virtue fix'd; 'tis fix'd as in a frost,
Contracted all, retiring to the breast;
But strength of mind is *exercise*, not *rest*: 95
The rising tempest puts in act the soul,
Parts it may ravage, but preserves the whole.
On Life's vast ocean diversely we sail,
Reason the card, but Passion is the gale:
Nor GOD alone in the still calm we find;
He mounts the storm, and *walks upon the Wind*. 100
  Passions, like Elements, tho' born to fight,
Yet mix'd and softned, in his work unite:
These, 'tis enough to *temper* and *employ*;
But what composes man, can man *destroy*?

VER. 83.] *The* PASSIONS, *and their* Use.
  B 4   Suffice

FIG. 105. Pope's changes, including substantives, in a copy of octavo *Works* II (1736), p. 23 (BL C. 122. e. 31; 182 × 110)

hand is similarly documented by Pope's letter to Bowyer. 'Mr Bowyer, I had this sheet sent before, as if it were printed off, but these Corrections of Mr Warburton's were not in it. I therfore return it you.' (*Correspondence*, iv. 514.)

Warburton's influence on the death-bed editions has been widely condemned, most memorably in the words of R. H. Griffith (with Warburton's own small

octavos, and any changes seem to be individual rather than due to a printer's general system. In these cases Bowyer would have been following the typography of Wright's edns., and if my judgement of Wright is sound, this would have been closely supervised by Pope and Warburton. Bowyer may have had more influence on the *Dunciad*, which he first printed in octavo and then in quarto.

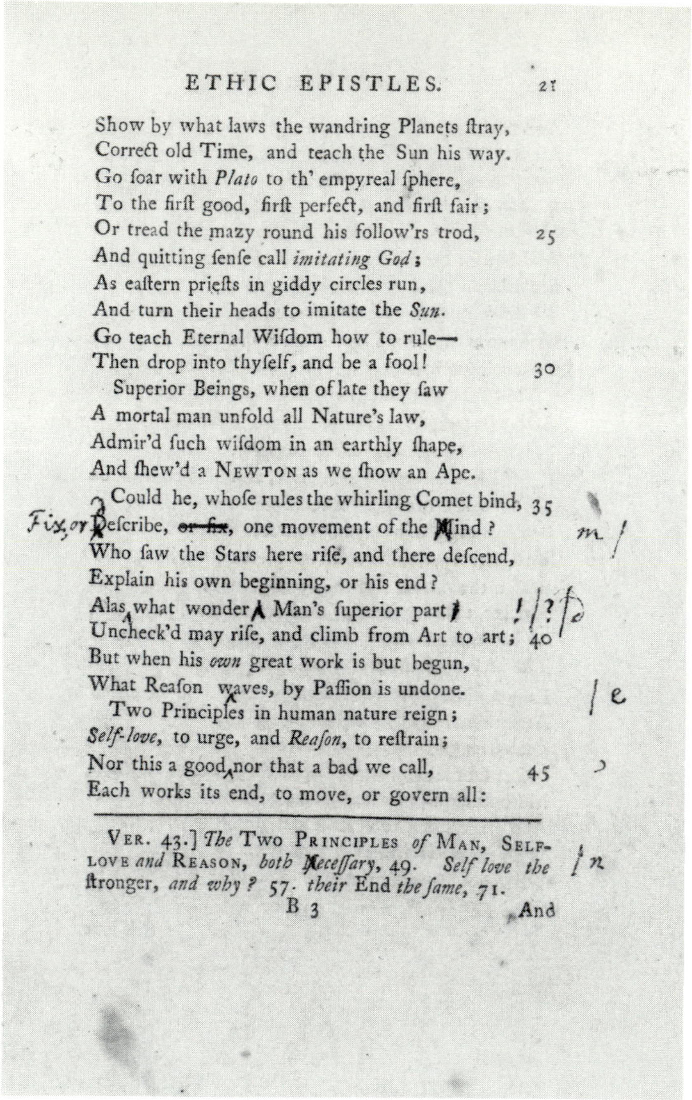

ETHIC EPISTLES. 21

Show by what laws the wandring Planets ſtray,
Correct old Time, and teach the Sun his way.
Go ſoar with *Plato* to th' empyreal ſphere,
To the firſt good, firſt perfect, and firſt fair;
Or tread the mazy round his follow'rs trod,
And quitting ſenſe call *imitating God*;
As eaſtern prieſts in giddy circles run,
And turn their heads to imitate the *Sun*.
Go teach Eternal Wiſdom how to rule—
Then drop into thyſelf, and be a fool!
Superior Beings, when of late they ſaw
A mortal man unfold all Nature's law,
Admir'd ſuch wiſdom in an earthly ſhape,
And ſhew'd a NEWTON as we ſhow an Ape.
Could he, whoſe rules the whirling Comet bind,
Deſcribe, ~~or fix~~, one movement of the Mind?
Who ſaw the Stars here riſe, and there deſcend,
Explain his own beginning, or his end?
Alas what wonder Man's ſuperior part
Uncheck'd may riſe, and climb from Art to art;
But when his *own* great work is but begun,
What Reaſon waves, by Paſſion is undone.
Two Principles in human nature reign;
Self-love, to urge, and *Reaſon*, to reſtrain;
Nor this a good nor that a bad we call,
Each works its end, to move, or govern all:

VER. 43.] *The* TWO PRINCIPLES *of* MAN, SELF-LOVE *and* REASON, *both Neceſſary*, 49. *Self love the* ſtronger, *and why?* 57. *their* End *the ſame*, 71.

B 3 And

FIG. 106. Pope's changes, including substantives, in a copy of octavo *Works* II (1736), p. 21 (BL C. 122. e. 31; 182 × 110)

caps.): 'as an EDITOR OF POPE Warburton, Early or Late, is to be lifted down from any pedestal of respect, and is to be treated with a very great deal of skepticism and very little of reverence.'[15] Yet however much we may be suspicious of Warburton's role in making changes in the last editions of Pope's lifetime, we must give him credit for the way he abandoned his own typographical prejudices in editing the 1751 edition of Pope. One of his own works, *The alliance between church and state*, 1736, will show his own preference for a complex typography,

[15] Griffith, 'Early Warburton? or late Warburton?', p. 131.

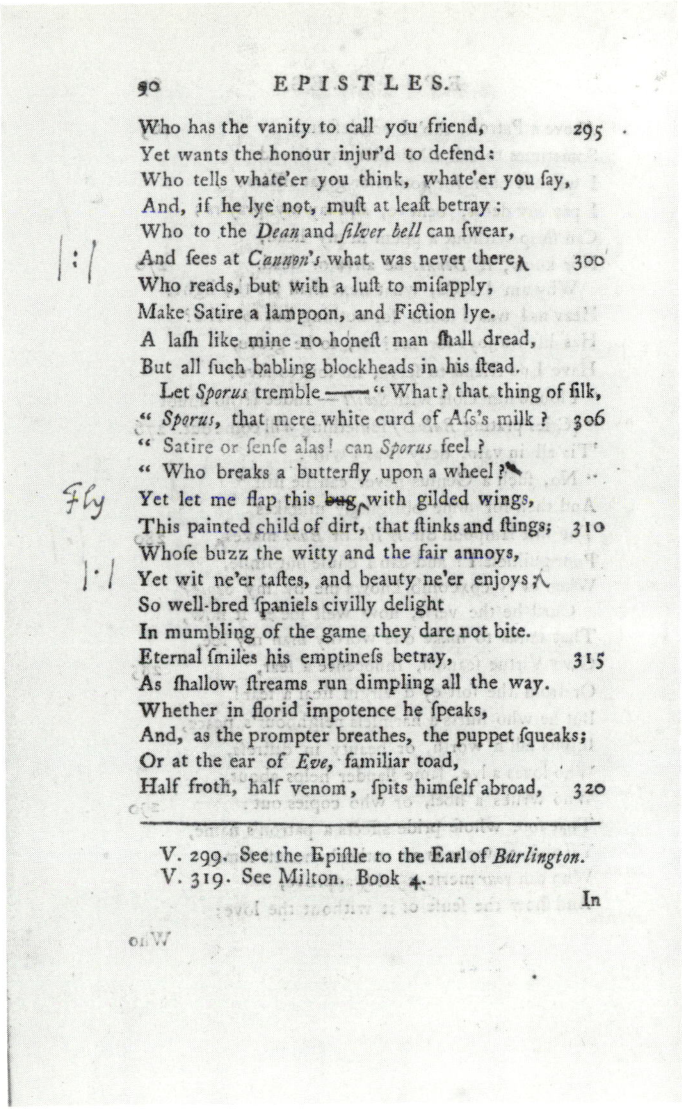

FIG. 107. Pope's changes, including substantives, in a copy of octavo *Works* II (1736), p. [2]90 (BL C. 122. e. 31; 182 × 110)

capitalizing nouns and using italics and caps. and small caps. (Figure 108); it is found also in his preface to the 1751 edition of Pope quoted above. Yet in the 1751 edition of Pope he accommodates himself entirely to Pope's style (Figure 109), removing those italics like '*Nile*' which had survived by error in the death-bed *Essay on criticism*, or in *Windsor Forest* (Figure 110) using the honorific caps. and small caps. for 'STUART' that Pope had used in his manuscript and in the first edition, though not in the *Works* of 1717. (Pope may well have made this change in the annotated octavo Warburton used, as he surely added italic to the dismissive '*and all that*' (Figure 111) in the *Rape of the lock*.) Though one may question

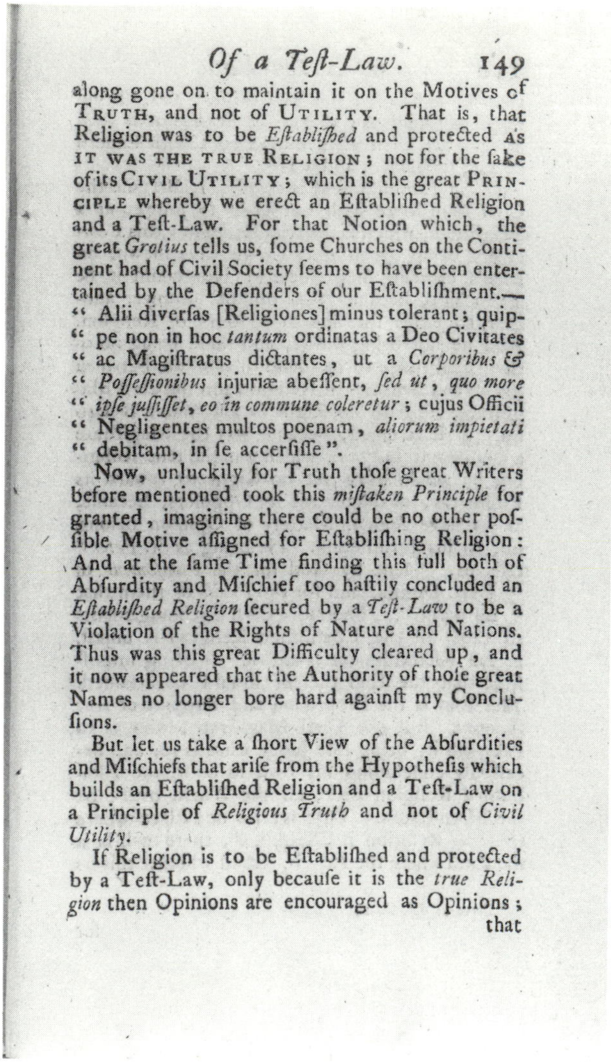

FIG. 108. Warburton's preference for a complex typography illustrated by his *The alliance between church and state* (1736), p. 149 (Bodl. G. Pamph. 7(5); 196 × 112)

Of a Teſt-Law.   149

along gone on to maintain it on the Motives of TRUTH, and not of UTILITY. That is, that Religion was to be *Eſtabliſhed* and protected as IT WAS THE TRUE RELIGION ; not for the ſake of its CIVIL UTILITY; which is the great PRINCIPLE whereby we erect an Eſtabliſhed Religion and a Teſt-Law. For that Notion which, the great *Grotius* tells us, ſome Churches on the Continent had of Civil Society ſeems to have been entertained by the Defenders of our Eſtabliſhment.——

" Alii diverſas [Religiones] minus tolerant; quip-
" pe non in hoc *tantum* ordinatas a Deo Civitates
" ac Magiſtratus dictantes, ut a *Corporibus &*
" *Poſſeſſionibus* injuriæ abeſſent, *ſed ut*, *quo more*
" *ipſe juſſiſſet, eo in commune coleretur* ; cujus Officii
" Negligentes multos poenam, *aliorum impietati*
" debitam, in ſe accerſiſſe ".

Now, unluckily for Truth thoſe great Writers before mentioned took this *miſtaken Principle* for granted, imagining there could be no other poſſible Motive aſſigned for Eſtabliſhing Religion : And at the ſame Time finding this full both of Abſurdity and Miſchief too haſtily concluded an *Eſtabliſhed Religion* ſecured by a *Teſt-Law* to be a Violation of the Rights of Nature and Nations. Thus was this great Difficulty cleared up, and it now appeared that the Authority of thoſe great Names no longer bore hard againſt my Concluſions.

But let us take a ſhort View of the Abſurdities and Miſchiefs that ariſe from the Hypotheſis which builds an Eſtabliſhed Religion and a Teſt-Law on a Principle of *Religious Truth* and not of *Civil Utility.*

If Religion is to be Eſtabliſhed and protected by a Teſt-Law, only becauſe it is the *true Religion* then Opinions are encouraged as Opinions ;
that

details of execution, Warburton's edition so far as italics and capitals are concerned is a self-effacing attempt to produce an edition in conformity with Pope's typographic intentions: and we must recognize that he probably had a better knowledge of these than anyone. So one solution for an editor is to follow Warburton's example in revising the typography of the octavos, which were the last editions to be revised by Pope alone, but to try to bring more delicacy of touch to the task.[16]

[16] Herbert Davis in his Oxford Standard Authors edn. of 1966 was aware of the basic situation, even if he did not appreciate the complexities of Pope's approach (some of his papers concerning the edition are deposited in Bodl. MS Eng. lett. 524–5). He followed the text of the death-bed quartos 'believing that

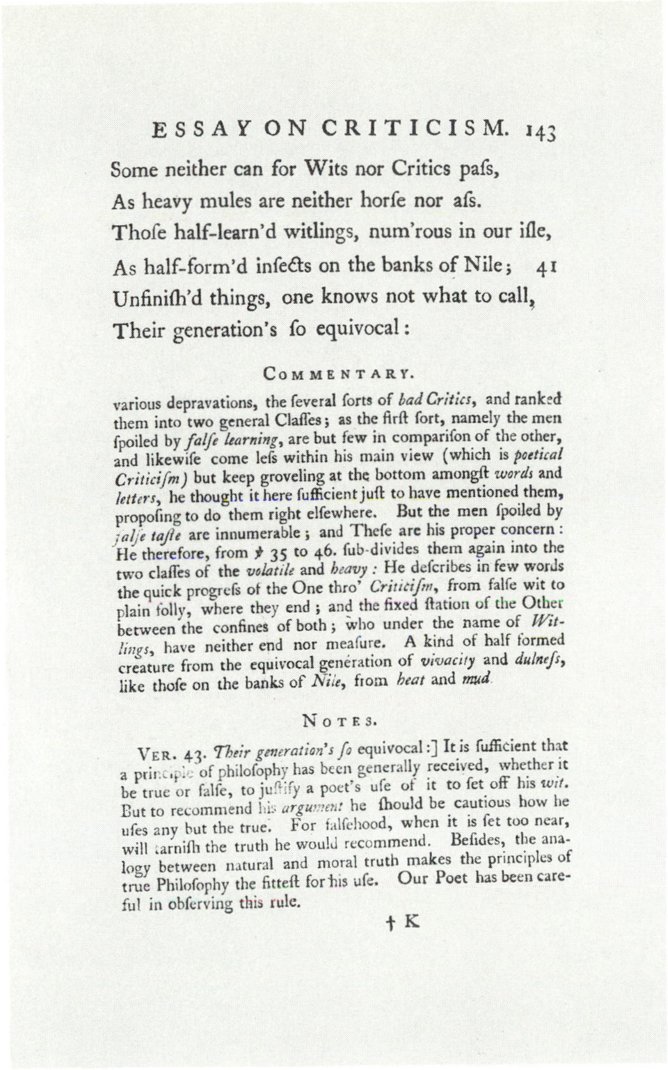

ESSAY ON CRITICISM. 143

Some neither can for Wits nor Critics pafs,
As heavy mules are neither horfe nor afs.
Thofe half-learn'd witlings, num'rous in our ifle,
As half-form'd infects on the banks of Nile; 41
Unfinifh'd things, one knows not what to call,
Their generation's fo equivocal:

COMMENTARY.

various depravations, the feveral forts of *bad Critics*, and ranked them into two general Claffes; as the firft fort, namely the men fpoiled by *falfe learning*, are but few in comparifon of the other, and likewife come lefs within his main view (which is *poetical Criticifm*) but keep groveling at the bottom amongft *words* and *letters*, he thought it here fufficient juft to have mentioned them, propofing to do them right elfewhere. But the men fpoiled by *falfe tafte* are innumerable; and Thefe are his proper concern: He therefore, from ℣ 35 to 46. fub-divides them again into the two claffes of the *volatile* and *heavy*: He defcribes in few words the quick progrefs of the One thro' *Criticifm*, from falfe wit to plain folly, where they end; and the fixed ftation of the Other between the confines of both; who under the name of *Witlings*, have neither end nor meafure. A kind of half formed creature from the equivocal generation of *vivacity* and *dulnefs*, like thofe on the banks of *Nile*, from *heat* and *mud*.

NOTES.

VER. 43. *Their generation's fo equivocal:*] It is fufficient that a principle of philofophy has been generally received, whether it be true or falfe, to juftify a poet's ufe of it to fet off his *wit*. But to recommend his *argument* he fhould be cautious how he ufes any but the true. For falfehood, when it is fet too near, will tarnifh the truth he would recommend. Befides, the analogy between natural and moral truth makes the principles of true Philofophy the fitteft for his ufe. Our Poet has been careful in obferving this rule.

† K

FIG. 109. Removal of italic in Warburton's edition of Pope's *Works* (1751), i. 143 (Keele University Library; 209 × 125)

If we reject this solution as relying on the subjective judgements of an editor and insist on accidentals that are demonstrably Pope's, then our only alternative is to turn to the last revised edition of each volume of the octavo *Works*. If we do this and reproduce the italics which Pope abandoned elsewhere, we could per-

they give not only in substance but in the accidentals what Pope finally wanted' (p. vi). While I agree in principle, I question whether in every detail the accidentals are those of Pope, quite apart from errors such as the use of italic for 'Nile' mentioned above. For the rest of the works Davis used Warburton's 1751 text, 'amended . . . to restore the punctuation in places where Warburton had tampered with it.' The only weakness here is that (as Mack showed in 'Pope's Horatian poems', pp. 43–4; *Collected in himself*, pp. 112–13) Warburton used an edn. of the octavo *Works*, II. ii which did not contain Pope's final revisions, and Davis follows him in this error. The overall result has a consistency of accidentals which is preferable to the *Twickenham* text.

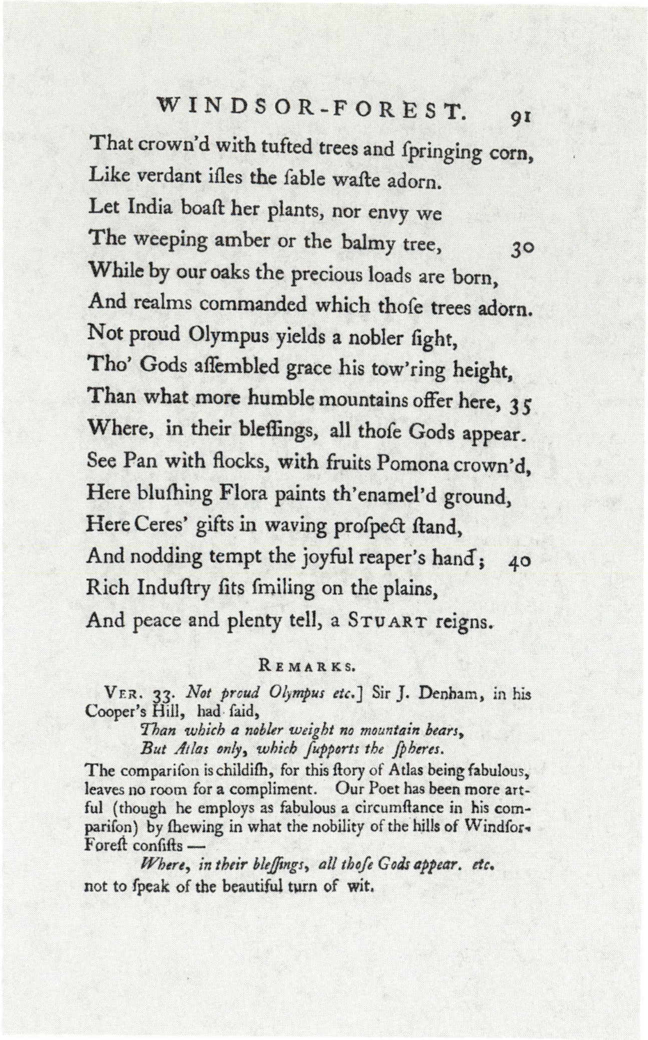

FIG. 110. Use of caps. and small caps. in Warburton's edition of Pope's *Works* (1751), i. 91 (Keele University Library; 209 × 125)

WINDSOR-FOREST. 91

That crown'd with tufted trees and springing corn,
Like verdant isles the sable waste adorn.
Let India boast her plants, nor envy we
The weeping amber or the balmy tree,   30
While by our oaks the precious loads are born,
And realms commanded which those trees adorn.
Not proud Olympus yields a nobler sight,
Tho' Gods assembled grace his tow'ring height,
Than what more humble mountains offer here,   35
Where, in their blessings, all those Gods appear.
See Pan with flocks, with fruits Pomona crown'd,
Here blushing Flora paints th'enamel'd ground,
Here Ceres' gifts in waving prospect stand,
And nodding tempt the joyful reaper's hand;   40
Rich Industry sits smiling on the plains,
And peace and plenty tell, a STUART reigns.

REMARKS.

VER. 33. *Not proud Olympus etc.*] Sir J. Denham, in his Cooper's Hill, had said,
Than which a nobler weight no mountain bears,
But Atlas only, which supports the spheres.
The comparison is childish, for this story of Atlas being fabulous, leaves no room for a compliment. Our Poet has been more artful (though he employs as fabulous a circumstance in his comparison) by shewing in what the nobility of the hills of Windsor-Forest consists —
Where, in their blessings, all those Gods appear. etc.
not to speak of the beautiful turn of wit.

haps argue that this is the form Pope chose for his popular editions, and we want to produce a popular edition. No doubt there will be inconsistencies caused by revisions made at different times, but any editor who seeks to reproduce an author's accidentals, even those of one as meticulous as Pope, is bound to accept that inconsistency is a fact of literary life.

Yet even this does not provide an easy solution, for there are still the two octavo volumes revised by Pope, that used by Warburton for the early poems and that, recently made available to an editor, with revisions to the epistles and satires. The first would have to be reconstructed by editorial discernment; while the examples I have given seem clear examples of Pope's hand at work, there will be many more borderline cases. The second presents us with a more clear-cut

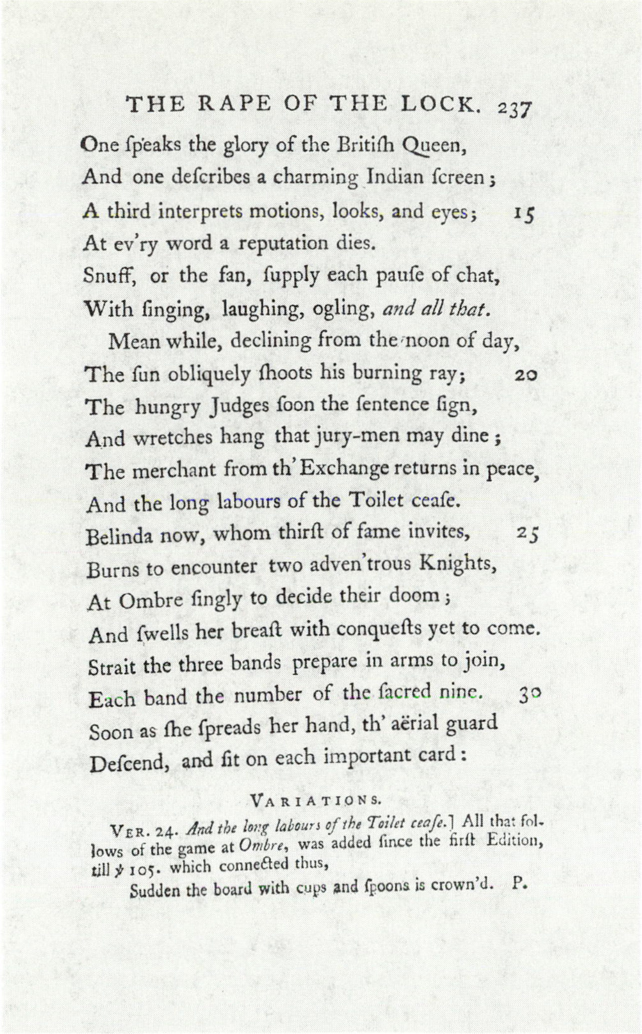

THE RAPE OF THE LOCK. 237

One speaks the glory of the Britiſh Queen,
And one deſcribes a charming Indian ſcreen;
A third interprets motions, looks, and eyes; 15
At ev'ry word a reputation dies.
Snuff, or the fan, ſupply each pauſe of chat,
With ſinging, laughing, ogling, *and all that*.
    Mean while, declining from the noon of day,
The ſun obliquely ſhoots his burning ray; 20
The hungry Judges ſoon the ſentence ſign,
And wretches hang that jury-men may dine;
The merchant from th'Exchange returns in peace,
And the long labours of the Toilet ceaſe.
Belinda now, whom thirſt of fame invites, 25
Burns to encounter two adven'trous Knights,
At Ombre ſingly to decide their doom;
And ſwells her breaſt with conqueſts yet to come.
Strait the three bands prepare in arms to join,
Each band the number of the ſacred nine. 30
Soon as ſhe ſpreads her hand, th'aërial guard
Deſcend, and ſit on each important card:

VARIATIONS.

VER. 24. *And the long labours of the Toilet ceaſe.*] All that follows of the game at *Ombre*, was added ſince the firſt Edition,
till ẏ 105. which connected thus,
    Sudden the board with cups and ſpoons is crown'd. P.

FIG. 111. Addition of italic for emphasis in Warburton's edition of Pope's *Works* (1751), i. 237 (Keele University Library; 209 × 125)

problem, but I can see no reason for rejecting the changes (unless they conflict with a change made in 1739 or later) since there is no reason to think they have any less validity than other changes made sporadically in the published octavos. Pope might have revised them yet again if time had permitted, but if we are to accept Pope's latest revisions, we must take them wherever we find them. Similarly the revisions to the preliminaries are clear improvements and must be accepted. The revisions to the text have made me hesitate because they have not been the object of Pope's second thoughts at the proof stage, but I can see no good reason why they should be rejected.

The problem of the revised octavos remains whichever policy an editor follows, and their treatment must be substantially the same. But the main conclusion must be that, whether an editor follows the difficult path of Warburton and tries to revise Pope's accidentals as he would have wished, or whether he plays it safe and follows the conventions of the octavos, it is the text found in the octavos and hitherto little regarded that must form the basis of his work.

Finally I would like to add two personal comments. I have always opposed the idea that bibliography and textual criticism should be considered arcane specialities, just as the Oxford English School has regarded them as a part of every graduate student's equipment. In one way my work on these lectures has confirmed me in this view, for once the facts have been assembled and looked at without preconceptions, the conclusions follow logically without much resort to specialized techniques. What does give me cause for concern, however, is the extent to which one needs a historical understanding of the trade and its practices before the facts can be seen in perspective. Until we have more studies of historical bibliography, the student will be dependent on the accumulated experience of men like Cyprian Blagden and Graham Pollard to whom I owe such a debt.

My other concern follows from what I have said of the need to study the practice of an author throughout his life before one can begin to formulate the principles on which any one of his works should be edited. Many of us must be aware of the problems in examining a thesis, reviewing a book, or reading a manuscript for a publisher, when bibliographical or textual matters are involved: a true judgement can only be formed by duplicating some, at least, of the author's or editor's work and seeing whether one gets the same answers. This is a lengthy and laborious process, and we often fall back on compromise solutions—or rather, expedients. If the case of Pope is in any way typical, reaching a judgement on an edition of one work by an author may well mean studying the whole publishing career of that author; which is the task of months if not years. This conclusion may well seem ludicrous in terms of the realities of daily life—or if we are devoted to the ideals of scholarship it may seem horrific. The fact remains that we need to know the weaknesses of a work like the *Twickenham* edition of Pope—to which I am still devoted—and if they can be cured before publication, so much the better.

APPENDIX A: POPE AND COPYRIGHT

This appendix draws together information on Pope's contracts, lawsuits, and plans for litigation. Without claiming to be comprehensive, it attempts to cover Pope's most important dealings in literary property.

The beginning of Pope's publishing career coincides neatly with the introduction of the first copyright legislation. The 'Pastorals' were published in Tonson's *Miscellany* on 2 May 1709 and the following year 'An act for the encouragement of learning' (8 Ann. c. 19) was introduced on 11 January and became law on 6 April.[1] At first this legislation merely formed a background to Pope's agreements with the book trade, which ran on conventional lines, but Pope increasingly saw the Queen Anne Act as a source of authorial power and began to exploit it in his own interest. He had powerful legal advisers in William Fortescue, an ally of Walpole who in 1741 was appointed Master of the Rolls, and William Murray, one of the century's most brilliant lawyers, who was appointed successively Solicitor-General, Attorney-General, and Lord Chief Justice; they helped Pope to devise ways of profiting from his work, and he was finally able to do so without giving up his rights to the booksellers. The 1709/10 Act is supposed to have brought a new alliance of author and bookseller, but from the 1720s Pope was using it to strengthen his own position and lessen his dependence on the book trade.

'An act for the encouragement of learning'

The first 'Copyright' Act was drawn up in the absence of any agreed concept of copyright. The *OED* has no record of the word before 1735, and the Act does not even use the two words 'copy right', which we first find in Pope's transactions with the book trade in 1727. Gay quotes Pope using 'Copyright' in a letter of 28 August 1732 (*Correspondence*, iii. 309), and this may be the first recorded occurrence of the word. The Act uses the terms 'copy', which at this time probably retained its primary meaning, 'a document which is to be copied', and 'property', a term which led to conceptual difficulties later in the century.[2] As John Feather has shown, the Act was passed in response to pressure from the book trade, which was hoping for the restoration of former privileges and the granting of perpetual copyright, but the resulting legislation did not simply serve the booksellers' interests, it conferred rights on authors as well. In the absence of documentary evidence,

[1] J. Feather in 'The book trade in politics: The making of the Copyright Act of 1710', *PubH* viii (1980), 19–44 and H. Ransom in *The first copyright statute* (Austin, 1956) give accounts of the passing of the Act. It is reprinted in O. Ruffhead (ed.), *Statutes at large*, rev. C. Runnington (1786), iv. 401–3; by Ransom; and by Wiles, *Serial publication in England before 1750*.

[2] See G. Walters, 'The booksellers in 1759 and 1774: The battle for literary property', *Library*, 5th ser. xxix (1974), 287–311.

we cannot be sure why the booksellers' hopes were disappointed, but it is probable the Statute of Monopolies stood in their way. This Statute of 1623/4 (21 Ja. I. c. 3) forbade monopolies except in the special case of grants of privilege to the inventors of new manufactures: existing grants of this sort were allowed to continue for twenty-one years, and new grants could be made for a period of fourteen years.[3] The Statute also made provision against consequent rises in the price of commodities. Although printing was specifically exempted in one of its clauses, this Statute must have provided the framework of regulation underlying the 1709/10 Act. All rights in the Queen Anne Act originate in the inventor, the author; existing rights extend for twenty-one years, and new ones for fourteen years in the first instance; and consequential attempts are made to control the price of the goods, books.

The 1709/10 legislation can be divided into six parts.

1. Authors or proprietors of books already printed were to retain the exclusive right of printing for 21 years from the date of the Act; and for new books there was to be a term of fourteen years. There were penalties for those offending against these rights: the books were forfeit to the proprietor to be damasked, and a penny was to be paid for each sheet (half to the treasury; half to be sued for).

2. No one could be subject to the penalties unless the book had been entered in the Register of the Stationers' Company 'in such manner as hath been usual'.

3. If book prices became unreasonable, appeal could be made to named authorities.

4. Nine copies 'upon the best Paper' had to be deposited at Stationers' Hall for the use of named libraries.

5. All lawsuits had to begin within three months of the offence.

6. At the end of the term of fourteen years the rights would return to the author for fourteen more. (This was a House of Lords amendment.)

In 1740 Pope made some detailed notes on copyright in response to a letter from Henry Lintot; they are preserved with his contracts in the British Library and are printed by Sherburn (*Correspondence*, iv. 223–4). These notes, which reveal a fiercely protective attitude to literary property and a close knowledge of the legislation, provide a helpful, if incomplete, guide to Pope's copyright career; they figure prominently in the early part of the following survey, which is essentially chronological.

Early Poems

In his notes Pope divides his early poems into three groups, according to the bookseller he dealt with: (1) Tonson senior; (2) Tonson junior; (3) Lintot.

(1) Tonson senior. Pope notes that his 'Pastorals', 'The Merchant's Tale', and the 'Episode of Sarpedon' fall under the section of the Act which granted a further 21 years to books already printed, 'Those 21 years expired in 1731. so that it is no man's Property, but common'. That was an accurate interpretation of the Act, but it is unlikely that it would have been acceptable to the Tonsons. Booksellers were already claiming a perpetual right to their copies in common law (in addition to their statutory rights), and the Tonsons, who claimed the rights to Shakespeare, were among those who had most to gain

---

[3] 'An act concerning monopolies and dispensations with penal laws, and the forfeitures thereof' is repr. in *Statutes at large*, iii. 90–2.

by pressing such a claim. None of Pope's early agreements, except for the major contracts for the *Iliad*, *Odyssey*, and *Works*, survives, so we have no idea what was originally agreed. Pope may simply have signed a receipt for payment, or he may, as the editors of the *Spectator* did, have signed away his rights 'for ever'.[4]

(2) Tonson junior. The other Tonson poems were published in the *Miscellanies* of 1712 and 1713, after the Act, and Tonson's rights would, as Pope notes, have expired after fourteen years, unless Pope had 'covenanted to the contrary'. In 1740 Pope could not remember the terms of his agreements, but thought the sums 'very triffling . . . no way proportiond to a Perpetuity'. As we shall see shortly, the Tonsons allowed all their poems to be included in the collected *Works* brought out by Lintot in 1717, but when Lintot and Pope wanted to print them again in 1736, as part of the set of octavo *Works* published in collaboration with Gilliver, there was a dispute. The Tonsons claimed a right to their poems, even though the period allowed by the Act was up, but Pope was not certain to allow it, as a note in the Stationers' Register from Lintot to its keeper, Simpson, makes clear: 'pray enter the property of this first Volume of Mr Popes Works in my name 'till Mr. Tonson can clear up his right to Mr. Pope's satisfaction.' If the matter was resolved then, Pope had forgotten by 1740.

(3) Lintot. Pope continues his notes, which show that the term 'copy right' is by now very familiar to him:

I never alienated, intentionally, any Copy for ever, without expressly giving a Deed in forms, to witness & that the Copy right was to subsist after the Expiration of the 14 years in Queen Ann's Act, which *then* was understood generally to be the Case, unless covenanted to the Contrary. In testimony of this, all my assignments to Mr Lintot for perpetuity, will be found expressly articled for.

It is possible that the early contracts with Lintot were unspecific about the term granted and that the understanding outlined in this note was one Pope succeeded in agreeing with Lintot at the time of the 1717 *Works*, when he did grant him rights in perpetuity. But there is insufficient evidence to make this claim with confidence.

Works, 1717

The arrangements for the *Works* draw together the strands of Pope's early career, and they have already been discussed in Chapter 1. Maynard Mack has shown what an important project this was for Pope,[5] and he may have had it in mind as early as 5 October 1713 when Tonson junior signed an agreement (BL MS Egerton 1951, fo. 1) which allowed Pope to use Tonson's poems in a collection, provided he was allowed 'books in proportion to the number of Sheets the said poems amount to in such volume and in proportion to the impression'. It is possible that Pope and Lintot were planning to use the Tonson poems in the second edition of *Miscellaneous poems and translations*, published on 4 December 1713, but whatever the immediate plan, this was an important agreement for

---

[4] The receipts for 10 guineas for the 'Tale of Chaucer' and 3 guineas for the 'Episode of Sarpedon' are in the Pierpont Morgan Library and are reproduced in *Early career*, facing p. 85. A receipt for Pope's contribution to Steele's miscellany (5 Oct. 1713) survives at the New York Public Library (not seen). The contract for the *Spectator* (BL Add. MS 21110) is reproduced in W. H. Arnold, *Ventures in book collecting* (New York and London, 1923), p. 186.

[5] Mack, 'Pope's 1717 Preface with a transcription of the manuscript text', in *Augustan worlds*, ed. J. C. Hilson *et al.* (Leicester, 1978), pp. 85–106 (*Collected in himself*, pp. 159–78).

Pope. It followed the normal practice for the trade when dealing in shared property, and Pope was to draw on this experience in his negotiations with Motte and Bathurst in 1737 and with Lintot in 1743.

The contract Pope signed with Lintot for the 1717 *Works* (BL MS Egerton Charter 129) is undoubtedly the strangest of his surviving agreements. The *Works* were published on 3 June 1717 but the indenture is dated 28 December 1717. In one way the delay is unsurprising because the rights to the poems, except for 'Eloisa to Abelard' and eight shorter poems which first appeared in this collection, were already held by Lintot and Tonson, and it is difficult to see at first glance why there was the need for a major contract at all. It is possible Lintot was made to feel such a need by having his attention drawn to the provisions of the Queen Anne Act. The beginning of the contract is unusual in that the recitals are not first concerned with plans to bring out the edition but with the provisions of the Act. It begins, 'Whereas by an Act of Parliament' and goes on to outline the provision for two terms of fourteen years, with reversion to the author at the end of the first. The reversion is the focus of the contract. Lintot is granted 'The Sole Liberty Right and priviledge of printing reprinting and selling the said Booke' for and during 'the Terme of Fourteen yeares next ensueing ... As for all and every other Terme and Termes as farr as [Pope] ... may have any Right Power and Authority by the said Act of Parliament or otherwise to grant and sell the same'. This would seem sufficient to secure the rights to Lintot for the second fourteen years, but the contract then goes on to provide that at the end of the first fourteen years Pope will not sell the rights to anyone other than Lintot and that he will make a further grant to Lintot for another term of fourteen years at no cost to Lintot other than that of drawing up the contract.

The articles of this agreement are so closely related to Pope's notes of 1740, where he argues that the second fourteen years are not included unless specifically convenanted for, that it may be possible to reconstruct something of the negotiations leading to the contract in 1717. Pope had already received his side of the bargain, 120 quartos on royal paper 'now actually delivered', but it is possible that Lintot was asking for some payment (either for print and paper or delivery, for there had been a dispute over the delivery of the subscribers' *Iliad*s), because the contract insists that 'the said Alexander Pope shall ... be from henceforth fully and absolutely acquitted and discharged of and from all manner of payments Costs or Charges whatsoever on the Account of the said One Hundred and twenty Books or any of them or for the printing or delivery of them or any of them'. Pope, who was advised by Fortescue, may have suggested to Lintot that unless he got a further assignment all his rights would expire at the end of fourteen years. As a result Pope got his copies free and in return sold his rights, including the second term, to Lintot for 5s. (a nominal sum that occurs frequently in Pope's contracts from now on). That this is a possible interpretation of the evidence will be confirmed by Pope's subsequent dealings with Motte.

A new light is cast on all these arrangements with Lintot and the Tonsons by Warburton's edition of Pope. An account for the large octavo second edition of 1752 survives in the British Library (MS Egerton 1959, fo. 29) which shows the ownership of the volumes in sheets and the consequent distribution of the profits between the booksellers. 'Messrs. Tonson & Co.' are allowed 12 sheets, 11 pages; Lintot 24 sheets, 10 pages; Bathurst 9 sheets. All the early poems would have been out of copyright by 1745 (adding 28 years to 1717), but Warburton and Knapton were still allowing the claim. It is a good indication of how in practice the booksellers operated a system of perpetual copyright, independent of the statutory provisions.

The *Iliad* and *Odyssey*

The main provisions of the Homer contracts are dealt with in Chapter 2, but some smaller matters may be dealt with here. There are interesting differences in the phrasing of the contracts, probably due to the influence of Fortescue on the *Odyssey* contract, which shows specific awareness of the Queen Anne Act. In the *Iliad* contract Pope assigns 'All and every the Copy and Copies ... and the sole and absolute property therof' to Lintot 'for and during all such time terme and termes of yeares and in as large ample and beneficiall manner' as he himself 'may can might or could have use or enjoy the Same'. The *Odyssey* contract, witnessed and probably drawn up by Fortescue, is much more specific: Lintot is to hold his rights 'frome the day of ... Publishing ... for and during the Term of fourteen Years fully to be Compleat and Ended; and for and during all and every Such farther time and times Term or Termes as He the Said Alexander Pope by any Act or Acts of Parliament or otherwise in Howsoever is Enabled to Grant and Assign the Same' (BL MS Egerton Charter 130). That satisfied the arguments underlying the contract for the *Works* in 1717.

An important difference between the two copies of the *Iliad* contract deserves remark here, even though there are problems in finding a satisfactory explanation for it. The Bodleian copy (MS Don. a. 6) says that the 750 subscribers' copies are to be printed 'on a Royall Paper of an Octavo size'; the British Library copy (MS Egerton Charter 128) reads 'on a Royall Paper of a Quarto size'; and, of course, the subscribers' copies were quartos; it would have been impossible to charge a guinea a volume otherwise. But the British Library copy also originally read 'octavo'—it is possible to see where the word has been scratched out before being overwritten 'quarto'—and as this is an unlikely mistake for a scrivener to make, there must be a strong probability that the original plan was for an octavo subscription. Certain features of the contract fit this explanation. The stress in this agreement is on the importance of the illustrations as the feature that differentiates the subscribers' copies from the rest; in the *Odyssey* contract they were differentiated by format as well. The *Iliad* contract forbids Lintot to use any illustrations in his trade copies, but in fact the duodecimos used those from Mme Dacier's translation, and the normally litigious Pope did nothing about it. The change in format for the subscribers' copies may have made it unimportant to him. A projected subscription for an octavo *Iliad* would throw new light on the octavo *Rape of the lock* which was issued with engraved plates, headpieces, and tailpieces on 4 March 1714, nineteen days before the signing of the *Iliad* contract. It could have been something of a trial run for the *Iliad*, designed as an example to potential subscribers of the sort of book they could expect. Perhaps its extraordinary success led Pope and Lintot to change their plans and invest in a quarto subscription. The only difficulty with this explanation is that Pope had been collecting subscriptions from as early as October 1713, though the printed proposals had not yet appeared, and subscribers must surely have been told what sort of book they would get— and how much it would cost. Only the discovery of documents relating to the early stages of subscription can settle the matter, but it is certainly possible the contract was drawn up well before October 1713.

The *Dunciad variorum* and Gay's *Polly*, 1729

Chapter 3 includes a discussion of the lawsuit Gilliver brought in Chancery against Watson, Astley, Clarke, and Stagg for pirating the *Dunciad variorum* (PRO C 11/2581/36).[6] The *Dunciad* was Pope's first independent publication and it is no coincidence that it was also the first lawsuit he became involved in. As Chapter 3 makes clear Gilliver was acting as Pope's agent; he obtained an injunction against Watson and the others but it was dissolved because he had not claimed that he had acquired the copy from the author. This was the first instance of what became a persistent problem for Pope; his close relations with printer and publisher seemed to promise easy anonymity in publication, but it then became difficult to demonstrate that the bookseller had a legitimate claim to the copy—particularly when he had none because Pope had decided to keep the property himself. Fortunately Pope's circle was able to take more effective action against Watson and Astley at this time through a suit over *Polly*.[7] *Polly* had been banned from the stage by the Lord Chamberlain, and Gay's defiant response was to print it. There was some danger attached to the enterprise and although Bowyer did some of the printing, possibly the engraved music, Pope's new printer, Wright, was given the responsibility of seeing it in print, paying Bowyer, and probably printing the rest of the volume himself.[8] Watson and Astley clearly found these controversial works printed for the author an irresistible temptation to piracy, and they pirated *Polly*, this time using 'T. Thomson' in the imprint.[9] But Gay did not share Pope's desire for anonymity, and *Polly* had been entered in the Register by James Roberts, the publisher, on 3 April 1729 as the property of 'John Gay Esqr'. As a consequence the action in Chancery (probably started a month before Pope's) had a different result. At the end of the report on the Gilliver case, Viner records: 'Afterwards in the same Term, an Injunction was granted in the **Case of Gay**, Author of the Sequel of the Beggar's Opera, against publishing and selling that Book, upon a Bill founded Stat. 8 Ann. cap. 19.' (*General abridgment of law and equity*, iv. 279.) Arbuthnot reports the same result informally to Swift on 9 June 1729:

Mr Pope is well. he had gott an injunction in chancery against the printers who had pyrated his dunciad; it was dissolv'd again because the printer could not prove any property nor did the Author appear. that is not Mr Gays case for he has own'd his book. (*Correspondence*, iii. 37.)

A perpetual injunction against the pirates of *Polly* was eventually obtained after Gay's death, on 6 December 1737.

The Lords' Assignment of the *Dunciad variorum* to Gilliver, 1729

This assignment (BL MS Egerton 1951, fo. 6), dated 16 October 1729, and possibly reflecting the nature of Pope's assignment to the lords, is a good example of an assignment

---

[6] See Sutherland, 'The Dunciad of 1729'. Richard Goulden of the BL generously made his notes on the Pope copyright cases available for consultation.

[7] See Sutherland, '*Polly* among the pirates', *MLR* xxxvii (1942), 291–303; Gay's bill is lost, but Sutherland is able to piece together an account from the answers, which he lists, p. 295 n.

[8] The reference to Wright's payment is in the summary accounts in the first printing ledger and is dated 19 Apr. 1729. There were 5,000 ordinary copies and 500 fine.

[9] The *Dunciad variorum* used 'Dob'. Dr Mervyn Jannetta of the British Library reports that books purporting to be printed by T. Johnson of The Hague (the original of T. Thomson) in this period merit close examination to determine whether they are Watson's; Watson's piracy of Pope's *Letters* (discussed below) is an example. There were three other piracies of *Polly* besides Watson's.

for unspecified duration. It merely says that the lords 'grant bargain sell assign and transferr unto the said Lawton Gilliver ... *The Book* intitled *The Dunciad* an *Heroick* poem and the Copy thereof and the sole right and liberty of printing the same'. This was to cause trouble later, when Pope became involved in a dispute with Henry Lintot over when the rights originally transferred to Gilliver expired.

Contract with Gilliver for Epistles, 1732

Chapter 3 gives an account of this agreement (BL MS Egerton 1951, fos. 8–9) in which Gilliver agreed to pay £50 for each epistle and thereby acquire the rights to print and publish for a year. One of the main purposes of this agreement was to get Pope's property entered in the Stationers' Register. Gilliver had to make the entry and he was to benefit from it for one year but 'after the determination of one year ... the same shall be In trust for the said Alexander Pope and to or for no other intent or purpose whatsoever' and Gilliver agreed that at the end of the year he would 'transferr and assign unto the said Alexander Pope ... All his right title and interest in every such respective Poem or Epistle for or by reason of the said Entry or otherwise'. This was an ingenious way of securing the property while, if necessary, retaining anonymity. One wonders whether there were end-of-year transfers and whether Pope paid 5s. for them.

Motte and the *Miscellanies*, 1727–1732

Pope's dealings with Motte straddle his move to publishing independence. They are remarkable for the first use of 'Copy right' in a Pope contract and of 'Copyright' in a letter quoting Pope. Motte published the first three volumes of the Pope–Swift *Miscellanies* and he regularly handled Swift's publications in England (with Pope sometimes as intermediary), but his slowness in paying for the third of these (called the 'last') gave Pope the chance to take the reins back into his own hands. In an agreement dated 29 March 1727,[10] Motte had undertaken to pay £50 for the first volume of *Miscellanies* and £4 a sheet octavo for the subsequent ones. In return the sole 'Copy right' was vested in him. The 'last' volume was published on 8 March 1728 and Motte should have completed his payments in four months, but in May 1729 he still owed £25 from a total of £250. This was a good opportunity for Pope to switch any further business to his new bookseller, Gilliver, and in an agreement of 1 July 1729 Motte gained the rights to the first three volumes for fourteen years—in spite of the missing £25—with the promise of being able to buy the second fourteen years for 5s. (on the model of the *Works* of 1717); but in return Motte had to give up all claim to the fourth volume of *Miscellanies*.

What seemed such a satisfactory agreement later gave Pope considerable trouble. He sold the rights of the fourth volume to Gilliver in 1732 only to find first that Motte complained to Swift that he should be the bookseller and then that William Bowyer claimed to have bought the rights from Matthew Pilkington. Pope was displeased. A letter from Gay and the Duchess of Queensberry to Swift of 28 August 1732 reports a letter from Pope saying,

---

[10] The agreements, in the Pierpont Morgan Library, are printed in Rogers, *Major satires of Pope*, p. 115. There is a good account of the *Miscellanies* affair by G. Sherburn, 'The Swift–Pope *Miscellanies* of 1732', *HLB* vi (1952), 387–90, with a correction at vii (1953), 248.

'Motte & another idle fellow I find have been writing to the Dean to get him to give them some Copyright which surely he will not be so indiscreet as to do when he knows my design . . . Surely I should be a properer person to trust the distribution of his works with than to a common Bookseller.' (*Correspondence*, iii. 309.)

The use of 'Copyright', even if it is a mistranscription by Gay, shows how deeply Pope was involved in the formation of the concept, but on this occasion he decided not to entangle himself further, leaving it to Gilliver to reach an accommodation with the others.

The *Letters* and the 'Booksellers' Bill', 1735

Pope first got his letters published by tricking Curll into buying sheets he had already had printed by Wright and Hughs. No sooner were they in Curll's hands than he was dragged off to the House of Lords, accused of printing the letters of some of its members. In his account of his transactions with Curll, *A narrative of the method by which the private letters of Mr. Pope have been procur'd and publish'd by Edmund Curll, bookseller*, Pope says, 'By THIS INCIDENT THE BOOKSELLERS BILL WAS THROWN OUT', and he closes with a plea for legislation to prevent the licence of the book trade, particularly the printing of letters without permission (*Correspondence*, iii. 464 and 467). The Booksellers' Bill Pope refers to was 'An act for the better encouragement of learning', a copy of which survives in Thomas Carte's papers in the Bodleian (MS Carte 207 (14)).[11] The bill proposed that rights to existing copies (it uses the term Copy-right) should extend for seven years after the date of enactment and that new works should have a copyright period of twenty-one years. The legal position of proprietors was to be strengthened by enabling them to sue pirates for 10s. a book (another 10s. to go to the Crown), with the possibility of recovering the costs of the suit. Minor provisions included: the requirement to deposit fourteen (rather than nine) copies at Stationers' Hall (something taken up again in 1737); power given to authorities to order reprints of out-of-print works; an obligation on publishers who brought out a new edition within twenty-one years of first publication to offer supplements to buyers of the first edition. From the author's point of view the major defect of the bill was that it reduced the author's copyright period (from twenty-eight to twenty-one years) while tending to increase that of the purchaser (from fourteen years to twenty-one).[12] The desire to influence the passage of this bill must have been at least one factor in Pope's planning of the episode of Curll and the *Letters*; it is impossible to believe the arrival of the books at Curll's shop on the day of the second reading was a coincidence. There are other cases where petitioners reinforced their case by drawing attention to offences that needed remedy, and that was surely the plan in this instance.[13]

Pope succeeded in securing protection for letter-writers in a subsequent suit against Curll in 1741.

[11] There are many drafts of the bill in MSS Carte 207 and 114; 207(14) seems to be the one published, for its details are those discussed in *A letter to a Member of Parliament concerning the Bill now depending in the House of Commons* in the same papers (piece 20). These papers are explored in J. Feather's valuable guide, 'The publishers and the pirates: British copyright law in theory and practice, 1710–1775', *PubH* xxii (1987), 5–32; the understanding of the 1709/10 Act in this Appendix differs slightly from his.

[12] The controversy over the bill is summarized by A. S. Collins, *Authorship in the days of Johnson* (1927), pp. 68–74.

[13] For an account of the episode, see McLaverty, 'The first printing and publication of Pope's letters'. For an example of drawing attention to an offence, see Feather, 'The book trade in politics', p. 25.

The *Letters* and the 'Copyright Bill', 1737

Pope's interest in contemporary copyright legislation is confirmed by his response to the bill of 1737. In the *London gazette* of 22 March he announced that the quarto edition of the *Letters* was ready for distribution, 'But a *Bill* being now depending in *Parliament*, to secure the *Property of Books*, it is presumed the Subscribers will admit of a short Delay in the Delivery of the same, till the Fate of the said Bill is determined.' (*Correspondence*, iv. 65 n.) A copy of the bill is preserved in the Carte papers in the Bodleian (MS Carte 207(2)).[14] For works published after 24 June 1737 copyright would extend during the author's life and for eleven years after death; if he died within ten years of publication the period after death would be extended to twenty-one years. Authors were protected by not being allowed to sell their rights for more than ten years at a time, and there were tougher penalties for pirates (5*s.* a sheet recoverable by the proprietor). Pope would have welcomed the extensions of the copyright period, but his thinking was also influenced by three other provisions of the bill. The first offered the possibility of renouncing the penalties under the Act and bringing a bill in a court of Equity to discover the profits made from the piracy and to sue for them. The second provided for an author's concealing his identity by arranging for someone to register a book in trust for him. The third provided (along the lines of the 1735 bill) that the publishers of subsequent editions of a book costing more than 5*s.* in sheets should make additions and alterations available separately so that purchasers of the first edition could buy and insert them. Writing to Buckley about *Works* II on 13 April 1737 Pope wanted to know how many quartos and folios Knapton had because he was planning 'New Works, & would print a number accordingly, in each Size: Thus I am fulfilling the act of Parliament before it commences' (*Correspondence*, iv. 66). Pope continued to print his new works in quarto and folio before issuing them in the octavo *Works*—but the law did not require him to do so; this bill also was lost.

The Watson Piracy of the *Letters*, 1737

This piracy is interesting because the surviving documents fill some gaps in our knowledge of Pope's dealings in this period[15] and because it shows the tide turning in Pope's favour in his dealings with the book trade's pirates. Under the influence of William Murray, then a promising barrister but later to be Lord Chief Justice, Pope developed an interest in new forms of legal action, and the last ten years of his life show this policy meeting with some success.

Watson has been encountered before as the pirate of the *Dunciad variorum* and *Polly*. In the autumn of 1737 he pirated the *Letters* under the imprint of Thomas Johnson of The Hague.[16] Pope, acting on Murray's advice, responded vigorously, first, on 18 November, asking Nathaniel Cole, solicitor to the Stationers' Company, to draft a bill of complaint, and subsequently sending Murray's draft to Cole for vetting. The first point

---

[14] The bill is summarized by Collins, *Authorship in the days of Johnson*, pp. 77–82 and discussed by Feather, 'The publishers and the pirates', pp. 11–12.

[15] The manuscripts were in the Guildhall Library, MSS 18778–9; generous extracts are printed by Mack, *Collected in himself*, pp. 491–501. At the time of writing, the documents have been withdrawn for sale by their owners, Janson Ltd., a descendant of the law practice of Nathaniel Cole, solicitor to the Stationers' Company in Pope's time.

[16] For the possibility of further such piracies see n. 9 above.

of interest, especially in view of the failure of the *Dunciad variorum* injunction in 1729, is the establishment of ownership of copy. Dodsley had used a special formula in the entry of the quarto *Letters* on 17 May 1737: 'I Robert Dodsley do Claim the Sole Property in and to the above book, by Virtue of an Assignment under the hand of Alexander Pope Esqr.' The same formula had first been used by Wright in entering the *Ode to Venus* to himself on 8 March 1737. Dodsley used it a second time on 17 May for the *Second book of the epistles of Horace*, but not for the octavo edition of the *Letters*, which he entered as *Works* V and VI on 31 October. Pope experimented with different forms of registration in this period and perhaps this formula was suggested by Murray; Pope tells Cole, 'this Mr Murray says was a sufficient entry and assignment which is ready to produce' (*Correspondence*, iv. 88). It is noteworthy that a few months later, on 14 January 1738, while this affair with Watson was still being pursued, Pope entered the *Sixth epistle of the first book of Horace* directly to himself in the Register.[17] This is the first work (other than the half share in *Works* II with Gilliver) entered to Pope in the Register and he went to Stationers' Hall in person and signed the book himself. Although the Queen Anne Act was understood to have enabled the author to enter a book himself, Pope may not have known that (though *Polly* was entered to Gay by Roberts); in advising him on the Watson case Cole may have told Pope that the formula Dodsley had used was unnecessary and that the author could secure his property directly.[18] At any rate Pope went on to have the dialogues of 1738, the *Prose works* II, and the *New Dunciad* entered to him in the Register. It was only when he returned to Bowyer as printer at the end of his career that we find Pope's works (the Cibber *Dunciad* and the *Essay on man*) again being entered to a member of the book trade, Mrs Cooper.

As Pope suspected, Dodsley would have had to establish his property in the 1737 *Letters* by producing his assignment. It is a surprising document.

I Alexander pope of Twickenham in Middlesex do hereby Assign & make over all my Right & Title to the Copy of a Book intituled (The Works of Alexr. pope in prose, being Letters of the sd A pope & Several of his friends) in Consideration of Five Shills paid & other valuable Consideration, Unto Robert Dodsley Bookseller in pall Mall, to have & hold the same, & so print Reprint & Vend, to him & his Assigns, for the space of Fourteen Years from the publication of the said Book; in pursuance of the Liberty granted & property vested in Authors by the Act of Parliament made in the Eighth Year of Queen Anne. In Witness whereof I have hereunto Set may hand & Seal, this twenty fourth of March one Thousand Seven hundred & Thirty *Seven* [as Cole notes this should be 'Six' in a legal document].

This looks like a sale for fourteen years, but there are two unusual features to take into account. First, the sum of 5*s*., used in other contracts to purchase a second term of fourteen years, is not a plausible payment for the rights to Pope's newly published *Letters*; the 'other valuable Consideration' must be the vital factor. Second, as the discussion in Chapter 3 points out, there is no sign that Dodsley subsequently owned the copyright. The other consideration might have been that Pope was to get the profits on all editions

---

[17] This epistle was addressed to Murray and the *Ode to Venus* concerns him. This may do no more than reflect the intimacy between the two men in this period, but it would be a brave pirate who would print a poet's epistle to his lawyer.

[18] For earlier regulations, see Blagden, *The Stationers' Company*, p. 110 n. and Brodowski, 'Literary piracy in England', p. 23. There are many examples of authors entering their own work after the Act of 1709/10; Thomas Carte and Henry Carey regularly did so.

during that period, with Dodsley being paid commission, or it might have been that Dodsley would sell the rights back to Pope for 5*s.* as soon as he had entered the work in the Register. The latter would be a refinement on the contract with Gilliver which sold the epistles for a year to get them registered, and it would explain why Pope was originally named as co-orator in the Chancery bill of complaint against Watson, only for his name to be crossed out; if Dodsley really held the copyright, there seems no reason anyone should have thought of naming Pope.

In the bill, Dodsley, on the advice of Murray, took a step envisaged in the proposed Act of Parliament which had been thrown out earlier in the year. He waived and disclaimed all the penalties allowed him by the Queen Anne Act and asked for a remedy in Equity instead. He asked to be told who was involved in the piracy, how many copies had been printed, and what profits had been made. Clearly the aim was to acquire full compensation. In the event, the dispute with Watson was settled out of court, being put in the hands of John Knapton for arbitration. Watson produced a flurry of defences, but he finally submitted to severe penalties: he gave up the whole impression to Dodsley, receiving only £25 compensation, and he gave a bond of £100 not to pirate Pope in the future. Watson had understood from Gilliver, who had been called in at an early stage to negotiate for the inexperienced Dodsley, that he might get as much as 9*d.* a book (Pope says 9¼*d.*) for the 1,646 books surrendered, which would have given him over £60 and covered his costs.[19] It is not difficult to guess who might have decided to drive a harder bargain.

Motte and Bathurst: *Works in prose*, 1737 and *Works in prose* II, 1741

The agreement Pope made with Motte and Bathurst over *Works in prose* (BL MS Egerton 1951, fo. 10) is transcribed and discussed in Chapter 3. It is unusually careless in wording, saying that Pope is 'willing' to include Motte and Bathurst's pieces in a volume or volumes of 'Letters and other Pieces in Prose' and that they 'covenant and agree to pay the price of the Print and Paper . . . for the same at the delivery of the Books'. It is not clear whether they were to take the whole of the edition or only a part proportionate to their contribution. It could hardly be the former because the contract goes on to stipulate the prices they were to charge, and that would have been pointless if they were handling the whole edition. The agreement imposes no obligation on Pope; it merely gives information about his state of mind: even though it was made under seal, Motte would probably have found little difficulty in setting this contract aside, had he so wished.

Later, probably on 4 August 1740, Bathurst asked Pope whether he should allow Lintot's claim to the *Key to the lock*: 'I desire the favour of a Line from you which I may shew him in justification of my Refusal.' Pope replied:

Mr Lintot has no right to the *Key to the Locke* these many Years. The term expired in the year 29 or 30. But till then I presume Mr Motte allowed it. If not, you may set agst it your Right to the small

---

[19] Watson estimated his paper cost at £30. 12*s.* 0*d.*, but the papers show the actual sum was £39. 10*s.* 6*d.* (62 reams at 12*s.* 9*d.* per ream), which suggests he was being honest in these negotiations.

poems in the End of Mr Lintots third volume of my Works 8⁰ to which Your Right from Mr Motte yet continues.[20]

The *Key to the lock*, purchased 31 April 1715, would have fallen into the same category as other early works sold to Lintot, but it was not included in the *Works* 1717 agreement. It was most unlikely, therefore, that Pope would recognize any continuing right or that Lintot, threatened with retaliation on his *Works* III, would pursue the matter.

Curll's Piracy of the Pope–Swift *Letters*, 1741

Pope had stopped Watson by starting legal action in 1737; in this case (PRO C 11/1569/29) he finally defeated Curll in the courts, and it must have given him particular pleasure that he did so over an edition of the letters. Murray was again his counsel. Curll claimed that he had printed only 500 copies of the *Letters* and sold only 16 of them, and he mounted a number of defences. The most important of these was that letters were not covered by the provision of the Queen Anne Act; the resulting judgment established copyright in letters and delimited the rights of recipients.[21]

Lintot and the *Iliad*, 1740–1742

At the time Pope made his copyright jottings in 1740 he was worried about being cheated over the *Iliad* and *Odyssey*. Thomas Osborne had advertised large-paper quartos of the *Iliad* in the *Daily gazetteer* in January 1740 and Pope knew that according to his contract the quartos should all be his. He therefore wrote to Lintot, querying the advertisements. Lintot replied that he had not sold any quartos to Osborne, but that he had sold odd volumes of the *Iliad*, including quartos, to Gilliver and given him 'Liberty to reprint the first Volume, in Quarto' (*Correspondence*, iv. 223). Pope was alarmed and 'interrogated' Gilliver, who said he had bought about 75 quarto sets lacking the first volume and had printed first volumes for them (there is a copy in the British Library, C. 130. e. 6). Pope suspected a major fraud by the Lintots. Noting the bookseller was not allowed by the contracts to print quarto *Iliad*s at all but was allowed to print quarto *Odyssey*s after ten years, he assumed Lintot had illicitly over-printed quarto *Iliad*s and had intended to print further copies of the *Odyssey* to make up sets. This picture of readers systematically collecting his quartos informed many of Pope's decisions in the latter part of his career, but here it misled him. When he questioned Bowyer he found that he had printed only 660 *Iliad*s and that Watts had printed 750 *Odyssey*s. Nevertheless Pope was prepared to take legal action, as two letters of the period suggest. The first, to Lintot on 31 January 1741, reminds him of the problem:

I Received Yours of this last post but it does not mention one that I wrote to you some time since which I Desird Mr Cole to deliver to you with a State of that Affair upon which I troubled you last

---

[20] Bathurst's letter and Pope's reply are printed by Mack, *Collected in himself*, p. 515; Mack's conventions have been adapted to Sherburn's.

[21] There are excellent accounts of the case by H. Ransom, 'The personal letter as literary property', *Studies in English*, xxx (1951), 116–31, and by P. Rogers, 'The case of Pope *v.* Curll', *Library*, 5th ser. xxvii (1972), 326–31.

summer at Mr Murrays and as to which I wonder you have given me no answer. I hope Mr Wright has returnd to you the 50 books in exchange for yours as he was directed to do some Weeks ago. (*Correspondence*, iv. 333.)

The second, to Cole on 17 March, suggests Lintot was still refusing to take any notice:

I was of opinion indeed that Mr L. would not be active to seek you on this affair, as he never answerd me upon it; but would require us to be pressing with Him . . . Perhaps you should tell him that unless he compromises, you must file a Bill directly: Or propose to leave it to a Reference. Judge Chappel was a Witness to the Articles, & a Friend to his Father: I'll leave it to him & Judge Fortescue—or (if he consents) to Yourself to determine what shall be my Reparation.   (*Correspondence*, iv. 336.)

Judge Chappel was a witness to the *Iliad* articles; he witnessed the British Library copy of the contract, but not the Bodleian copy. The references are, therefore, to the *Iliad* dispute and Sherburn's suggestion that Lintot was claiming rights to the *Key to the lock* falls in consequence.[22]

Henry Lintot may not have appreciated the terms of his father's agreements when he gave Gilliver permission to print, but, as Pope reports in a letter to George Arbuthnot, he later found what he claimed to be 'a Discharge in full for all Books, remaining of my Homers' (*Correspondence*, iv. 394). This must be a reference to 'Mr. Pope's Assignment for the Royal Paper that were then left of his Homer' that Nichols records as belonging to Lintot's papers (*Anecdotes*, viii. 300). Pope told Arbuthnot he would not 'proceed in our Bill till you see what this is', and I suspect this assignment settled the matter. As Sherburn points out, a reference to *Joseph Andrews* enables us to date this letter some time after February 1742, which means the dispute had dragged on for two years.

Lintot and the *Dunciad*, 1742–1743

Gilliver, on the road to bankruptcy, had sold his rights to the *Dunciad*: one-third in 1739 to John Clarke (probably his former partner to whom he later owed £400), who sold it to John Osborne, who sold it to Henry Lintot on 18 January 1740, and the remaining two-thirds directly to Lintot on 15 December 1740.[23] Lintot had an edition of 4,000 copies (plus 100 fine) printed by Woodfall by 4 July 1741 and Lintot claimed in the resulting lawsuit that this was done 'with the Consent of the Complainant who corrected the Sheets of that Edition or Impression as they came from the press'.[24] The questions at issue between Pope and Lintot which took them to Chancery on 16 February 1743 (PRO C 11/549/39) seem to have been: (1) when did Lintot's rights expire? (2) could Lintot go on selling his edition after his rights expired? The answer was of some importance to

---

[22] Sherburn's reference here is to the supplementary agreement, not the *Iliad* articles, and that may be why he does not find Chappel's name. In any case, Pope would hardly expect to receive reparation for printing the *Key to the lock* himself, whether it was Lintot's property or his own. The 50 copies were probably being exchanged for the 55 copies of *Works* I. ii printed on fine paper by Woodfall by Dec. 1740 (P.T.P., 'Pope and Woodfall', p. 378); Wright would be sending the *Miscellanies* or the forthcoming *Works in prose* in exchange. The exchange could have been connected with the claim to the *Key to the lock*, amicably settled, but that seems less likely.

[23] A summary of this and the following case is given by Vincent, 'Some *Dunciad* litigation'; Richard Goulden has generously provided additional information.

[24] Vincent, 'Some *Dunciad* litigation', p. 287.

Pope because he was deep in preparation of his new Warburton–Cibber edition and did not want a rival on the scene. The first news we have of a settlement is a letter of 21 May 1743 in which Pope tells Warburton his suit with Lintot is at an end (*Correspondence*, iv. 455). The terms imposed or agreed are not known, but it was probably decided that Pope should wait until fourteen years had elapsed from the date of the Lords' assignment to Gilliver; that would have been 16 October 1743. Mrs Cooper took her first batch of 500 copies of the new quarto *Dunciad* from Bowyer's warehouse on 14 October and it was published on 29 October. It is unlikely that Lintot was officially allowed to go on selling copies after that date; the Act specifies that after the first fourteen years 'the sole Right of printing and disposing of Copies shall return to the Authors thereof'. But there are clearly many practical difficulties in enforcement, and a compromise may have been reached.

Jacob Ilive's Piracy of *Dunciad* IV, 1742–1743

On the same day Pope began his suit against Lintot, 16 February 1743, he began an action against Ilive (PRO C 11/837/14); once more Murray seems to have been the instigator. Ilive, who represented himself as the servant or agent of Ellen Ackers—his sister—had printed an edition of 1,000 copies (only 975 turned out to be perfect) of the *New Dunciad* at the request of 'David' [Daniel?] Lynch. The printing had to be of the 'same Letter and of the same Size with a Book or Books then lately printed call'd Popes and Swift's last Letters'. The details of the printing costs are given, but the identification of the volume remains doubtful. Pope announced his initial success in a letter to Allen of 27 December 1742: 'My coming to Town has put a Stop to the Pyraters of my book, who have surrendered all their Copies to get free from further prosecution, but had sold 5 or 600, for which they can't refund, in my wrong.' (*Correspondence*, iv. 433.) Ilive's evidence includes a reference to the 400 copies sent to Pope's counsel, which confirms this is the same case. Pope must have begun the Chancery action in the hope of compensation.

Bickham's Piracy of the *Essay on man*, 1744

This Chancery case (PRO C 11/626/30) has recently been the subect of an admirable discussion by David Hunter.[25] It was unusual because Bickham was an engraver. In a bill of charges drawn up by Murray, dated 9 January 1743/4, Pope claimed that Bickham had printed a large edition of the first book of the *Essay on man*. Bickham replied with the customary defences, but added a new one—that as he was printing from engraved plates rather than from the usual types he could cast off as many copies as he pleased. Pope would have been in difficulty if the case had come to a hearing and he had sought the penalties laid down in the Act. The poem had been entered in the Register by Wilford, to himself, beginning on 10 March 1733, but Gilliver probably held the copyright at the time. As Hunter points out, Murray skirts round this problem in the charges, but a hearing might have tested the efficacy of Pope's complex agreement with Gilliver of 1732. But Bickham did not contest and the court granted an injunction on 23 January.

Once again Pope was acting to protect an immediate interest: he was about to bring out

[25] Hunter, '*Pope* v. *Bickham*: An infringement of *An essay on man* alleged', *Library*, 6th ser. ix (1987), 268–73.

the quarto edition of the *Essay on man* with Warburton's notes; Mrs Cooper took her first 250 copies on 20 January and entered the work in the Register on 17 February 1744.

Lintot and the *Essay on criticism*, 1744

There is a good argument that the *Essay on criticism* was out of copyright by 1744. It was first published in 1711 and two terms of fourteen years had already expired. Lintot had bought the copyright, from Pope or Lewis, on 17 July 1716 and Pope may have been allowing 28 years from that date or, or more likely, from the *Works* of 1717. He may even by this date have been assuming that he had granted Lintot a perpetual right in common law. Whatever his reasons, he was scrupulous in recognizing Lintot's rights to the poem in 1744. On 23 February he wrote to Bowyer:

I hope you have Enterd the Essay on Man & the Essay on Criticism with the Commentary and Notes of W. Warburton, Printed for Wm Bowyer in the Hall book—I desire you to remember exactly, & minute down what Mr Lintot has said to you of printing any thing of mine, &c. There may be Occasion for it, if ever he ventures at it, and I must beg you to be particularly watchful; if it can be found done at his Press, I would have you write him word, & keep a Copy of the Letter, that 'you have publishd but *so many* [Bowyer was to enter the number] books to try the Tast of the Town: that the Proportion of Sheets belonging to Him, being the whole Text of the Essay on Crit. makes 4 sheets a 6th part of the book; that you inclose him a Bill of the Costs of paper & print of *so many* books as you have publishd, and have them ready to be deliverd him, on payment. Or if he would pay no money, to deduct it out of the Number, without asking him Ever to allow for any Books more than as they shall sell: 'but that he may either take his proportion, whenever Mrs Cooper takes a number from time to time, or I will allow it to him & account with him for it as they are sold, without ever charging him for the remainder.'   (*Correspondence*, iv. 501–2.)

Pope thought it safe to enter the *Essay on criticism* in the Register because of the new property in Warburton's notes, hence the specific instructions on wording. In fact Mrs Cooper's entry for the book of 17 February had been more cautious, naming only the *Essay on man*. It seems from this letter that Lintot was angry and had threatened to print some of Pope's copies in retaliation. The offer Pope tells Bowyer to make grows out of the arrangement with Tonson in 1713 and the similar arrangements offered to Motte and Bathurst in 1737, but perhaps finally rejected by them (see Chapter 3 above). Lintot is offered the proportion of the edition his text amounts to in the volume—as Sherburn points out, a deduction is made to take account of Warburton's notes—but Pope is at pains to avoid foisting a proportion of the edition on Lintot whether he wants it or not. He need only take as many as he can sell, or a proportion of each batch will be sold for him and the profits handed over. This arrangement, which indemnifies Lintot against loss, would have met the objections Motte and Bathurst may have had in 1737; possibly it represents the solution which was eventually arrived at and enabled Pope to go ahead with his *Works in prose* II.

APPENDIX B: GAY AND CAPITALS AND ITALICS

Chapter 4 discusses Gay's abandonment in 1716 of the capitalization of nouns in three re-prints of earlier works (*The shepherd's week*, *Trivia*, and *The what d'ye call it*); the very fact that they are reprints, where a compositor would naturally follow copy, marks the significance of the change.[1] At the same time these works abandon the use of italic in the text for all purposes (except for the street cries, 'clean your Shoes' and 'stop thief' in *Tri-via*), thus prefiguring Pope's practice twelve years later in the *Dunciad variorum*. *The what d'ye call it* goes even further in starting broken speeches with lower case, and also after stops where one would expect a capital. Even 'christians' starts with a lower-case 'c'. We cannot say how far Pope and Gay influenced one another in these changes, but Gay like Pope continued to employ heavy capitalization in his manuscripts after his change in typographic style.

Perhaps for this reason, or because Gay did not press his wishes on his printers, his next single works (*Three hours after marriage* (1717), *Two epistles* (1717), and *A panegyri-cal epistle* (1721)) revert to the old use of capitals and italics, but his subscription quarto, *Poems on several occasions* (1720) adopts the new style as far as capitals go. Its use continues between 1722 and 1727 in the works printed for Tonson by Watts, but the first two editions of the *Beggar's opera* (1728), which are printed by Watts on his own account, revert to the old style; I think Watts was old-fashioned in his style when he was printing for himself and not for Tonson. The third edition, however, is changed to the new style, and in doing so matches *Polly*, printed for the author by Bowyer and Wright in 1729.

As with Pope, I think an editor should disregard the accidentals of the manuscripts and first editions in accordance with the author's change of typography; and in this respect the old Oxford edition of his works by G. C. Faber seems to me preferable to the new by Vinton Dearing, despite the greater technical sophistication of the latter.

---

[1] See Table 17 and pp. 184–5.

LIST OF SELECTED SOURCES
AND WORKS CONSULTED

Manuscripts are listed under the name of the writer or the chief person concerned. The listing of documents in the PRO is especially selective. Pope's poetical manuscripts are not listed; guides to them will be found in Butt's article and Mack's *Collected in himself*, listed below. Books are published in London unless otherwise indicated.

Manuscripts

ADDISON, JOSEPH, and STEELE, RICHARD, *Spectator*: Agreement with Tonson, 10 November 1712, BL Add. MS 21110.

BOWYER, WILLIAM (elder and younger), Ledgers 1710–81, Bodl. MSS Dep. b. 243–4, Dep. c. 718–23.

—— Paper stock ledger 1717–73, Bodl. MS Don. b. 4.

DAVIS, HERBERT, Correspondence concerning Oxford Standard Authors edition of Pope, 1965–6, Bodl. MS Eng. lett. c. 524–5.

FOXON, DAVID F. 'Pope and the early eighteenth-century book-trade', Lyell Lectures 1975–6 [copies at Bodl. (MS Johnson c. 12), BL, Beinecke, William Andrews Clark].

—— 'The Stamp Act of 1712', Sandars Lectures 1978 [copies at Cambridge University Library, BL].

Grub Street journal, Minute book of the partners, The Queen's College, Oxford, MS 450.

McKENZIE, D. F., 'The London book trade in the later seventeenth century', Sandars Lectures 1976 [copies at Cambridge University Library, BL].

POPE, ALEXANDER, *Dunciad*: Complaint against James Watson *et al.*, 6 May 1729, and their answer, PRO C 11/2581/36.

—— *Dunciad*: Assignment to Lawton Gilliver by the three Lords, 16 October 1729, BL MS Egerton 1951, fo. 6.

—— *Dunciad*: Complaint against Henry Lintot, 16 February 1742/3, and Lintot's answer, PRO C 11/549/39.

—— *Dunciad*: Complaint against Jacob Ilive, 16 February 1742/3, and his answer, PRO C 11/837/14.

—— Epistles: Agreement with Gilliver, 1December 1732, BL MS Egerton 1951, fos. 8–9.

—— *Essay on man*: Complaint against George Bickham, 9 January 1743/4, PRO C/11/626/30.

—— *Iliad*: Agreement with Bernard Lintot, 23 March 1713/14, BL MS Egerton Charter 128, Bodl. MS Don. a. 6.

—— *Iliad*: Supplementary agreement with Lintot, 10 February 1715/16, BL MS Egerton 1951, fos. 2–3.

—— *Letters*: Papers for proceedings against Watson, 24 November 1737–Spring 1738, Guildhall MSS 18778–9.

—— *Letters*: Bond from Watson not to pirate works, 26 April 1738, BL MS Egerton 1951, fo. 13.

—— *Letters*: Complaint against Edmund Curll, 4 June 1741, PRO C/11/1569/29.

—— *Odyssey*: Agreement with Bernard Lintot, 18 February 1723/4, BL Egerton Charter 130, Houghton MS Eng. 232.2.

—— *Odyssey*: Draft proposals, BL Add. MS 4809, fo. 87ᵛ.

—— *Prose*: Agreement with Benjamin Motte and Charles Bathurst, 24 February 1736/7, BL MS Egerton 1951, fo. 10.

—— Shakespeare: Agreement with Jacob Tonson, 22 May 1721, Houghton MS Eng. 233.13.

—— Shakespeare: Draper's record of payments from Tonson's accounts, Folger MS S. a. 163.

—— *Works*: Agreement with Tonson, 5 October 1713, BL MS Egerton 1951, fo. 1.

—— *Works* I: Agreement with Bernard Lintot, 28 December 1717, BL MS Egerton Charter 129.

—— *Works* II: Declaration to Gilliver, BL MS Egerton 1951, fo. 12.

TONSON, JACOB, jun., Contract with Henry Huddle and Thomas Perry for repairing house in Bow Street, 29 October 1717, BL Add. MS 28275, fos. 40–1.

—— Letter from John Hughes, 17 August 1719, BL Add. MS 28275, fo. 61.

—— Payment of rents and refitting of house, 1736–7, Bodl. MS Eng. misc. b. 45.

TONSON, JACOB, sen., Payments in overseer's accounts and rate collector's book of St Paul's, Covent Garden, Victoria Library, City of Westminster, MSS H482–95, H5–25.

WARBURTON, WILLIAM, Account for Pope's *Works*, 17 January 1753, BL MS Egerton 1959, fo. 29.

—— Letter to John Knapton, BL MS Egerton 1954, fo. 5.

—— Letter to Robert Dodsley, 26 December 1755, Edinburgh University Library, MS La. II. 153.

Ward trade sales catalogues, Bodl. John Johnson Collection.

WATTS, JOHN, Sales of copyrights to Thomas Lownds, 30 June 1758, Bodl. MS Eng. misc. c. 297, fo. 50ʳ.

Books, articles, and theses

ACKERS, CHARLES, *A ledger of Charles Ackers*, ed. D. F. McKenzie and J. C. Ross, Oxford Bibliographical Society Publications, NS xv (Oxford, 1968).

ADDISON, JOSEPH, *Letters*, ed. W. Graham (Oxford, 1941).

ARNOLD, WILLIAM HARRIS, *Ventures in book collecting* (New York and London, 1923).

AULT, NORMAN, *New light on Pope* (1949).

—— (ed.), *Pope's own miscellany* (1935).

BARBER, GILES, 'Bolingbroke, Pope, and the *Patriot king*', *Library*, 5th ser. xix (1964), 67–89.

BARKER, NICOLAS, 'Pope and his publishers', *TLS*, 3 September 1976, p. 1085.

BARNARD, JOHN, 'Dryden, Tonson, and subscriptions for the 1697 *Virgil*', *PBSA* lvii (1963), 129–51.

BEDFORD, EMMETT G., and DILLIGAN, ROBERT J., *A concordance to the poems of Alexander Pope*, 2 vols. (Detroit, 1974).

BELANGER, TERRY, 'Booksellers' sales of copyright' (unpublished Ph.D. dissertation, University of Columbia, 1970).

—— 'Booksellers' trade sales, 1718–1768', *Library*, 5th ser. xxx (1975), 281–302.

BENNET, THOMAS, and CLEMENTS, HENRY, *The notebook of Thomas Bennet and Henry Clements*, ed. Norma Hodgson and Cyprian Blagden, Oxford Bibliographical Society Publications, NS vi (Oxford, 1956).

BLAGDEN, CYPRIAN, 'The memorandum book of Henry Rhodes, 1695–1720 II', *BC* iii (1954), 103–16.

—— *The Stationers' Company: A history, 1403–1959* (1960).

BODDY, MARGARET, 'Tonson's "loss of Rowe"', *NQ* ccxi (1966), 213–14.

BOND, RICHMOND P., 'The pirate and the *Tatler*', *Library*, 5th ser. xviii (1963), 257–74.

BOSWELL, JAMES, *Life of Johnson*, ed. George Birkbeck Hill, rev. L. F. Powell, 6 vols. (Oxford, 1934–50).

BOWERS, FREDSON, 'Scholarship and editing', *PBSA* lxx (1976), 161–88.

BOYCE, BENJAMIN, 'Baroque into satire: Pope's frontispiece for the "Essay on man"', *Criticism*, iv (1962–3), 14–27.

BRADY, FRANK, 'The history and structure of Pope's *To a lady*', *SELit* ix (1969), 439–62.

BRODOWSKI, JOYCE HELENE, 'Literary piracy in England from the Restoration to the early eighteenth century' (unpublished doctoral dissertation, Columbia University, 1973).

BRONSON, BERTRAND H., 'Printing as an index of taste', in his *Facets of the Enlightenment* (Berkeley and Los Angeles, 1968), pp. 326–65.

BROWNELL, MORRIS R., *Alexander Pope and the arts of Georgian England* (Oxford, 1978).

BUTT, JOHN, 'Pope's poetical manuscripts', Warton Lecture on English poetry, *Proceedings of the British Academy*, xl (1954), 23–39.

CALLAN, NORMAN, 'Pope's *Iliad*: A new document', *RES* NS iv (1953), 109–21.

CARTER, HARRY, *A history of the Oxford University Press* (Oxford, 1975).

CASE, A. E., *A bibliography of English poetical miscellanies, 1521–1750* (Oxford, 1935).

CLEMENTS, HENRY. See Bennet, Thomas.

COCHRANE, J. A., *Dr. Johnson's printer: The life of William Strahan* (1964).

COLLINS, A. S., *Authorship in the days of Johnson* (1927).

'Common-place notes', *Gentleman's magazine*, lvii (1787), 76.

CROTTET, E., *Supplément à la 5me édition du Guide de l'amateur de livres à figures du XVIIIᵉ siècle* (Amsterdam, 1890).

DAVIES, DAVID WILLIAM, *The world of the Elseviers, 1580–1712* (The Hague, 1954).

DAVIS, HERBERT, 'Bowyer's paper stock ledger', *Library*, 5th ser. vi (1951), 73–87.

DEARING, VINTON A., 'New light on the first printing of the letters of Pope and Swift', *Library*, 4th ser. xxiv (1944), 74–80.

—— 'The Prince of Wales's set of Pope's works', *HLB* iv (1950), 320–38.

—— 'The 1737 editions of Alexander Pope's letters', in *Essays critical and historical dedicated to Lily B. Campbell*, ed. Louis B. Wright (Berkeley and Los Angeles, 1950), pp. 185–97.

—— 'Pope, Theobald, and Wycherley's *Posthumous works*', *PMLA* lxviii (1953), 223–36.

—— 'Two notes on the copy for Pope's letters', *PBSA* li (1957), 327–33.

DUNTON, JOHN, *The life and errors of John Dunton* (1705).

EADE, J. C., 'Lewis Theobald's translation rates: A hard bargain', *Library*, 6th ser. i (1979), 168–70.

EEGHEN, ISABELLA HENRIETTA VAN, *De Amsterdamse boekhandel 1680–1725*, 5 vols. in 6 parts (Amsterdam, 1960–78).

ELIAS, A. C., jun., 'The Pope–Swift *Letters* (1740–41): Notes on the first state of the first impression', *PBSA* lxix (1975), 323–43.

FEATHER, JOHN, 'The book trade in politics: The making of the Copyright Act of 1710', *PubH* viii (1980), 19–44.

—— 'The commerce of letters: The study of the eighteenth-century book trade', *ECS* xvii (1984), 405–24.

—— *The provincial book trade in eighteenth-century England* (Cambridge, 1985).

—— 'The publishers and the pirates: British copyright law in theory and practice, 1710–1775', *PubH* xxii (1987), 5–32.

FIELDING, HENRY, *The complete works*, intro. William Ernest Henley, 16 vols. (1903).

FLEEMAN, J. D., '18th-century printing ledgers', *TLS*, 19 December 1963, p. 1056.

FOXON, DAVID F., 'Concealed Pope editions', *BC* v (1956), 277–9.

—— 'Two cruces in Pope bibliography', *TLS*, 24 January 1958, p. 52.

—— *English verse 1701–1750*, 2 vols. (Cambridge, 1975).

—— 'Greg's "Rationale" and the editing of Pope', *Library*, 5th ser. xxxiii (1978), 119–24.

FRANKLIN, BENJAMIN, *The autobiography*, ed. L. W. Labaree *et al.* (New Haven and London, 1964).

GASKELL, PHILIP, 'Notes on eighteenth-century British paper', *Library*, 5th ser. xii (1957), 34–42.

—— *A bibliography of the Foulis Press* (1964).

GAY, JOHN, *Dramatic works*, ed. John Fuller, 2 vols. (Oxford, 1983).

—— *The letters*, ed. C. F. Burgess (Oxford, 1966).

—— *The poetical works*, ed. G. C. Faber (Oxford, 1926).

—— *Poetry and prose*, ed. Vinton Dearing and Charles E. Beckwith, 2 vols. (Oxford, 1974).

GEDULD, HARRY M., *Prince of publishers* (Bloomington and London, 1969).

GENT, THOMAS, *The life of Thomas Gent*, ed. J. Hunter (1832).

GOLDGAR, BERTRAND A., 'Pope and the *Grub-street journal*', *MP* lxxiv (1977), 366–80.

GOULDEN, R. J., 'Auction catalogue printing 1715–1730', *Factotum*, x (1980), 23–7.

—— *The ornament stock of Henry Woodfall 1719–1747*, Occasional Papers of the Bibliographical Society, iii (1988).

GREENWOOD, JEREMY, *Newspapers and the Post Office 1635–1834*, Postal History Society Special Series Publication, xxvi (1971).

GREG, W. W., *The collected papers*, ed. J. C. Maxwell (Oxford, 1966).

GRIFFITH, REGINALD H., *Alexander Pope: A bibliography*, 2 vols. [strictly vol. i, parts 1 and 2] (Austin, 1922–7).

—— 'A piracy of Pope's *Iliad*', *SP* xxviii (1931), 737–41.

—— 'Early Warburton? or late Warburton?', *Studies in English* (University of Texas), xx (1940), 123–31.

—— 'Pope editing Pope', *Studies in English* (University of Texas), xxiv (1944), 5–108.

—— Review of *The Dunciad*, ed. James Sutherland, *PQ* xxiv (1945), 152–7.

GUERINOT, J. V., *Pamphlet attacks on Alexander Pope 1711–1744* (1969).

HALSBAND, ROBERT, '*The rape of the lock*' and its illustrations, *1714–1896* (Oxford, 1980).

HAMMELMAN, HANS ANDREAS, *Book illustrators in eighteenth century England*, ed. and completed T. S. R. Boase (New Haven, 1975).

HANSON, LAURENCE, *Government and the press 1695–1763* (Oxford, 1936).

HARDMAN, PHILLIPA, 'An addition to Griffith's bibliography of Pope', *Library*, 5th ser. xxxiii (1978), 326–8.

HART, JOHN A., 'Pope as a scholar-editor', *SB* xxiii (1970), 45–59.

HAZEN, A. T., 'The meaning of the imprint', *Library*, 5th ser. vi (1951), 120–3.

HELLINGA, WYTZE Gs, *Copy and print in the Netherlands* (Amsterdam, 1962).

HESSE, ALFRED W., 'Pope's role in Tonson's "loss of Rowe"', *NQ* ccxxii (1977), 234–5.

HEWARD, EDWARD, *Lord Mansfield* (Chichester, 1979).

HODGART, MATTHEW J. C., 'The subscription list for Pope's *Iliad*, 1715', in *The dress of words: Essays on Restoration and eighteenth century literature in honour of Richmond P. Bond*, ed. Robert B. White, jun. (Lawrence, Kansas, 1978), pp. 25–34.

HUNTER, DAVID, '*Pope* v. *Bickham*: An infringement of *An essay on man* alleged', *Library*, 6th ser. ix (1987), 268–73.

An impartial history of the life, character, amours, travels, and transactions of Mr. John Barber, city-printer, common-councilman, alderman, and Lord Mayor of London (1741).

JACK, IAN, *The poet and his audience* (Cambridge, 1984).

JOHNSON, SAMUEL, *The letters of Samuel Johnson*, ed. R. W. Chapman, 3 vols. (Oxford, 1952).

—— *The lives of the English poets*, ed. G. Birkbeck Hill, 3 vols. (Oxford, 1905).

KINSLEY, WILLIAM, 'The *Dunciad* as mock-book', *HLQ* xxxv (1971–2), 29–47.

KNAPP, ELISE F., 'Community property: The case for Warburton's 1751 edition of Pope', *SELit* xxvi (1986), 455–68.

KOON, HELENE, 'Pope's first editors', *HLQ* xxxv (1971–2), 19–27.

KORSHIN, PAUL, 'Types of eighteenth-century literary patronage', *ECS* vii (1973–4), 453–73.

KUPERSMITH, WILLIAM, 'Asses, adages, and the illustrations to Pope's *Dunciad*', *ECS* viii (1974–5), 206–11.

LAUFER, ROGER, *Introduction à la textologie* (Paris, 1972).

LE GAL, SIMONNE, 'En marge de l'exposition du "livre anglais"; l'Homère de Pope', *Bulletin du bibliophile et du bibliothècaire*, NS i (1952), 49–54.

LERANBAUM, MIRIAM, *Alexander Pope's 'Opus magnum' 1729–1744* (Oxford, 1977).

LONGMAN, C. J., *The house of Longman 1724–1800* (1936).

LYNCH, KATHLEEN M., *Jacob Tonson, Kit-Cat publisher* (Knoxville, 1971).

MACDONALD, HUGH, *John Dryden: A bibliography of early editions and of Drydeniana* (Oxford, 1939).

MACK, MAYNARD, 'The first printing of the letters of Pope and Swift', *Library*, 4th ser. xix (1938–9), 465–85.

—— 'Pope's Horatian poems: Problems of bibliography and text', *MP* xli (1943–4), 33–44.

—— Review of *Correspondence of Alexander Pope*, ed. George Sherburn, *PQ* xxxvi (1957), 389–99.

—— 'Some annotations in the second Earl of Oxford's copies of Pope's *Epistle to Dr. Arbuthnot* and *Sober advice from Horace*', *RES* NS viii (1957), 416–20.

—— 'Two variant copies of Pope's *Works . . . Volume II*: Further light on some problems of authorship, bibliography, and text', *Library*, 5th ser. xii (1957), 48–53.

—— 'Pope's 1717 Preface with a transcription of the manuscript text' in *Augustan worlds*, ed. J. C. Hilson *et al.* (Leicester, 1978).

—— *Collected in himself* (Newark, London, and Toronto, 1982).

—— *Alexander Pope: A life* (New Haven, 1985).

—— '"In affectionate touch": Letters from and to Pope', *Scriblerian*, xx (1987–8), 1–7.

—— (ed.), *The last and greatest art: Some unpublished poetical manuscripts of Alexander Pope* (Newark, London, and Toronto, 1984).

McKENZIE, ALAN T., 'The solemn owl and the laden ass: The iconography of the frontispieces to *The Dunciad*', *HLB* xiv (1976), 25–39.

McKENZIE, D. F. *The Cambridge University Press*, 2 vols. (Cambridge, 1966).

—— 'The London book trade in 1668', *Words*, iv (1974), 75–92.

—— (ed.), *Stationers' Company apprentices 1641–1700*, Oxford Bibliographical Society Publications, NS xvii (Oxford, 1974).

—— (ed.), *Stationers' Company apprentices 1701–1800*, Oxford Bibliographical Society Publications, NS xix (Oxford, 1978).

McLAVERTY, JAMES, 'John Wright and Lawton Gilliver: Pope's printer and bookseller' (unpublished B.Litt. dissertation, University of Oxford, 1974).

—— *Pope's printer, John Wright*, Oxford Bibliographical Society Occasional Publications, xi (Oxford, 1977).

—— 'Lawton Gilliver: Pope's bookseller', *SB* xxxii (1979), 101–24.

—— 'Pope's Horatian poems: A new variant state', *MP* lxxvi (1979–80), 304–6.

—— 'The first printing and publication of Pope's letters', *Library*, 6th ser. ii (1980), 264–80.

—— 'The mode of existence of literary works of art: The case of the *Dunciad variorum*', *SB* xxxvii (1984), 95–105.

MASLEN, KEITH I. D., 'New editions of Pope's *Essay on man 1745–48*', *PBSA* lxii (1968), 177–88.

—— 'Printing for the author: From the Bowyer printing ledgers, 1710–75', *Library*, 5th ser. xxvii (1972), 302–9.

MEANS, JAMES, 'Sir Richard Blackmore and the frontispiece to *The Dunciad variorum* (1729)', *Scriblerian*, vi (1973–4), 101–2.

MENGEL, ELIAS F., jun, 'The *Dunciad* illustrations', *ECS* vii (1973–4), 161–78.

MONTAGU, MARY WORTLEY, *Essays and poems*, ed. Robert Halband and Isobel Grundy (Oxford, 1977).

MORES, E. R., *A dissertation upon English typographical founders and foundries (1778) with a catalogue and specimen of the typefoundry of John James (1782)*, ed. Harry Carter and Christopher Ricks, Oxford Bibliographical Society Publications, NS ix (Oxford, 1961).

MOXON, JOSEPH, *Mechanick exercises on the whole art of printing*, ed. Herbert Davis and Harry Carter, 2nd ed. (Oxford, 1962; first printed in parts 1683).

MYERS, ROBIN, and HARRIS, MICHAEL (eds.), *Development of the English book trade, 1700–1899* (Oxford, 1981).

——— (eds.), *Author/publisher relations during the eighteenth and nineteenth centuries* (Oxford, 1983).

NICHOL, DONALD W., 'Pope, Warburton and the Knaptons: Problems of literary legacy (with a book-trade correspondence)' (unpublished Ph.D dissertation, University of Edinburgh, 1984).

—— '"So proper for that constant pocket use": Posthumous editions of Pope's *Works* (1751–1754)', *Man and nature*, vi (1987), 81–92.

NICHOLS, JOHN, *Literary anecdotes of the eighteenth century*, 8 vols. (1812–14).

P., P.T., 'Pope and Woodfall', *NQ* xi (1855), 377–8, 418–20; xii (1855), 197, 217–19.

PAPALI, G., *Jacob Tonson, publisher* (Auckland, 1968).

PHILIP, IAN, *The Bodleian Library in the seventeenth and eighteenth centuries* (Oxford, 1983).

POPE, ALEXANDER, *The correspondence*, ed. George Sherburn, 5 vols. (Oxford, 1956).

—— *Poetical works*, ed. Herbert Davis (1966).

—— *The prose works* I, ed. Norman Ault (Oxford, 1936).

—— *The prose works* II, ed. Rosemary Cowler (Oxford, 1986).

—— *The Twickenham edition of the poems*, ed. John Butt *et al.*, 11 vols. (London and New Haven, 1939–69).

—— *The works*, ed. Joseph Warton, 9 vols. (1822).

—— *The works*, ed. W. Elwin and W. J. Courthope, 10 vols. (1871–89).

—— For editions of manuscripts, see under Mack, Schmitz, and Wasserman.

RANSOM, HARRY, *The first copyright statute* (Austin, 1956).

—— 'The personal letter as literary property', *Studies in English*, xxx (1951), 116–31.

REED, TALBOT BAINES, *A history of the old English letter foundries*, rev. and enlarged A. F. Johnson (1952).

RIVERS, ISABEL (ed.), *Books and their readers in eighteenth-century England* (Leicester, 1982).

RIVINGTON, SEPTIMUS, *The publishing family of Rivington* (1919).

ROBINSON, F. J. G., and WALLIS, P. J., *Book subscription lists: A revised guide* (Newcastle, 1975).

ROGERS, PAT, 'The case of Pope v. Curll', *Library*, 5th ser. xxvii (1972), 326–31.

—— 'Pope and his subscribers', *PubH* iii (1978), 7–36.

—— 'Not in Sherburn', *BJECS* viii (1985), 59–64.

ROGERS, ROBERT W., 'Notes on Pope's collaboration with Warburton in preparing a final edition of the *Essay on man*', *PQ* xxvi (1947), 358–66.

—— *The major satires of Alexander Pope* (Urbana, 1955).

ROOT, ROBERT K., 'Pope's contributions to the Lintot miscellanies of 1712 and 1714', *ELH* vii (1940), 265–71.

RUFFHEAD, OWEN, *The life of Alexander Pope* (1769).

—— (ed.), *Statutes at large*, rev. Charles Runnington, 10 vols. (1786).

RYSKAMP, CHARLES, '"Epigrams I more especially delight in": The receipts for Pope's *Iliad*', *PULC* xxiv (1962–3), 36–8.

SAMBROOK, JAMES, 'The *Dunciad* illustrations', *ECS* viii (1974–5), 211–12.

SAYCE, R. A., 'Compositorial practices and the localization of printed books, 1530–1800', *Library*, 5th ser. xxi (1966), 1–45.

SCHMITZ, ROBERT M., 'The "Arsenal" proof sheets of Pope's *Iliad*: A third report', *MLN* lxxiv (1959), 486–9.

—— (ed.), *Pope's Windsor Forest 1712* (St Louis 1952).

—— (ed.), *Pope's Essay on criticism 1709* (St Louis, 1962).

SHAW, BERNARD, *Collected letters*, ed. Dan H. Lawrence, 3 vols. (1965, 1972, 1985).

SHERBURN, GEORGE, *The early career of Alexander Pope* (Oxford, 1934).

—— 'The Swift–Pope *Miscellanies* of 1732', *HLB* vi (1952), 387–90; with a 'Corrigendum', *HLB* vii (1953), 248.

—— 'Letters of Alexander Pope, chiefly to Sir William Trumbull', *RES* NS ix (1958), 388–406.

SIMPSON, PERCY, *Proof-reading in the sixteenth, seventeenth, and eighteenth centuries* (Oxford, 1935).

SNYDER, HENRY L., 'The reports of a press spy for Robert Harley', *Library*, 5th ser. xxii (1967), 326–45.

SPENCE, JOSEPH, *Observations, anecdotes, and characters of books and men*, ed. James M. Osborn, 2 vols. (Oxford, 1966).

STEELE, RICHARD, *The correspondence*, ed. Rae Blanchard (Oxford, 1941).

STRAUS, RALPH, *Robert Dodsley, poet, publisher & playwright* (1919).

SUTHERLAND, JAMES R., 'The circulation of newspapers and literary periodicals, 1700–30', *Library*, 4th ser. xv (1934–5), 110–24.

—— 'The Dunciad of 1729', *MLR* xxxi (1936), 347–53.

—— '*Polly* among the pirates', *MLR* xxxvii (1942), 291–303.

—— *The Restoration newspaper and its development* (Cambridge, 1986).

SWIFT, JONATHAN, *The correspondence*, ed. Harold Williams, 5 vols. (Oxford, 1963–5).

TANSELLE, G. THOMAS, 'Greg's theory of copy-text and the editing of American literature', *SB* xxviii (1975), 167–229.

—— 'Recent editorial discussion and the central questions of editing', *SB* xxxiv (1981), 23–65.

—— 'Historicism and critical editing', *SB* xxxix (1986), 1–46.

TODD, WILLIAM B., 'Concealed Pope editions', *BC* v (1956), 48–52.

TREADWELL, MICHAEL, Letter on subscription in *TLS*, 7 July 1972, p. 777.

—— 'Congreve, Tonson, and Rowe's "Reconcilement"', *NQ* ccxx (1975), 265–9.

—— 'London trade publishers 1675–1750', *Library*, 6th ser. iv (1982), 99–134.

TURNER, MICHAEL L., 'The minute book of the partners in the *Grub Street journal* [with transcription]', *PubH* iv (1978), 49–94.

UPDIKE, DANIEL BERKELEY, *Printing types*, 2nd ed., 2 vols. (Cambridge, 1937).

VANDER MEULEN, DAVID L., 'A descriptive bibliography of Alexander Pope's *Dunciad*, 1728–1751' (unpublished dissertation, University of Wisconsin-Madison, 1981).

—— 'Pope's revisions during printing: A variant section in *The Dunciad*', *MP* lxxviii (1981), 393–8.

—— 'The printing of Pope's *Dunciad* 1728', *SB* xxxv (1982), 271–85.

—— 'The identification of paper without watermarks: The example of Pope's *Dunciad*', *SB* xxxvii (1984), 58–81.

VARNEY, A. J., 'The composition of Pope's *Windsor Forest*', *DUJ* NS xxxvi (1974), 57–67.

VINCENT, HOWARD P., 'Some *Dunciad* litigation', *PQ* xviii (1939), 285–9.

VINER, CHARLES (ed.), *General abridgment of law and equity*, 24 vols. (1751–8).

VOET, LEON B. *The golden compasses*, 2 vols. (Amsterdam, 1972).

WALLIS, P. J., 'Book subscription lists', *Library*, 5th ser. xxix (1974), 255–86.

WALTERS, GWYN, 'The booksellers in 1759 and 1774: The battle for literary property', *Library*, 5th ser. xxix (1974), 287–311.

WASSERMAN, EARL R. (ed.), *Pope's 'Epistle to Bathurst'* (Baltimore, 1960).

WATSON, GEORGE (ed.), *New Cambridge bibliography of English literature*, 5 vols. (Cambridge, 1969–77).

WENDORF, RICHARD, 'Robert Dodsley as editor', *SB* xxxi (1978), 235–48.

WILES, ROY, *Serial publication in England before 1750* (Cambridge, 1957).

WIMSATT, W. K., *The portraits of Alexander Pope* (1965).

WINN, JAMES A., 'On Pope, printers and publishers', *ECL* vi (1981), 93–101.

WISE, T. J., *A Pope library* (1931; reprinted with intro. K. I. D. Maslen, 1973).

WOODMAN, CHARLES BATHURST, 'Pope's arrangements with Mr. Benj. Motte', *Gentleman's magazine*, NS xliv (1855), 363–6.

INDEX

Primary entries for peers are under their titles rather than their family names. Illustrations and tables are represented by italic page numbers at the end of entries.

KING ALFRED'S COLLEGE
LIBRARY